Dust Explosion and Fire Prevention Handbook

Scrivener Publishing
100 Cummings Center, Suite 541J
Beverly, MA 01915-6106

Publishers at Scrivener
Martin Scrivener (martin@scrivenerpublishing.com)
Phillip Carmical (pcarmical@scrivenerpublishing.com)

Dust Explosion and Fire Prevention Handbook

A Guide to Good Industry Practices

Nicholas P. Cheremisinoff, Ph.D.

Scrivener Publishing
WILEY

Co-published by John Wiley & Sons, Inc. Hoboken, New Jersey, and Scrivener Publishing LLC, Salem, Massachusetts.
Published simultaneously in Canada.

For general information on our other products and services or for technical support, please contact our Customer Care Department within the United States at (800) 762-2974, outside the United States at (317) 572-3993 or fax (317) 572-4002.

Wiley also publishes its books in a variety of electronic formats. Some content that appears in print may not be available in electronic formats. For more information about Wiley products, visit our web site at www.wiley.com.

For more information about Scrivener products please visit www.scrivenerpublishing.com.

Cover design by Kris Hackerott

Library of Congress Cataloging-in-Publication Data:

ISBN 978-1-118-77350-5

Printed in the United States of America

10 9 8 7 6 5 4 3 2 1

Contents

About the Author

Nicholas P. Cheremisinoff is a graduate of Clarkson College of Technology, where he received his B.Sc., M.Sc. and Ph.D. degrees in chemical engineering. He has nearly 40 years of industry, applied research and international business experience, and is the author of numerous engineering reference textbooks concerning good industry practices in the management of dangerous and hazardous materials. He is the Principal of No Pollution Enterprises, which is a firm specializing in environmental and worker safety litigation support.

Preface

Airborne dust created by the handling of many industrial materials can combine in an air/dust mixture that could result in a violent, damaging explosion. A combustible dust is defined by the NFPA (Standards 68 and 654) as "any finely divided solid material 420 microns or smaller in diameter which presents a hazard when dispersed or ignited in air." ISO is even more conservative and reports any finely divided solid material smaller in diameter 500 microns may present an explosion hazard. Most organic (carbon containing) and metallic dust will exhibit some combustibility characteristics. Therefore, if dust is present in any form within a working environment efforts should be taken to assess whether the potential for a hazard exists or not, and to devise appropriate practices and safeguards to mitigate the risks.

Preventing dust explosions has gained increased attention in recent years. In the United States the Chemical Safety Board has proposed new regulations to reduce the dangers of combustible dust. The European Community has already implemented two directives for that same purpose. Directive 94/9/EC, often referred to as ATEX-95 (Atmosphères Explosives), defines the safety requirements concerning equipment and protective systems intended for use in potentially explosive atmospheres. The other EC directive, 1999/92/EC (ATEX 137), outlines the minimum requirements for the protection and safety of workers at risk from explosive atmospheres.

Dust explosions can result when a flame propagates through combustible particles that have dispersed in the air and formed a flammable dust cloud. Whether an explosion happens or not depends on the supply of oxygen to the fire and the concentration of the fuel. If the concentration of the oxygen or the fuel is too high or low, then an explosion is very unlikely.

Consider the combustion engine in your car – there are three combustion components (fuel, air/oxygen and the ignition spark) which work together in a controlled manner to produce an explosion inside the enclosed cylinder. For the explosion to take place, the ratio of fuel to air must be in the proper proportion. If the fuel tank is empty, the air source is blocked or if the ignition does not work, then any one of these components is considered controlled, combustion cannot occur and the motor will not start.

Industrial dust explosions can be instigated by many sources, including static sparks, friction and glowing or smoldering materials. But before dust can explode, the following factors need to be present:

- The dust must be combustible.
- The dust must be capable of becoming airborne.
- The dust must have a size distribution capable of flame propagation.
- The dust concentration must be within the explosion limits.
- An ignition source must be present.
- The atmosphere must contain sufficient oxygen to support and sustain combustion.

When all of these factors are present, a dust explosion can occur. Eliminating just one of these requirements would make a dust explosion very unlikely. This then is the overall objective of the volume – to examine the causes of dust explosions and to provide readers with an overall understanding of good industry practices to prevent such events from occurring.

Explosions are defined as sudden reactions involving a rapid physical or chemical oxidation reaction, or decay generating an increase in temperature or pressure, or both simultaneously. When the flame speed exceeds the speed of sound, the event is referred to as a detonation. Otherwise, the explosion is known as a deflagration. Detonations are much more destructive than deflagrations. Typically, dust explosions are relatively slow combustion processes. If ignition occurs in a dust cloud in an open area, then little or no overpressure results and the primary hazard is a fireball. But if a deflagration occurs in a confined space such as a piece of equipment or constricted ductwork or tubes, the results may be devastating, causing substantial damage to operations, injury to operating personnel and even fatalities. The reader will find a number of case studies documented in this volume which testify to the devastating results of industrial dust explosions.

The volume begins with a glossary of terms that are commonly applied to the safe handling of dusts as they relate to fire and explosion issues. The reader may wish to spend a few moments familiarizing him or herself with some of the terms if this subject is relatively new to them.

Chapter 1 examines the physical and thermodynamic properties of particles which comprise dust. Properties such as size, shape, particle size distribution and the combustible nature of some materials are examined, thereby orienting the reader to more in-depth discussions to follow in later chapters.

Chapter 2 provides a general overview of the characteristics and parameters that can be the cause of dust explosions. The basic ingredients that are required for a dust explosion to occur are discussed and important terms and concepts relevant to dust instability are examined.

Chapter 3 provides discussions on the various factors that influence dust explosibility, including but not limited to particle size and particle size distribution, dust concentration, oxidant concentration, ignition temperature, turbulence of the dust cloud, maximum rate of pressure rise, admixed inert dust concentration and the presence of flammable gases.

Chapter 4 delves into the topics of explosions in grain dust elevators, the causes, and good industry practices for prevention. The first recorded incident of a dust explosion was in 1785, in a flour mill in Turin, Italy. A series of accidents during World War I led to a flurry of scientific activity culminating in the publication of numerous pamphlets and bulletins by the U.S. Department of Agriculture. This work identified grain dust as the specific ingredient common to all accidents, and recommended best practices were made in order to prevent the occurrence of these accidents. Despite early industry recognition and statements of good practices, there have continued to be numerous incidents over the decades leading to enormous human and financial losses.

Chapter 5 is titled Coal Dust Explosibility and Coal Mining Operations. This chapter provides an examination of coal dust explosions, safe handling operations, and coal mine safety practices. There are three necessary elements which must occur simultaneously to cause a fire: fuel, heat, and oxygen (known as the fire triangle). Removing any one of these elements eliminates the possibility of fire. But for an explosion to occur, there are five essential elements: fuel, heat, oxygen, suspension, and confinement. These form the five legs of the so-called explosion pentagon. Like the fire triangle, removing any one of these requirements would prevent an explosion from propagating. When a burning fuel is placed in suspension by a sudden blast of air, all five sides of the explosion pentagon are satisfied and an explosion would be imminent. The reader will find pertinent information in this chapter for both good practices for dust management in mining operations and for general processing operations.

Chapter 6 provides information on combustible metals, their properties and some common sense guidelines for safe handling of metal dusts. Most metals are combustible to a varying degree, depending on their physical conditions. Many will undergo dangerous reactions with water, acids, and certain other chemicals; and some metals are subject to spontaneous heating and ignition. The hazard of an individual metal or alloy varies depending on the particle size and shape that is present. The reader will find a wide variety of data and useful information on the safe handling of these materials, plus general guidelines for management of dusts for fire and explosion prevention that are relevant to all materials.

Chapter 7 covers phlegmatization, the use of diluents and the application of inert gases. Each of these practices can reduce the risks of explosible dusts.

Chapter 8 addresses Leak Detection and Repair (LDAR) programs. Because of the possibility of flammable vapors being present in many operations, LDAR should be considered a critical part of the dust management program.

Appendix A is an assembly of general guidelines on safe work practices. Dust explosion and fire safety management programs should be carefully integrated with the overall safe work practices and procedures of the facility. This appendix provides useful general information and good industry practices for safe work ethics and handling of dangerous chemicals.

The author wishes to thank the staff of No Pollution Enterprises for assisting in research, styling and proofreading the manuscript. A heartfelt thank you is also extended to the publisher for their fine production efforts.

Nicholas P. Cheremisinoff, Ph.D.

1

Combustible Dusts

1.1 Introduction

According to the National Safety Council[1], dust is defined as "solid particles generated by handling, crushing, grinding, rapid impact, detonation, and decrepitation of organic or inorganic materials, such as rock, ore, metal, coal, wood, and grain." Dust is a by-product of different processes that include dry and powdery material conveying, solids crushing and screening, sanding, trimming of excess material, tank and bin feeding and storing of granular materials, and a number of other processes. The creation of dust during material handling and processing operations may pose the obvious problem of inhalation risks to workers, often characterized as chronic or long term worker exposures. However, when combustible dust is produced and allowed to accumulate, risks can create immediate danger to life and health from explosions. Combustible dust explosions have

---

[1] *Fundamentals of Industrial Hygiene*, 3rd Edition, National Safety Council. Chicago, Ill., 1988.

resulted in the loss of life, multiple injuries and substantial property and business damage. A few examples[2] are:

- In 2002, an explosion at Rouse Polymerics International, a rubber fabricating plant in Vicksburg, Miss., resulted in injuring eleven employees, five of whom later died of severe burns. The explosion occurred with the ignition of an accumulation of a highly combustible rubber.
- In 2003 an explosion and fire occurred at the West Pharmaceutical Services plant in Kinston, N.C., resulting in the death of six workers, injuries to dozens of employees, and hundreds of job losses due to the destruction of the plant. The facility produced rubber stoppers and other products for medical use. The fuel for the explosion was a fine plastic powder that had accumulated unnoticed above a suspended ceiling over the manufacturing area.
- In 2003 an explosion and fire damaged the CTA Acoustics manufacturing plant in Corbin, Ky., fatally injuring seven employees. The facility produced fiberglass insulation for the automotive industry. The combustible dust associated with the explosion was a phenolic resin binder used in producing fiberglass mats.
- In 2003, a series of explosions severely burned three employees, one fatally, and caused property damage to the Hayes Lemmerz manufacturing plant in Huntington, Ind. The Hayes Lemmerz plant manufactured cast aluminum automotive wheels. The explosions were fueled by aluminum dust, a combustible by-product of the manufacturing process.
- In 2008 combustible sugar dust was the fuel for a massive explosion and fire at the Imperial Sugar Co. plant in Port Wentworth, Ga., resulting in 13 deaths and the hospitalization of 40 more workers, some of whom received severe burns.

These are only a few examples of dust explosions in which there was loss of life and the substantial destruction of assets and properties.

---

[2] Berry, C., A. McNeely, K. Beauregard, and J. E. Geddie, A Guide to Combustible Dusts, Occupational Safety and Health Division, N.C. Department of Labor, Feb. 2009.

Before we can understand the causes of dust explosions and ways to prevent them, we need to understand what dust is. The physical, chemical and thermodynamic properties of dust are important for a myriad of reasons ranging from the protection of workers from inhalation hazards, explosions and fire, and the overall safe and economic handling of materials that are prone to creating dust.

In this chapter we focus on the physical and thermodynamic properties of particles which comprise dust. Properties such as size, shape, particle size distribution and the combustible nature of some materials are discussed, orienting the reader to more in-depth discussions to follow in later chapters.

1.2 Metrics

Dusts are generated from solid or granular materials and can exist over a wide range of particle sizes depending on the material handling and processing operation. They may also form through the processes of sublimation and thermal oxidation as well as from combustion-related processes. Particles that are too large to remain airborne settle out due to gravity, while the smallest particles can remain suspended in air almost indefinitely as colloidal suspensions.

The unit of measure used to characterize dust particle size is the 'micrometer', more commonly known as a micron or μm. The micrometer is a unit of length equal to 10^{-4} (0.0001) centimeter or approximately 1/25,000 of an inch, or another way of stating this – there are *25,400 microns* in one inch. In metric units a micron represents one-millionth of a meter. By way of physical comparisons:

- Red blood cells are typically 8 μm (0.0008 cm) in size
- Human hair is 50 – 600 μm in diameter
- Cotton fiber, 15 – 30 μm

The human eye can see particles to as low as 40 microns. Table 1.1 provides some typical dimensions for materials the reader may relate to.

However, the term *particle size* requires some thought. What do we really mean by particle size? Certainly when a particle is spherical, size equates with the diameter of a sphere. But particles not only come in different sizes, they exist in different shapes.

Table 1.1 Typical particle size comparisons.

Particle Descriptor	**Particle Size (microns)**	
	Low Range	**Upper Range**
Oxygen	0.00050	
Carbon Dioxide	0.00065	
Atmospheric Dust	0.001	40
Viruses	0.005	0.3
Rosin Smoke	0.01	1
Tobacco Smoke	0.01	4
Oil Smoke	0.03	1
Smoldering or Flaming Cooking Oil	0.03	0.9
Sea Salt	0.035	0.5
Coal Flue Gas	0.08	0.2
Clay	0.1	50
Corn Starch	0.1	0.8
Paint Pigments	0.1	5
Radioactive Fallout	0.1	10
Face Powder	0.1	30
Metallurgical Dust	0.1	1,000
Metallurgical Fumes	0.1	1,000
Burning Wood	0.2	3
Carbon Black Dust	0.2	10
Combustion-related - motor vehicles, wood burning, open burning, industrial processes	2.5	
Bacteria	0.3	60
Copier Toner	0.5	15
Insecticide Dusts	0.5	10
Talcum Dust	0.5	50
Asbestos	0.7	90

Particle Descriptor	Particle Size (microns)	
	Low Range	Upper Range
Calcium Zinc Dust	0.7	20
Anthrax	1	5
Smoke from Synthetic Materials	1	50
Yeast Cells	1	50
Milled Flour, Milled Corn	1	100
Auto and Car Emission	1	150
Coal Dust	1	100
Fiberglass Insulation	1	1,000
Fly Ash	1	1,000
Lead Dust	2	
Spider web	2	3
Mold	3	12
Spores	3	40
Cement Dust	3	100
Starches	3	100
Bone Dust	3	300
Iron Dust	4	20
Red Blood Cells	5	10
Gelatin	5	90
Coffee	5	400
Grain Dusts	5	1,000
Antiperspirant	6	10
Mustard	6	10
Textile Dust	6	20
Tea Dust	8	300
Mold Spores	10	30
Fertilizer	10	1,000

(Continued)

Table 1.1 (*Cont.*)

Particle Descriptor	Particle Size (microns)	
	Low Range	Upper Range
Ground Limestone	10	1,000
Pollens	10	1,000
Textile Fibers	10	1,000
Cayenne Pepper	15	1,000
Ginger	25	40
Saw Dust	30	600
Human Hair	40	600
Mist	70	350
Dust Mites	100	300
Beach Sand	100	10,000
Spanish Moss Pollen	150	750
dot (.)	615	
Glass Wool	1,000	
Eye of a Needle	1,230	
one inch	25,400	

1.3 Size and Shape

One of the most important physical properties of particulates is size. Particle size measurement is routinely carried out across a wide range of industries and is often a critical parameter in the manufacture of many different products. Size has a direct influence on material properties including reactivity or dissolution rate e.g. catalysts, tablets; in the stability in suspension e.g. sediments and paints; for efficacy of delivery e.g. asthma inhalers; in the texture and feel e.g. food ingredients; in product appearance e.g. powder coatings and inks; in terms of flowability and handling e.g. granules; in viscosity e.g. nasal sprays; in the packing density and porosity of a product, e.g. ceramics.

Understanding how particle size affects products and processes is critical to many manufacturing operations. It is also important to the safe handling of materials especially in terms of inhalation risks to workers that come into contact with dusty materials, and as discussed later on, in terms of explosions and fires.

We begin by recognizing that particles are 3-dimensional objects and unless they are perfect spheres (e.g. emulsions or bubbles), they cannot be fully described by a single dimension such as a radius or diameter. To simplify both the measurement and characterization of particles, it is convenient to define particle size using the concept of equivalent spheres. In this way particle size is defined by the diameter of an equivalent sphere having the same property as the actual particle like volume or mass for example. Different measurement techniques and reference definitions use different equivalent sphere models and therefore will not necessarily give exactly the same result for the particle diameter. Examples include:

- Sphere with the same maximum or minimum length of a particle
- Sphere with the same weight of a particle
- Sphere with the same volume as a particle
- Sphere with the same surface area of a particle
- Sphere capable of passing through the same sieve aperture as a particle
- Sphere having the same settling or sedimentation rate as a particle

See figure 1.1 for reference. It is important that any size relied on should be carefully referenced to a specific measurement technique and/or reference definition.

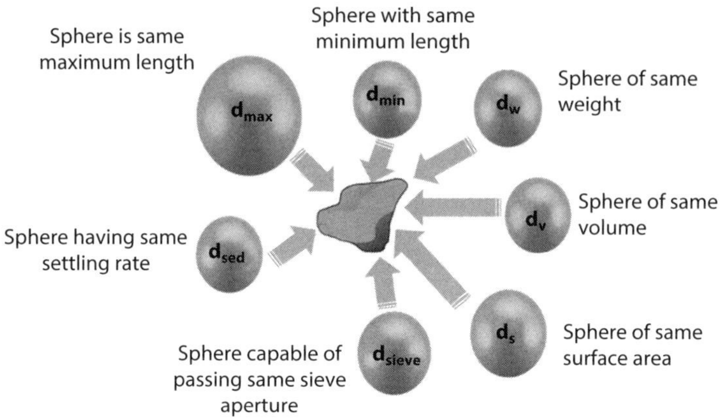

Figure 1.1 Examples of particle size definitions.

The concept of equivalent spheres is useful in terms of a convenient metric for the characterization of particles, however, surface area is more relevant to the subject of this book. Particle-gas interfacial area is a critical property of a two-phase gas-solid system that has direct relevance to the property of ignition as we will see from later discussions. For now, however, we shall continue to dwell on the definitions of particle size.

While the concept of equivalent sphere is reasonable for regular shaped particles, it may not always be appropriate for irregular shaped particles, such as needles or plate-like particulates, where the size in at least one dimension can differ significantly from that of the other dimensions. See figure 1.2 as an example.

Figure 1.2 illustrates a rod shaped particle for which a volume equivalent sphere would give a particle diameter of 198μm. This is not an accurate description of the particle's true dimensions. One option then is to define the particle as a cylinder with the same volume which has a length of 360μm and a width of 120μm. This definition more accurately describes the size of the particle and may provide a better understanding of the behavior of this particle during processing or handling.

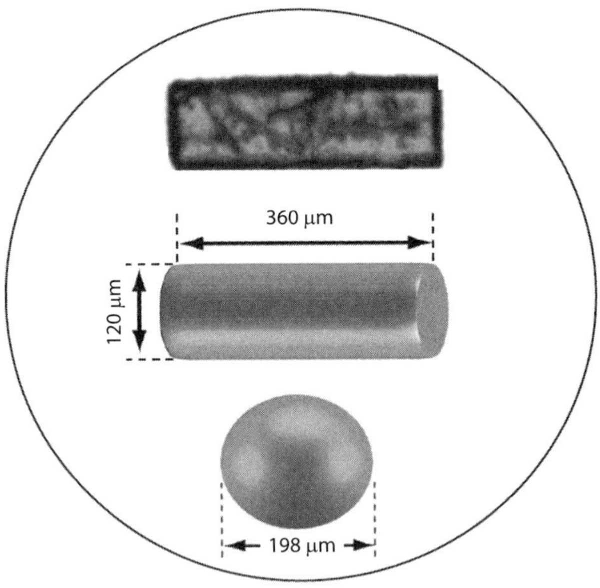

Figure 1.2 Example of volume equivalent rod and sphere of a needle-shaped particle (Source: after A Basic Guide to Particle Characterization, Malvern Instruments Worldwide, 2012 Malvern Instruments Ltd., wwwmalvern.com).

1.4 Size Distribution

Let us consider what we really mean by the term *particle size*. The term alone refers to a single metric or measurement. But is this truly an accurate way to describe dust?

To answer this we really must consider dust to be comprised of a particle system which is made up of many particles. Consider a special case where all of the particles making up the particle system have the same or almost same particle size. In this example the particle system is defined as being Monodisperse. However, in a case where the particle system is made up of particles different in size, the system is defined as Polydisperse. It is the size of the particle system or particle diameter distribution which reflects the regularity or irregularity of the sizes of all the particles.

The term *particle size distribution* actually refers to an index, which is a means of expression indicating the sizes of particles that are present in specified proportions – i.e., the relative particle amount is expressed as a percentage where the total amount of particles is 100 % in the sample particle group measured. Volume, area, length, and quantity are used as standards (i.e., metrics) to define particle amount.

The term *frequency distribution* is applied to define in percentage the amounts of particles existing in respective particle size intervals after the range of target particle sizes is divided into separate intervals or bins. One may further characterize particle sizes by cumulative distribution (for particles passing a sieve size) whereby we can express the percentage of the amounts of particles of a specific particle size or below a size. Alternatively, cumulative distribution (for particles remaining on the sieve) expresses the percentage of the amounts of particles of a specific particle size or above.

The concept of *particle size distribution* depends very much on the *particle size* definition used. The shape of almost all particles cannot be simply and quantitatively expressed as spheres. Rather, particles are complex systems comprised of irregular shapes, and in some instances are not individual particle entities but made up of aggregates. This is why the indirect definition of a *sphere-equivalent diameter* is most useful. Under this definition, when a certain particle is measured based on a certain principle of measurement, the particle size of the measured particle is expressed by the diameter of a spherical body that displays the same result (i.e. measurement quantity or pattern). Referring back to figure 1.1 as an example, consider an equivalent sphere definition based on particle settling rate. In this example the particle size is based on a measurement method known as the *precipitation method*. Here the particle size of the particle (i.e., its diameter) is calculated assuming a sphere having the same settling velocity and density as a sphere is the actual particle. However, if another measurement technique such as the *laser diffraction/scattering method* is used – i.e., a method in which the particle size of the particle to be measured assumes the same diffracted/scattered light pattern as a 1 μm-diameter sphere is 1 μm regardless of the shape of the particle – then we have a very different size definition and hence a very different size distribution.

If the principle of measurement differs, the definition of particle size, in other words, the scale itself used as the measurement standard differs. In which case, completely different measurement results will be obtained even if the term "particle size distribution" is the same. Accordingly, we really have no choice but to consider the principle of measurement itself to be a scale or a standard. For this reason, it is meaningless to scientifically rank precision or accuracy when comparing various principles of measurement.

In selecting the principle of measurement or an analyzer, one must clearly understand and state the objectives. Only by doing so can the properties and specifications of the analyzer (e.g. measuring range, resolution and sample state during measurement) be determined as appropriate for a specific measurement. Noting these considerations let's turn our attention back to the definitions of distribution.

Unless the particle network is mono disperse, i.e. every single particle has exactly the same dimensions, there must exist a statistical distribution of particles of different sizes. It is common practice to represent this distribution in the form of either a frequency distribution curve, or a cumulative (undersize) distribution curve. There are several terms that are used to define distributions:

- Weighted distributions
- Number weighted distributions
- Volume weighted distributions
- Intensity weighted distributions

1.4.1 Weighted Distributions

A particle size distribution can be represented in different ways with respect to the weighting of individual particles. The weighting mechanism will depend upon the measuring principle applied. In all cases, the statistical representation of the size distribution is based on a weighted basis.

1.4.2 Number Weighted Distributions

A counting technique such as image analysis will provide a numerical weighted distribution where each particle is given equal weight irrespective of its size. This is most often useful when knowing the absolute number of particles is important, in foreign particle detection for example, or where high resolution (particle by particle) is required.

1.4.3 Volume Weighted Distributions

Static light scattering techniques such as laser diffraction provide a volume weighted distribution. Here the contribution of each particle in the distribution relates to the volume of that particle (equivalent to mass if the density is uniform), i.e. the relative contribution will be proportional to (size). This is often extremely useful from a commercial perspective as the distribution represents the composition of the sample in terms of its volume/mass, and is sometimes related to dollar value for each size fraction.

1.4.4 Intensity Weighted Distributions

Dynamic light scattering techniques provide an intensity weighted distribution, where the contribution of each particle in the distribution relates to the intensity of light scattered by the particle. For example, using a technique known as the Rayleigh Approximation, the relative contribution for very small particles will be proportional to (size).

When comparing particle size data for the same sample measured by different techniques, it is important to recognize that the types of distribution being measured and reported can produce very different particle size results.

Let us now consider the statistical parameters defining the size distribution.

1.4.5 Size Distribution Statistics

Statistics is defined as the study of how to collect, organize, analyze, and interpret numerical information from data. This is by no means a simple topic and one where there are significant levels of interpretation and disagreement among engineers, scientists and mathematicians.

Descriptive statistics concerns methods of organizing, picturing and summarizing information from data.

Inferential statistics involves methods of using information from a sample to draw conclusions about the data population. We must always be cognizant of the fact that statistical inferences are no more accurate than the data they are based on , i.e., the weakest link in an analysis.

An important key to applying statistical definitions is the understanding of the variables which define a parameter of interest. A variable is the characteristic of the individual to be measured or observed, in this case size. Taking a more general view, if we wanted to do a study about the people who have climbed a particular tall mountain, then the individuals in the study would be the actual people who made it to the top. The variables to measure or record observations about might be the height, weight, race, gender, income, etc. of the individuals that made it to the top of the mountain.

Variables can fall into two general categories, quantitative and qualitative. A quantitative variable has a value or numerical measurement for which operations such as addition or averaging make sense. This is suitable for sizes. In contrast, a qualitative variable describes an individual by placing the individual into a category or group such as male or female, or perhaps irregular and needle-like particles found in a particle population.

We may also describe a population in terms of *Nominal Level* (in name only). These are *qualities* of a population with no ranking/ordering and no numerical or quantitative value. Data generally consist of names, labels and categories.

We may also describe qualities of a population in terms of *Ordinal Level*, i.e., data may be arranged in some order, but the differences between the data values are meaningless. For example, of 20 particle size samples taken for analysis, 15 were rated good quality, 4 were rated better quality, and 1 was rated best quality.

The term *Interval Level* also has relevance. Data values can be ranked and the differences between data values are meaningful. However, there is no intrinsic zero, or starting point, and the ratio of data values are meaningless. Some simple examples are calendar dates and celsius & fahrenheit temperature readings have no meaningful zero and ratios are therefore meaningless. And so it goes with particle sizes, e.g., the smallest particle bin size in an analysis was 0 to 2.5μm. The lowest value is not absolute but rather established by the lower limit of resolution of a measurement technique or more often, an arbitrary definition applied by the investigator.

There are also the following *Levels of Measurement* that we should be cognizant of:

- Nominal Level (in name only): Refers to qualities with no ranking/ordering, no numerical or quantitative value.
- Ordinal Level: These qualities can be arranged in some numerical order, but the differences between the data values are meaningless.
- Interval Level: Data values can be ranked and the differences between data values are meaningful. However, there is no intrinsic zero, or starting point, and the ratio of data values are meaningless.
- Ratio Level: A term that is similar to interval, except there is an inherent zero, or starting point, and the ratios of data values have meaning.

To simplify the interpretation of particle size distribution data, a choice of statistical parameters can be calculated and used for reporting and analyses purposes. The choice of the most appropriate statistical parameter for any given sample will depend upon how that data will be applied and what it will be compared against. For example, if we wanted to report the most common particle size in a sample population, we could choose between the following parameters:

- A mean, representing the 'average' size of the population
- The median, representing the size where 50% of the population is below or above the midsize particle size
- The mode, representing the size with the highest frequency

If the shape of the particle size distribution is asymmetric, as is often the case, we would not expect these three values to be exactly equivalent.

The most often relied on reference value for a particle size distribution is the D_{p50}, also known as the median diameter or the medium value of the particle size distribution. It represents the value of the particle diameter at 50% of the cumulative distribution. See as an example, figure 1.3 which reports the particle size distribution for a sample of beach sand. The graph shows the cumulative particle size distribution graph (curve), where the ordinate represents the cumulative particle size distribution from 0% to 100%, and the abscissa represents the particle size.

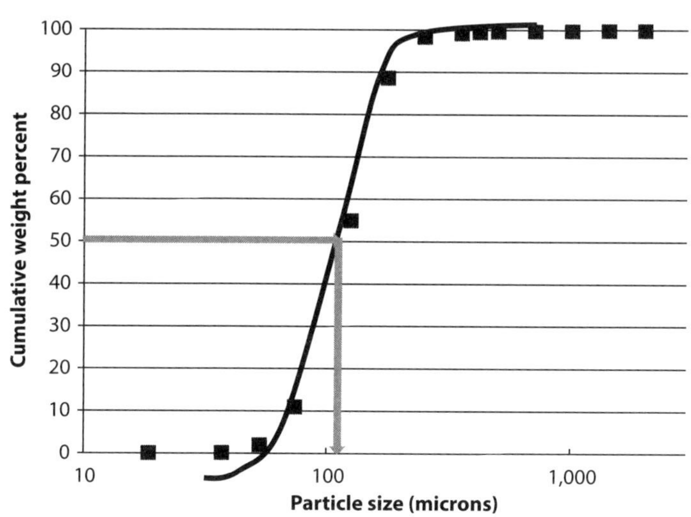

Figure 1.3 Cumulative particle size distribution for a sample of beach sand.

Located on the graph is the cumulative distribution 50% on the ordinate, along with its corresponding particle size value in ordinate (i.e., the D_{p50}). In this example, the D_{p50} value is about 125 μm. In other words, 50% of the particles in the sample are larger than 125 μm, and 50% smaller than 125 μm. D_{p50} is usually used to represent the particle size of group of particles.

One should think of a particle size distribution as the number of particles that fall into each of the various size ranges given as a percentage of the total number of all sizes in the sample of interest.

D_{p50} may also be expressed as D_{v50} (particle size represented by volume), D_{w50} (particle size represented by weigh or mass), and D_{n50} (particle size represented by number of particles). The specific definition depends on convention, the investigator or the application to a specific engineering calculation.

1.5 Why Some Dusts are Combustible

The National Fire Protection Association (NFPA) defines a *combustible dust* as "a combustible particulate solid that presents a fire or deflagration hazard when suspended in air or some other oxidizing medium over a range of concentrations, regardless of particle size or shape."[3]

As a general rule of thumb, combustible particulates having an effective diameter of 420 μm or smaller, as determined by passing through a U.S. No. 40 Standard Sieve, are generally considered to be combustible dusts. However, agglomerates of combustible materials that have lengths that are large compared to their diameter (and will not usually pass through a 420 μm sieve) can still pose a deflagration hazard. Therefore, any particle that has a surface area to volume ratio greater than that of a 420 μm diameter sphere should also be considered a combustible dust according to the NFPA definition.

The vast majority of natural and synthetic organic materials, as well as some metals, can form combustible dusts. The NFPA's Industrial Fire Hazards Handbook states, "any industrial process that reduces a combustible material and some normally noncombustible materials to a finely divided state presents a potential for a serious fire or explosion."

[3] The 2013 *NFPA 654: Standard for the Prevention of Fire and Dust Explosions from the Manufacturing, Processing, and Handling of Combustible Particulate Solids*

Examples of natural and synthetic organic materials that can form combustible dusts include:

- Food products (e.g., grain, cellulose, powdered milk, sugar, flour, starch, cocoa, maltodextrin)
- Pharmaceuticals (e.g., vitamins; cosmetic powders)
- Wood (e.g., wood dust, wood flour)
- Textiles (e.g., cotton dust, nylon dust)
- Plastics (e.g., phenolics, polypropylene)
- Resins (e.g., lacquer, phenol-formaldehyde)
- Biosolids (dried wastes from sewage treatment plants)
- Coal and other carbon dusts

Examples of inorganic materials and metals that can form combustible dusts include:

- Aluminum
- Iron
- Magnesium powder
- Manganese
- Sulfur

There are a myriad of industries where combustible dust explosion hazards exist; examples include:

- Agriculture – pesticides, herbicides, fertilizer manufacturing
- Chemicals manufacturing
- Food (e.g., candy, sugar, spice, starch, flour, feed)
- Grain elevators, bins and silos
- Tobacco
- Plastics
- Wood processing and storage
- Furniture
- Paper
- Tire and rubber manufacturing
- Textiles
- Pharmaceuticals
- Metal powder processing or storage (especially magnesium and aluminum)

While there are unique challenges in each of these industry sectors, there are many common threads that all sectors should follow in order to mitigate risks of dust explosions and fires.

1.6 Common Causes of Dust Explosions and Risk Mitigation

1.6.1 General

Dust explosions occur when combustible dust is airborne in a room or inside equipment, oxygen is present and there is a source of ignition. The confinement of the room, vessel or the internals of a piece of equipment causes pressure to build up as the dust burns, leading to an explosion. The general concept of a dust-related explosion is often represented by the so-called Dust Explosion Pentagon, illustrated in figure 1.4.

As most often reported in an industrial dust explosion incident, an initial (primary) explosion raises settled or bulk dust into the air producing a second, larger explosion that may follow the first. These can lead to deadly explosions with such great force as to cause great physical destruction to the surroundings, injuries and death.

The dual explosion scenario is illustrated in figure 1.5. The primary explosion (deflagration) results in the formation of a dust cloud. The heat from the primary explosion may provide sufficient energy as an ignition source to ignite the dust cloud resulting in the secondary explosion.

Dust explosions constitute a high risk to many industry sectors. The degree of risk is often overlooked and inadequate attention is paid to defining risks and mitigating them. In 2006 the U.S. Chemical Safety and

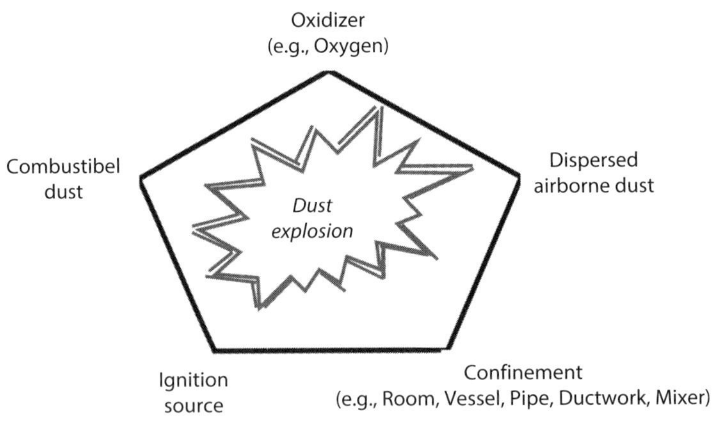

Figure 1.4 The Dust Explosion Pentagon.

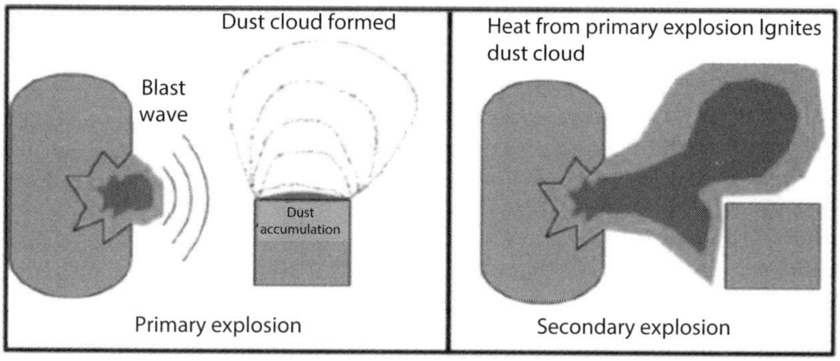

Figure 1.5 Primary and secondary explosions occur according to OSHA (source: www.osha.gov).

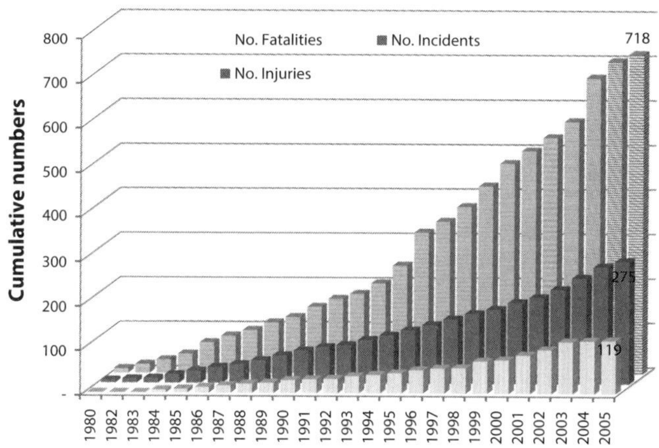

Figure 1.6 Cumulative numbers of dust explosion incidents, numbers of injuries and number of fatalities from dust explosions (source: Appendix A, U.S. Chemical Safety and Hazard Investigation Board, Investigation Report: Combustible Dust Hazard Study, Report No. 2006-H-1, November 2006).

Hazard Investigation Board published an extensive study[4] on combustible dust hazards. The study identified numerous dust explosions between 1980 and 2005 across many industry sectors. Figure 1.6 and table 1.2 summarize some of their findings. The board's research reported at least 281 combustible dust fires and explosions occurred in general industry between

[4] U.S. Chemical Safety and Hazard Investigation Board, Investigation Report: Combustible Dust Hazard Study, Report No. 2006-H-1, November 2006

Table 1.2 Industry demographics of dust explosions (source: Appendix A, U.S. Chemical Safety and Hazard Investigation Board, Investigation Report: Combustible Dust Hazard Study, Report No. 2006-H-1, November 2006).

Year	Food Products	Chemicals Mfg.	Furniture, Lumber, Wood Products	Primary Metals Industry	Electrical Services	Fabricated Metal Products	Equipment Mfg.	Misc.
1980	6	–	–					–
1982	1	2	–					–
1983		–	2					–
1984	1	1	1	2	1		1	1
1985	1	1	2		2			2
1986	3	2	2		1			1
1987	1	1	1	1		2		–
1988	3	2	2	1			1	–
1989	6	1	–	1	1	2		–
1990	2	1	1	4	1			2
1991	2	3	2				1	–
1992	1	–	2	1			1	–
1993	3	3	2			1		2

Year	Food Products	Chemicals Mfg.	Furniture, Lumber, Wood Products	Primary Metals Industry	Electrical Services	Fabricated Metal Products	Equipment Mfg.	Misc.
1994	3	1	3		1		1	1
1995	4	2	1	1		2	1	1
1996	2	1	1	3	1		3	1
1997	5	3	-			3	1	1
1998	3	1	4		1	1	2	1
1999		3	2	2	2	1		-
2000	4	4	2	2	1	2		-
2001	4	3	3		1	1		-
2002	4	6	3	2	2			1
2003	2	6	10	1	2	4	1	-
2004	4	6	6	2	4		2	1
2005		5	4				2	1
Totals	**65**	**58**	**56**	**23**	**21**	**19**	**17**	**16**

1980 and 2005, causing at least 119 fatalities and 718 injuries in the United States. These statistics include seven catastrophic dust explosions within the past decade alone, involving multiple fatalities and significant community economic impacts.

One of the critical findings of the board was that no Occupational Safety and Health Administration (OSHA) standard comprehensively addresses combustible dust explosion hazards in general industry. They note that OSHA's Grain Facilities Standard has successfully reduced the risk of dust explosions in the grain industry, but many of these practices have not been adopted across other industry sectors. Secondary dust explosions, due to inadequate housekeeping and excessive dust accumulations, caused much of the damage and casualties in the more recent catastrophic incidents.

Among some of the disturbing findings of the board is that while consensus standards developed by the National Fire Protection Association (NFPA) provide detailed guidance for preventing and mitigating dust fires and explosions and are widely considered to be effective, these standards are voluntary unless adopted as part of a fire code by a state or local jurisdiction. As such, the NFPA standards have not been adopted in many states and local jurisdictions or have been modified inconsistently. The board also noted that the OSHA Hazard Communication Standard (HCS) inadequately addresses dust explosion hazards, or safe work practices and guidance documents, in Material Safety Data Sheets (MSDSs). Also, training programs for OSHA compliance officers and fire code inspectors generally do not address recognizing combustible dust hazards. Even more disturbing statements by the board are that 41 % of the 140 combustible powder MSDSs the CSB surveyed did not warn users about explosion hazards, and only 7 referenced appropriate NFPA dust standards to prevent dust explosions.

1.6.2 Explosion Hazard Zones Classification

Hazard zones are defined so appropriate explosion protection measures can be implemented. For the European Community (EC), zone definitions are described in Directive 1999/92/EC. Explosion hazard zones are classified depending on the frequency and duration of the potentially explosive atmosphere. This classification provides the scope of measures to be taken according to Annex II Section A in Directive 1999/92/EC in conjunction with Annex I of the Directive 94/9/EC.

In the workplace, the explosion hazard areas are normally classified as zone 21 and 22. Zone 20 is restricted to very small and inaccessible areas in workplaces, or the inside of technical equipment. The following are the definitions applied by the EC:

- Zone 20 - A place in which an explosive atmosphere in the form of a cloud of combustible dust in air is present continuously, or for long periods or frequently.
- Zone 21 - A place in which an explosive atmosphere in the form of a cloud of combustible dust in air is likely to occur in normal operation occasionally.
- Zone 22 - place in which an explosive atmosphere in the form of a cloud of combustible dust in air is not likely to occur in normal operation but, if it does occur, will persist for a short period only.

Layers, deposits and heaps of combustibles must be considered as any other source that can form an explosive atmosphere. The term 'normal operation' means the situation when installations are used within their design parameters.

1.7 Closing Remarks and Definitions

In the chapters that follow, we shall explore in greater detail the causes of dust explosions and ways to reduce the risks of fires and explosions. It is important to note that the greatest authority in the world on dust explosions is the National Fire Protection Association (NFPA). This volume is only a primer and intended to provide basic information and general awareness of a complex industry problem for which not enough attention and diligence is paid to. The NFPA references cited in this volume should be sought for technical details and practical guidance.

In the chapters to follow, we shall examine recognized industry best practices for the risk management for dusts, fumes and fibers. Operations, which have the potential to form explosive dusts, need to apply diligence in the identification of the hazards. This includes but is not limited to applying proven methods and practices that identify dust, fume and fibre hazards including:

- Reviewing chemical/material specific information such as material safety data sheets (MSDS) and product information sheets
- Performing regular workplace inspections

- Consulting with employees who are performing operations and who have firsthand knowledge of dust formation and operating conditions that lead to dusty situations
- Reviewing past incidents, health concerns, hazard reports and learning lessons from these. This includes the careful examination and consideration of industry association information, e.g., newsletters, websites; incident reports from WorkSafe or other work health and safety regulators.
- Conducting specialist audits

Because lives and assets are at risk from dust explosion situations, those responsible for safety within an organization need to assess the risks that are specific to the operations and jobs performed by employees. For health risks, one may refer to the *Guidance Note for the Assessment of Health Risks arising from The Use of Hazardous Substances in the Workplace, NOHSC: 3127 (1994)* available at www.ascc.gov.au, as a starting place for basic information.

Risk assessments include a review of information and MSDS, and a review of current practices. These actions will help identify problem or high-risk areas. For explosion risks, consider whether the elements of the dust fire pentagon could occur and what damage or injury would result in the event of an explosion. Consult with employees on risk assessments, and keep a record of the assessments. Complex risk assessments should be done by, or with the assistance of a competent person such as an occupational hygienist and/or a fire prevention specialist.

Risk mitigation means the practices and technologies that are intended to control the hazard. The general hierarchy of controls (*listed from most to least effective*) are:

- Elimination
- Substitution
- Isolation
- Engineering (preferably at source)
- Administrative
- Personal protective equipment

Example of specific controls include:

- Separation of hazardous processes from other work areas
- Ventilation systems (especially local exhaust ventilation near the source)

- Use of dust collection equipment
- Placing vents on equipment where a dust explosion could occur
- Use of dust extraction on hand held tools
- Separation of heat and ignition sources from combustible dusts
- Using spark detection systems
- Using wet or damp work methods
- Adopting an aggressive cleaning program (including areas where dust may be unseen) –i.e., good housekeeping
- Using vacuum (using high efficiency "HEPA" filters) or wet cleaning instead of dry sweeping or compressed air cleaning
- Employee training and supervision
- Supply and use of appropriate personal protective equipment

Any controls that are relied on should be well maintained and regularly inspected to make sure they keep working well.

Good industry practices include having systems and equipment in place to control an emergency, which include:

- Monitoring and health surveillance - Air monitoring can help an organization determine whether controls are working properly, or if hazardous dust/fume/fibre levels are present. Air monitoring may also be used as part of risk assessments.
- In addition, health surveillance of employees is required for some hazardous dusts, fumes and fibers. Where there is a health risk and where it is possible to identify specific health outcomes special risk mitigation practices should be applied. Health surveillance is not a control, however in some cases it can help to identify problems early and improve controls.

Table 1.3 provides definitions of technical safety parameters of combustible dusts.

The following are some general resources that the reader can obtain more information from:

1. Cancer classifications: International Agency for Research on Cancer, *http://monographs.iarc.frCode of practice for the management and control of asbestos in workplaces [NOHSC: 2018(2005)]*,available at www.ascc.gov.au under Publications.

Table 1.3 General Technical Safety Parameters of Dusts.

Parameter	Definition/Description
Size of Particle	Dust particles larger than 400 μm are typically not combustible. Dust particles are combustible when they measure less than 400 μm and up to 20 μm. It is important to note that the transportation and processing of large-particle dust materials often results in abrasion, producing finer particles.
Combustible Dust Concentrations	As with gases, dust is combustible within certain concentration parameters: Lower combustibility limit: approx. 20–60 g/m^3 air; Upper combustibility limit: approx. 2–6 kg/m^3 air. These parameters vary widely across the spectrum. Highly combustible dusts can form a flammable mix with less than 15 g/m^3.
Maximum Explosive Pressure	In simple closed containers, flammable dusts can cause an explosive pressure between 6 and 10 bar. In exceptional cases, such as with light metal dusts, explosive pressure of up to 20 bar is possible.
K_{St}-Value	This is a classifying parameter that describes the volatility of the combustion. It equals the figure for the maximum speed of pressure build-up during the explosion of a dust/air mix in a container measuring 1 m^3.
Moisture	The moisture of a dust is an important factor for potential ignitions and explosions. Although no exact parameters exist, it is known that dust containing higher amounts of moisture require a higher ignition energy and are less likely to be swirled up.
Minimum Ignition Energy	The minimum energy of an electrical spark, which, under defined conditions, is able to ignite the explosive dust/air mix. Not every spark is capable of causing ignition. The decisive factor is whether sufficient energy is introduced into the dust/air mix to trigger an independent combustion of the entire mix. A modified Hartmann tube is used to determine the minimum ignition energy.
Ignition Temperature	The lowest temperature of a heated wall that ignites the dust/air mix upon brief contact. The shape of the vessel in which the ignition temperature is measured has proven to be especially critical. It may be assumed that ignition on differently shaped surfaces is, in practice, only possible at much higher temperatures. In the case of dust from food products and animal feed, this figure is between 410 and 500 °C depending on type.

Parameter	Definition/Description
Smoldering Temperature	The lowest temperature of a hot surface on which a 5 mm dust deposit is ignited. The smoldering temperature describes the ignition characteristics of thin dust layers. If the layer is thicker, or if the ignition source is completely buried in dust, the thermal insulation provided by the dust layer increases, which changes the smoldering temperature entirely, sometimes lowering it considerably, which could trigger an exothermic reaction. Experiments have shown that the smoldering temperature decreases nearly linearly as the thickness increases. Smoldering temperature is sometimes considerably lower than ignition temperature for an airborne mix of the same dust. The estimated maximum permissible surface temperature for electrical equipment may be higher, depending on the dust's thermal conductivity. Unnoticed smolder spots can be present for long periods in thick layers of dust and can, if the dust is swirled up, become effective ignition sources.

2. Controlling wood dust hazards at work http://www.commerce.wa.gov.au/WorkSafe/PDF/Guidance_notes/Guide_wood_dust.pdf
3. Flour dust http://www.commerce.wa.gov.au/WorkSafe/PDF/Factsheets/flour_dust.pdf
4. Grain movement and storage: http://www.commerce.wa.gov.au/WorkSafe/Content/Industries/Agriculture,_forestry_and_fish/Further_information_/Agriculture_workbook/Grain_movement_and_storage.html
5. OSHA, 2005, 'Combustible Dust in Industry: Preventing and Mitigating the Effects of Fire and Explosions'. (*Includes description of major explosions in the US and detailed assessment and control strategies*). http://www.osha.gov/dts/shib/shib073105.html
6. OSHA, 2008 Hazard Alert: Combustible Dust Explosions http://www.osha.gov/OshDoc/data_General_Facts/OSHAcombustibledust.pdf
7. Welding: http://www.commerce.wa.gov.au/WorkSafe/Content/Safety_Topics/Plant_and_machines/Additional_in formation/Welding.html

8. Welding Institute of Australia: *Technical Note 7: Health and Safety in Welding*; available for purchase from http://www.wtia.com.au/catalog.htm
9. Working with fiberglass http://www.commerce.wa.gov.au/WorkSafe/Content/Safety_Topics/Hazardous_substances/Additional _resources/Working_with_fibreglass.html

The following are some general definitions that supplement dust particle characterizations.

Equivalent circle diameter - Diameter of circle with equivalent projected area as a particle. See example below:

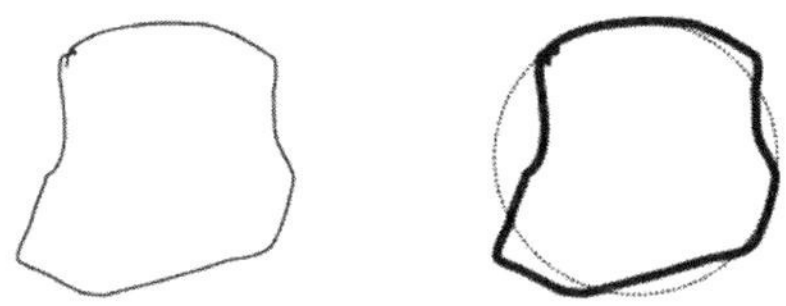

Enclosing circle diameter - Diameter of circle containing projected area. See example below:

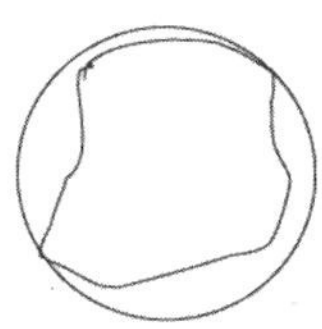

Frequency distributions, Cumulative frequency distribution – Defining F_N as the fraction of number of particles with diameter (F_v for volume, F_m for mass, F_s for surface area) less than or equal to a given diameter. One can obtain the cumulative frequency distribution from discrete data. The derivative of cumulative frequency distribution with respect to particle diameter is equal to the differential frequency distribution. Differential frequency distribution is a normalized particle size distribution function:

$$d\,F_N/d(dp) = f_N(dp) = \frac{1}{N}\frac{dN}{d(dp)}$$

$$d\,F_N/d(dp) = f_N(dp) = \frac{1}{V}\frac{dV}{d(dp)} = \frac{1}{M}\frac{dM}{d(dp)}$$

The following is an example of a cumulative frequency distribution:

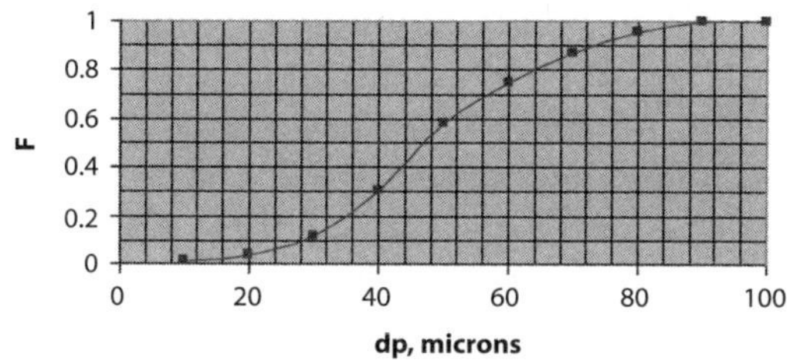

Martin's diameter - Length of line bisecting projected area (a given particle could have a range). See example below:

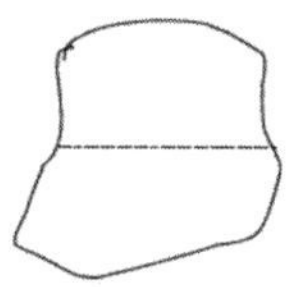

Particle concentration - There are different ways to describe concentration. Low concentrations of suspended particles: usually number, mass or volume concentrations are used. Number concentration = number of particles/ unit volume of gas. See illustrations below for concepts.

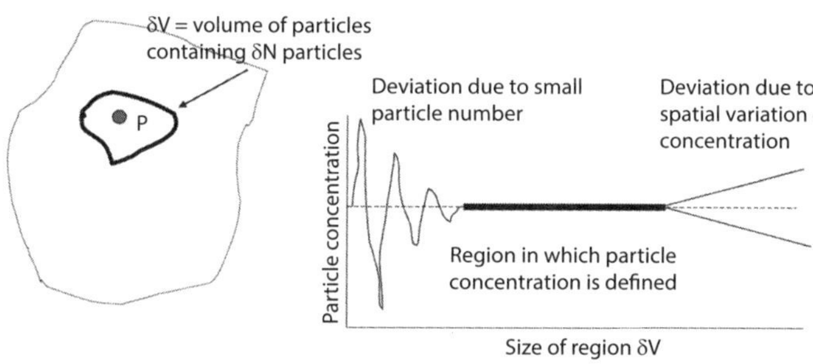

Along this same discussion we consider *Mass* and *Volume Concentrations*. Mass concentration refers to particle mass per unit volume of gas. Volume concentration refers to particle volume per unit volume of gas. Volume concentration can be related to ppm by volume, or it can be dimensionless.

Additional definitions are necessary, especially when considering powders.

- Bed or bulk density - mass of particles in a bed or other sample volume occupied by particles and voids between them.

- Tap density - density after being "packed", mass/volume.
- Void fraction - volume of voids between particles volume occupied by particles and voids between them.

Particle size- equivalent diameters – There are several definitions that one may apply. Examples of equivalent diameter definitions are as follows:

- *Sieve equivalent diameter* – diameter equal to the diameter of a sphere passing through the same sieve aperture as a particle.
- *Surface area equivalent diameter* – diameter equal to the diameter of a sphere with the same surface area as a particle.
- *Aerodynamic diameter* – diameter of a unit density sphere having the same terminal settling velocity as the particle being measured. This diameter is very important for describing particle motion in impactors and cyclone separators. However, in shear flows, describing the motion of irregular particles is a complex problem and it may not be possible to describe their motion by modeling their aerodynamic spherical equivalents.
- *Volume diameter* – diameter of a sphere having the same volume. Obtained from Coulter counter techniques.
- *Surface volume diameter* – diameter of sphere having same surface to volume ratio. Obtained from permeametry (measuring pressure drop with flow through a packed bed).
- *Mobility diameter* – diameter equal to the diameter of a sphere having the same mobility in an electric field as a particle.

Shear diameter – Defines how far one must move the particle so that it is not overlapping its former position. See example below:

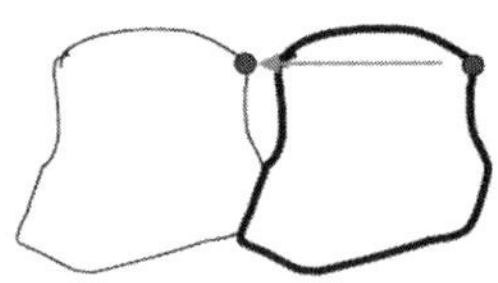

Radius of Gyration - The Radius of Gyration of an area about a given axis is a distance k from the axis. At this distance k an equivalent area is thought of as a line area parallel to the original axis. The moment of inertia of this line area about the original axis is unchanged.

2

The Basics of Dust Explosions

2.1 Conditions for Dust Fires and Explosions

2.1.1 Explosion Limits

Eckhoff[1] notes that most solid organic materials, as well as many metals and some nonmetallic inorganic materials, will burn or explode if finely divided and dispersed in sufficient concentrations.

Combustible dusts can be intentionally manufactured powders. Examples include corn starch or aluminum powder coatings. They may also be generated by handling and processing solid combustible materials such as wood and plastic pellets. Work activities such as polishing, grinding, transporting, and shaping many of these materials can produce very small particulates, which become airborne and settle on surfaces, crevices, dust collectors, and various equipment surfaces. When these dusts are disturbed, they can generate potentially explosive dust clouds.

---

[1] Ignitability and Explosibility of Dusts, Table A.1, Appendix A.2 (Eckhoff, 2003).

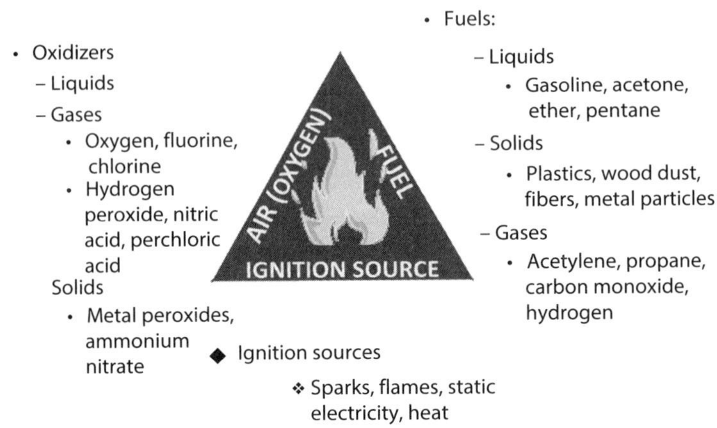

Figure 2.1 The Fire Triangle.

Relatively small amounts of accumulated dust can cause explosions and fires. The Chemical Safety Board[2] has reported that the explosion that devastated a pharmaceutical plant in 2003 and killed six employees was caused by dust accumulations mainly under 0.25 inches deep. The NFPA states in its standard[3] that more than 1/32 of an inch of dust over 5 percent of a room's surface area poses a significant explosion hazard.

Like all fires, a dust fire occurs when fuel (i.e., the combustible dust) is exposed to an ignition source in the presence of an oxidizer such as oxygen in air. Most people are familiar with the fire triangle, depicted in figure 2.1. The removal of any one of the legs of the triangle eliminates the possibility of a fire and explosion. As an aside, it may be impossible to eliminate all sources of ignition in an industrial setting, and as such, removing or controlling the fuel and the oxidizer become critical to risk mitigation.

Another important concept for all fires is illustrated by the slide in figure 2.2. From a plot of vapor pressure (or concentration since concentration is proportional to vapor pressure through Raoult's Law), the reader may recall from basic chemistry that the vapor pressure of an ideal solution is directly dependent on the vapor pressure of each chemical component and the mole fraction of the component present in the solution.

---

[2] U.S. Chemical Safety and Hazard Investigation Board, Investigation Report: Combustible Dust Hazard Study, Report No. 2006-H-1, November 2006

[3] NFPA Standard 654 (2006).

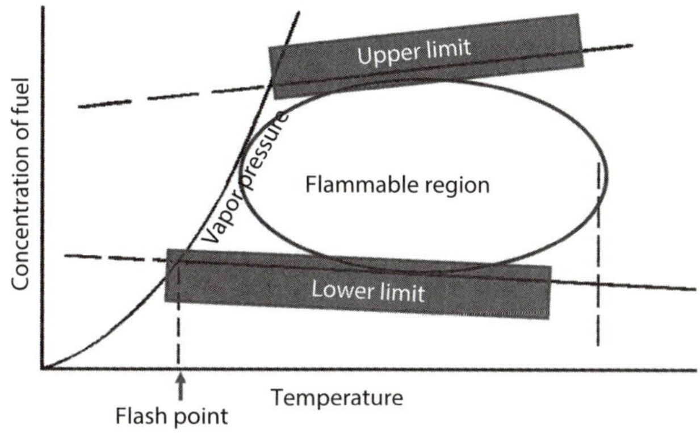

Figure 2.2 Illustration of the concept of flammability.

Before a fire or explosion can occur, three conditions must be met simultaneously. A fuel (ie. combustible gas) and oxygen (air) must exist in certain proportions, along with an ignition source, such as a spark or flame (i.e., the fire triangle, figure 2.1). The ratio of fuel and oxygen that is required varies with each combustible gas or vapor.

The minimum concentration of a particular combustible gas or vapor necessary to support its combustion in air is defined as the lower explosive limit (LEL) for that gas. Below this level, the mixture is too "lean" to burn. The maximum concentration of a gas or vapor that will burn in air is defined as the upper explosive limit (UEL). Above this level, the mixture is too "rich" to burn. The range between the LEL and UEL is known as the flammable range for that gas or vapor.

The term, flash point, refers to the lowest temperature at which a volatile material can vaporize, forming an ignitable mixture in air. Measuring a flash point requires an ignition source. At the flash point, the vapor may cease to burn when the source of ignition is removed.

The flash point is not to be confused with the autoignition temperature, which does not require an ignition source, or the fire point, the temperature at which the vapor continues to burn after being ignited. Neither the flash point nor the fire point is dependent on the temperature of the ignition source, which is much higher.

The flash point is often used as a descriptive characteristic of liquid fuels, and it is also used to help characterize the fire hazards of liquids. "Flash point" refers to both flammable liquids and combustible liquids. There are various standards for defining each term. Liquids with a flash point less than 60.5 or 37.8°C (141 or 100°F), depending upon the standard being applied, are considered flammable, while liquids with a flash point above those temperatures are considered combustible.

These are the standard definitions used to describe fuels but for combustible dusts the explanation of the conditions responsible for fires is a little more complex. A dust explosion requires the simultaneous presence of two additional elements to the ones shown in figure 2.1; namely both dust suspension and confinement are the two additional elements required – thus giving rise to the Dust Explosion Pentagon (figure 1.4 in chapter 1). Suspended dust burns more rapidly and confinement allows for pressure buildup. The removal of either the suspension or the confinement elements prevents an explosion, although a fire may still occur.

In an analogous way of the flammability range commonly used for vapors, the concentration of suspended dust must be within an explosible range in order for an explosion to occur. Dust explosions can be very energetic, creating powerful waves of pressure that can destroy buildings.

When all of the elements of the dust explosion pentagon are in place, rapid combustion known as *deflagration* can occur. Deflagration is defined as a rapid burning slower than the speed of sound. When this occurs within a confined space such as an enclosure (e.g., building, room, vessel or process equipment) the resulting pressure rise can cause an *explosion (a rapid burning faster than the speed of sound).* People caught in dust explosions are often either burned by the intense heat within the burning dust cloud or injured by flying objects or falling structures, or may be thrown great distances.

Important points to remember are the 5 basic ingredients that are required for a dust explosion to occur:

1. Combustible particulates sufficiently small to burn rapidly when ignited
2. A suspended cloud of these combustible particulates at a concentration above the minimum explosible concentration (MEC)
3. Confinement of the dust cloud by an enclosure or partial enclosure
4. Oxygen concentration greater than the limiting oxygen concentration (LOC) for the suspended dust cloud
5. Delayed ignition source of adequate energy or temperature to ignite the suspended cloud

Some important NFPA definitions to bear in mind:

NFPA 654 - "A combustible particulate solid that presents a fire or deflagration hazard when suspended in air or some other oxidizing medium over a range of concentrations, regardless of particle size or shape."

"Any finely divided solid material that is 420 microns or smaller in diameter (material passing a U.S. No. 40 Standard Sieve) and presents a fire or explosion hazard when dispersed in air."

Many combustible fiber segments, flat platelets, and agglomerates do not readily pass through a No. 40 sieve, but they can be dispersed to form a combustible dust cloud.

2.1.1.1 Minimum Explosible Concentration

MEC values are determined in the U.S. per the ASTM E1515 test procedure involving tests with various dust concentrations and a pyrotechnic igniter in a 20-liter sphere. The MEC corresponds to the smallest concentration that produces a pressure at least twice as large as the initial pressure at ignition. MEC values are not very sensitive to particle diameter for diameters less than about 60 μm, but increase significantly with increasing diameter above this approximate threshold.

The majority of the materials known to cause explosions have MEC values in the range 30 to 125 g/m^3. These concentrations are sufficiently high that a 2 m thick cloud can prevent seeing a 25 watt bulb on the other side of the cloud.

2.1.1.2 Confinement

The confinement needed for a dust explosion is usually from the process equipment or storage vessel for the powder or dust. In the case of fugitive dust released from equipment and containers, the room or building itself can represent the confinement.

Often, the dust cloud occupies only a fraction of the equipment or building volume. The resulting explosion hazard is called a *partial volume deflagration hazard.*

Pressures produced from partial volume deflagrations and the corresponding deflagration venting design bases are described in NFPA 68. Example applications include dust collectors and spray driers.

2.1.2 Ignition Sources

The ignition criteria for a dust explosion can take several forms: e.g., hot temperatures, burning embers and agglomerates, self-heating, impact/friction, electrical equipment and electrostatic discharges.

2.1.2.1 Hot Surfaces

An example of a hot surface is a dust cloud accidentally entering a hot oven or furnace. In a plastics plant a phenolic resin dust explosion incident occurred because a resin dust cloud was generated during the cleaning of fugitive dust from the area around the oven. In an application like this it is important to know the minimum dust cloud oven ignition temperature, which can be determined by oven tests described in ASTM E1491.

2.1.2.2 Burning Embers and Agglomerates

Burning embers and agglomerates are common ignition sources in dust explosions. Smoldering or flaming particulate embers or agglomerates (also called smoldering nests) are produced by frictional heating, e.g. during sanding or cutting, local heating associated with hot work on equipment and ducts containing dust deposits, hot powder accumulations on drier walls and the result of small heat sources, e.g. a portable lamp, accidentally embedded in a particulate pile. The following are some typical examples of dust explosions caused by burning embers and agglomerates:

- A dust explosion incident occurred because of a hot nest generated by some bolts falling into a hammermill used for pharmaceuticals production.
- Ignitions in many reported incidents are due to flaming, rather than smoldering nests/agglomerates.
- Burning agglomerate transport experiments show that glowing agglomerates could be transported large distances through otherwise empty piping with air transport velocities of 10–20 m/s, but the glowing was extinguished rapidly when non-burning dust was added to the flow. The glowing particles were not able to ignite the flowing dust.
- Other tests showed that burning nests did not ignite fine sawdust in the transport duct, but did ignite the sawdust cloud when it reached the filter media dust collector at the end of the duct.

Note that some vendors provide spark/ember detection and extinguishing systems to prevent ignitions by burning agglomerates transported through ducting. Optical detectors sense the radiant energy from the burning embers or agglomerates, and the control module triggers water spray through nozzles situated at an appropriate distance downstream of the detector.

2.1.2.3 Self-Heating

In the case of self-heating, certain particulate materials are prone to self-heating. These conditions can potentially lead to spontaneous ignition. The predominant chemical reaction is low level oxidation. Self-heating is manifested as smoldering in the interior of a large storage pile of particulates or in an accumulated layer in a dryer. If the smoldering particulates in the pile or dryer are subsequently disturbed and exposed to air, the smoldering can evolve into flaming. When the flaming nest or agglomerate is then transported to a hopper or dust collector, it can ignite the suspended dust cloud. Various laboratory tests have been developed to determine self-heating onset temperatures for different sample sizes and configurations. These include particulate basket tests in an isothermal oven, heated air flow tests with a slow rate of air temperature rise, and material in a package test to determine the self-accelerating decomposition temperature. Examples of materials that are prone to self-heating are:

- ABS resin powder
- Activated carbon
- Coal
- Various chemical intermediates
- Freshly manufactured/dried wood chips, anhydrous calcium hypochlorite, and hops
- Organic peroxides and other potentially unstable chemicals can self-heat by exothermic decomposition
- Various agricultural materials, such as bagasse and soybeans, start self-heating by microbiological processes

2.1.2.4 *Impact and Frictional Heating*

Impact and Friction heating can be the result of several different scenarios, common ones being:

- During combustible powder processing and during maintenance/repairs involving cutting and grinding operations;
- Equipment such as grinders, hammermills, and other size reduction equipment are prone to causing sparks and ignitions;
- Blenders with rotating element tip speeds greater than 1 m/s are vulnerable to creating hot spots and high shear conditions;
- Tramp metal stuck in a screw conveyor or a particle classifier is a source of frictional ignition.

The vulnerability of a combustible dust to impact/friction ignition is characterized in terms of the material spark minimum ignition energy

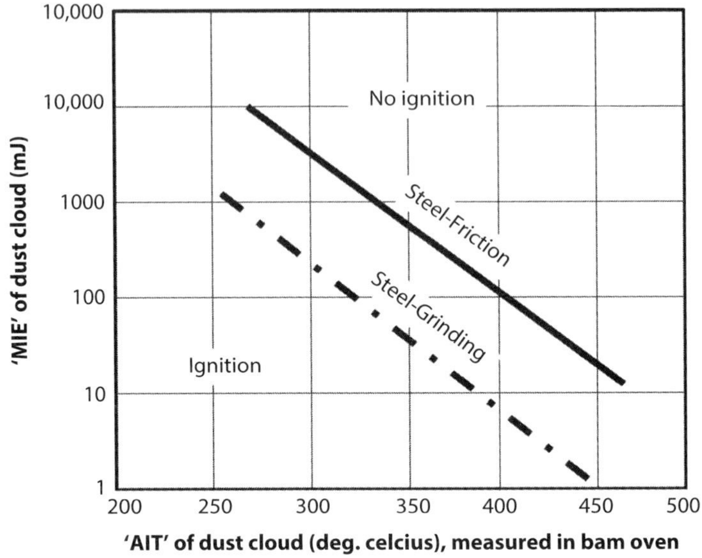

Figure 2.3 Shows generalized relationship between minimum ignition energy (MIE) and cloud auto-ignition temperature (AIT) as per testing by ASTM E2019.

(MIE) and cloud auto-ignition temperature (AIT). Testing to measure MIE values is described in ASTM E2019. Figure 2.3 shows the generalized relationship between MIE and AIT values to determine which dusts can be ignited by impact or frictional contact between steel surfaces. For example, a dust with an MIE of 10 J should be immune to steel-steel frictional or impact ignitions as long as its AIT is greater than 275°C. Dusts with lower MIE values but have larger AIT values still may not be prone to steel impact or frictional ignitions per figure 2.3.

2.1.2.5 *Electrical Equipment*

Electrical equipment and wiring can potentially cause ignition dust clouds by sparks, arcs, or heated surfaces. Dust ignition-proof equipment is enclosed in a manner that excludes dusts and does not permit arcs, sparks, or heat otherwise generated or liberated inside of the enclosure to cause ignition of exterior accumulations or atmospheric suspensions of a specified dust on or in the vicinity of the enclosure. When electrical equipment and wiring is used in locations in which combustible dusts can be present, there is a need to establish the Class II hazardous location classification of the area.

Per NFPA 70, a Class II Division 1 location is one in which combustible dust is in the air *under normal operating conditions* in quantities to produce explosive or ignitable mixtures, or where mechanical failure or abnormal operation of machinery or equipment might cause such explosive or ignatible mixtures to be produced, and might also provide a source of ignition through simultaneous failure of electrical equipment (NFPA 70 definition). The following are possible conditions for the existence of a Class II Division 2 location:

- A location in which combustible dust due to abnormal operations may be present in the air in quantities sufficient to produce explosive or ignitable mixtures.
- Dust accumulations that could be either suspended or ignited during equipment malfunctions or abnormal operations.
- Class II locations are further classified as Group E, F, or G depending on the type of dust material.
- NFPA 499 provides guidance and examples for the assignment of appropriate Class II Division 1 and 2 classifications for combustible powder and dust processing and handling operations.
- NFPA 70 Article 500.7 permits Dust Ignition-proof electrical equipment in Class II Division 1 and 2 areas.
- Intrinsically safe electrical equipment (in which all circuits cannot produce a spark or thermal effect capable of igniting a dust cloud per UL 913) is also allowed.
- Dust-tight equipment is permitted in Class II Division 2 areas. Article 502 of NFPA 70 describes the types of acceptable wiring in Class II Division 1 and 2 locations. Threaded metal conduit together with dust tight boxes and fittings is an acceptable method commonly used. The use of electrical sealing putty at boundaries of Class II areas is also described in Article 502.

2.1.2.6 Electrostatic Discharges

Electrostatic discharges are preceded by charge accumulation on insulated surfaces, ungrounded conductors (including human bodies), or particulate materials with high resistivities. A subsequent electrostatic discharge is only an ignition threat if it is sufficiently energetic in comparison to the minimum ignition energy of the pertinent dust cloud. Various types of electrostatic charges are listed in Table 2.1.

Table 2.1 Types of electrostatic charges.

Type of Discharge	Maximum Energy (mJ)	Typical Examples
Corona	0.1	Wires, Type D Bulk Bags
Brush	1–3	Flexible boots and socks
Bulking Brush	1–10	Piles of powders with resistivities > 10′Ω-m in hopper or silo
Propagating Brush	1000–3000	Boots, plastic pipe or ductwork
Spark	>10,000	Ungrounded conductor - e.g., baghouse cage, a person
Corona	0.1	Wires, Type D Bulk Bags
Brush	1–3	Flexible boots and socks
Bulking Brush	1–10	Piles of powders with resistivities > 109 ′Ω-m in hopper or silo
Propagating Brush	1000–3000	Boots, plastic pipe or ductwork
Spark	>10,000	Ungrounded conductor - e.g., baghouse cage, a person

Since combustible dust MIE values are substantially greater than 0.1 mJ, corona discharges are not an ignition threat. Type D bulk bags are deliberately designed and fabricated to safely dissipate accumulated charges via corona discharges.

Propagating Brush Discharge – This is a special case we need to be aware of. A propagating brush discharge can occur when a charged non-conductor is in direct contact with a conductive surface, such as a metal surface coated with a plastic film or a layer of high resistivity powder. The propagating brush discharge occurs when the surface charge density is sufficiently large to cause electrostatic breakdown at the nonconductor surface. Streamers carry the surface charge to a central region where it intensifies (see figure 2.4). Propagating brush discharges can ignite dusts with MIE values less than about 3 J. Since combustible dust MIE values are substantially greater than 0.1 mJ, corona discharges are not an ignition threat. Type D bulk bags are deliberately designed and fabricated to safely dissipate accumulated charges via corona discharges. But again, propagating brush discharges are a different matter.

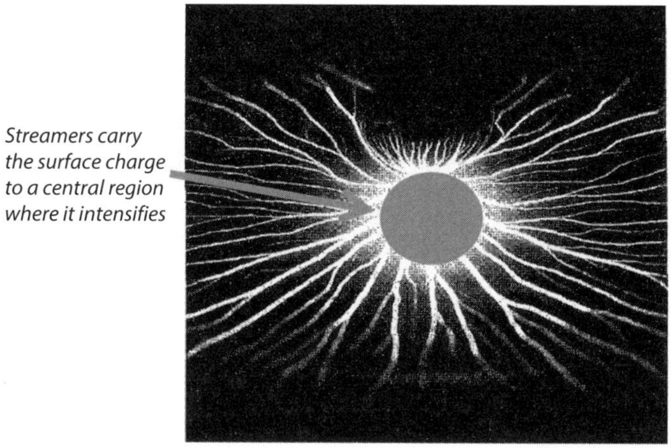

Figure 2.4 Propagating brush discharge.

2.2 Primary and Secondary Dust Explosions

Dust explosions can either be *primary* or *secondary*. A primary dust explosion occurs when a dust suspension within a container, room, or piece of equipment is ignited and explodes.

A secondary explosion occurs when dust accumulated on floors or other surfaces is lofted and ignited by a primary explosion (refer to figures 2.5 and 2.6). The blast wave from the secondary explosion can cause accumulated dust in other areas to become suspended in air, which may generate additional dust explosions. Depending on the extent of the dust deposits, a weak primary explosion may cause very powerful secondary dust explosions. The initiating event for a secondary dust explosion is not necessarily a dust explosion itself. There may very well be a flammable vapor source as an example which can result in the second explosion.

The simple illustration shown in figure 2.6 helps to emphasize the need for good housekeeping practices. The best way to prevent secondary dust explosions is to minimize dust accumulations on surfaces. Good industry practice to prevent secondary dust explosions includes good housekeeping, designing and maintaining equipment thus preventing dust leaks, the application of well-maintained dust collectors, minimizing flat surfaces where dusts may accumulate, and sealing hard-to-clean areas. Along with these practices proper equipment and techniques to clean combustible dust accumulations must be applied. In all situations, diligence is needed to prevent or minimize dust clouds. Hence only vacuum

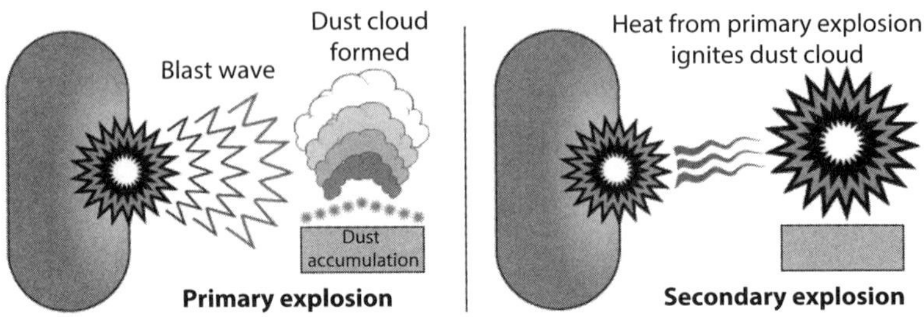

Figure 2.5 Primary and secondary explosion mechanism.

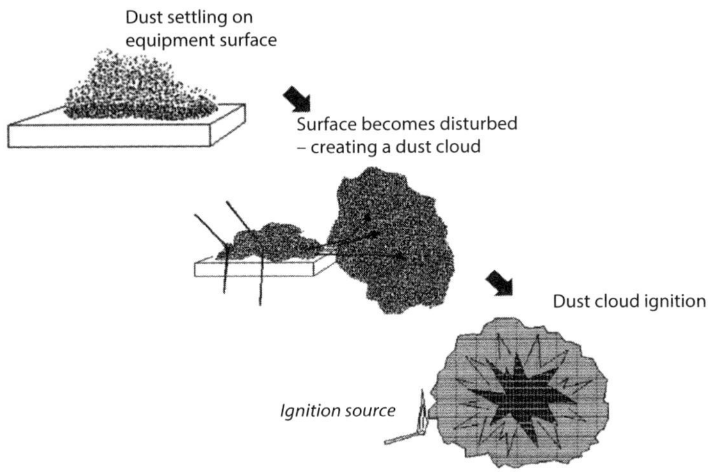

Figure 2.6 Illustrates the sequence of events typically leading to a dust explosion.

cleaning devices that are approved for combustible dust areas should be used.

2.3 Explosions within Process Equipment

Dust explosions can occur in process equipment when there is a particulate concentration between the Minimum Explosible Concentration (MEC) and the Upper Explosive Concentration (UEC) and then an ignition source develops or reaches the combustible cloud. Both the MEC and UEC depend on the oxidant that is present within the enclosure of the equipment.

Some of the types of equipment that are used in dust handling, and which have been involved in dust explosions include:

- bag openers (slitters)
- blenders and mixers
- dryers and ovens
- dust collectors
- pneumatic conveyors
- size reduction equipment (grinders)
- silos and hoppers
- hoses
- loading spouts
- flexible boots

There are many instances reported in the literature over the years concerning explosions in dust collectors. Common explosion scenarios which help to explain the possible causes for these events include the following:

- First we should recognize that these types of equipment are almost omnipresent in all particulate-handling facilities.
- They inherently concentrate the smaller particles, which are easier to ignite than the mostly larger particles in other equipment.
- Dust collectors are often structurally weaker than other process equipment, and therefore more prone to explosion damage. We may expect therefore more situations where secondary explosions may occur.

2.3.1 Baghouse Dust Explosion Case Study

Let's consider a case study involving a baghouse dust explosion. In this example we examine the conditions leading to an explosion event in a baghouse dust collector that was used to collect a pharmaceutical product from a hammer mill/flash drying operation. Figure 2.7 is an illustration of a typical baghouse (fabric filter) unit.

A post incident investigation revealed the following sequence of events which explain the causes of the explosion.

1. The impact hammer mill had been operating for several minutes when the operator heard unusual grinding sounds coming from inside the hammer mill.

Figure 2.7 Typical baghouse unit.

2. Upon heating the grinding sounds, the operator immediately shut down the mill but just as an explosion occurred within the dust collector, located inside the building on the second floor.
3. The pressure wave from the explosion caused the explosion vent (a hinged panel) of the dust collector to open, and the explosion products and unburned powder were directed outside the building via a vent duct. This might have been the end of the story, but it was determined that a screen had been securely fastened at the end of the duct to prevent birds from entering, and as the vent panel swung upward and outward, it struck the screen and opened no further. It is estimated that the screen prevented the explosion vent panel from opening to no more than 50 percent of the vent area.
4. With the vent partially obstructed, the access door to the dust collector failed under pressure and released a dust cloud into the building, which then ignited. The flame front went through the vent duct and followed the dust cloud through the access door, resulting in a fireball at both locations. On the first floor, a fireball was seen exiting the vicinity of the rotary valve outlet at the bottom of the dust collector, which feeds a sifter. There was no secondary explosion on the first or second floors. However, windows were blown out at both floors. The ensuing fire

> in the dust collector engulfed the wool filter bags (which were burned up) and the remaining powder in the collector hopper, but the fire was quickly extinguished by the automatic sprinkler system inside the dust collector.

Upon further investigation, the root cause of the incident was determined to be a carbon steel bolt from the inside of the feeder (which feeds wet powder to the hammer mill/flash dryer) had become loose and fell into the hammer mill. The bolt became trapped inside the 3600 rpm mill, where it became heated from friction to above the autoignition temperature of the powder. The hot metal ignited some of the powder in the mill, which was pneumatically conveyed into the dust collector. In the collector, a dust cloud created by the blow ring (pulse jet on the bag house), was ignited by the hot powder conveyed in from the hammer mill. An inspection of the feeder revealed that six 3/8-inch carbon steel bolts were missing.

The underlying culprit in this case study is a lack of preventive maintenance, both on the hammer mill, and in the vent which was a crucial part of the intended safety device. We can see this as a reoccurring theme in many equipment explosion incidents. Often, preventive maintenance is either lacking or simply given low priority. Careful consideration of the importance of even the simplest of safety devices intended to minimize the risk of an explosion or fire is often not done.

Collective data on dust explosions in the United States, United Kingdom, and Germany indicate that more than one fourth of all dust explosions occur in dust collectors. Within the United States, the ratio is higher at more than 40 percent. Zalosh, *et al.*[4] states that three possible reasons for the high occurrence of dust collector explosions are: 1) they [dust collectors] are almost omnipresent in particulate handling facilities, 2) they inherently concentrate the smaller particles, which are easier to ignite than the mostly larger particles in other equipment, and 3) they often employ pulse jet cleaning, which by nature, periodically generates dust clouds inside the collector. Data from the German compilation of dust explosions indicate that the most frequent ignition sources have been mechanical sparks (41%), smoldering nests (11 percent), electrostatic discharges (10%) and mechanical heating via friction (7%) (See Zalosh, et.al 2005).

---

[4] Zalosh, Robert, Stanley Grossel, Russell Kahn, and Daniel Sliva, 2005. "Dust Explosion Scenarios and Case Histories in the CCPS Guidelines for Safe Handling of Powders and Bulk Solids", 39th AIChE Loss Prevention Symposium Session on Dust Explosions, Atlanta, April 2005.

2.3.2 Blender and Grinder Dust Explosions

Blenders, like those shown in figure 2.8 by way of examples, are common pieces of equipment used to mix particulates of two or more different compositions. These operations involve inter-particulate friction and particulate-wall friction which in turn cause electrostatic charge generation. If the particulate resistivity is sufficiently high, the electrostatic charge can continue to accumulate with correspondingly increasing voltage differences.

If the blender wall is not well grounded, charge and associated high voltages can accumulate on the blender wall. If the particulate minimum ignition energy (MIE) is sufficiently low, and if the eventual electrostatic discharge occurs in a location where combustible concentrations exist, the result is a dust explosion.

In one case study examined, an explosion occurred in a plastics manufacturing plant in a blender that was used to mix the primary polymer with various additives. The primary polymer had a resistivity of 2 x 10^{16} ohm-cm, a MIE of about 7 mJ, and a MEC of 20 g/m^3. The latter two values are lower than those of most organic powders. Even with significantly

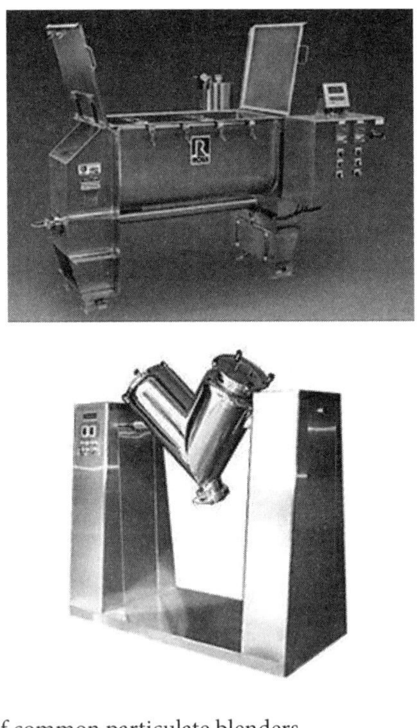

Figure 2.8 Examples of common particulate blenders.

larger MEC values, concentrations > MEC should be anticipated toward the top of the blender during normal operation and throughout most of the blender volume during batch loading and unloading.

Good industry practices to prevent blender explosions include blender construction and grounding considerations to minimize electrostatic charge generation, and blender design and operational considerations to reduce the hazards of frictional heating and tramp metal entry into the blender.

Basically, any kind of operation that lends itself to impact/friction ignition needs to be carefully considered. During mixing and blending operations, if the impeller is misaligned or is deformed or has inadequate clearances, or tramp metal enters the mixer, there can be a recipe for disaster. Even tramp metal in a particle classifies or conveyor may lead to a dust explosion.

Grinders, pulverizers, and other size reduction equipment inherently dissipate large energy inputs required to break up the particles. This energy dissipation inevitably causes heating of the particles and metal surfaces. Particles accumulating in the grinder can easily overheat, smolder, and ignite a dust explosion during grinder loading or unloading.

Frictional heating of the housing of a grinder being used to produce 50 μm (300-mesh) silicon powder caused one explosion. Tramp metal was the apparent ignition source in another explosion that occurred in a hammermill being used for an intermediate stage powder with a low MIE and autoignition temperature (AIT).

Evidence of the frictional or impact heating of hammers in a hammermill used to produce powdered sugar is shown in figure 2.9. The heating of sugar and metal in the vicinity of the mill ignited a sugar dust explosion

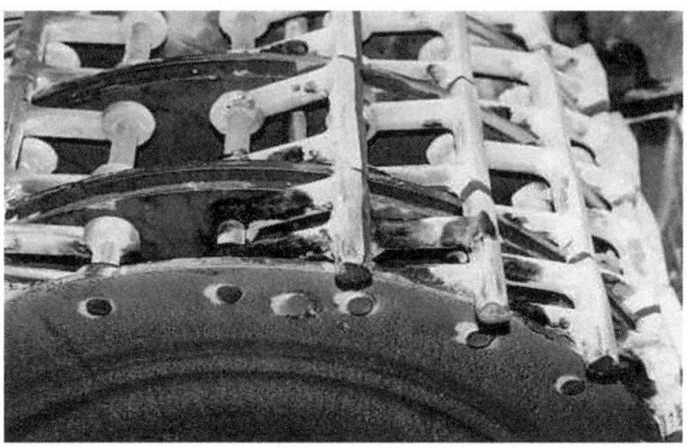

Figure 2.9 Shows wear on the hammers of a hammer mill suggesting frictional heating cause the explosion.

that burned the hammermill operator who was responding to the sound of a severe vibration due to tramp metal or a broken hammer in the mill.

Good industry practices to prevent fires and explosions in grinders and pulverizers include:

- Monitoring the mill motor current
- Incorporating an interlock shutdown upon high current draw
- Using magnetic separators to find and remove tramp metal before it enters the mill
- Using special enclosed mills to allow inerting of powders with extremely low MIE and AIT values

2.3.3 Dryer Dust Explosion Scenarios

The overheating of particulates on a hot surface is a very common ignition source. Although particulates near the dryer inlet may be too wet to be readily ignited, particulates exiting the dryer are both dry and often suspended in concentrations above the MEC. Particulate matter can accumulate on the hot surface and form a smoldering nest. The hot surface temperature can be sufficiently high to directly ignite a suspended dust cloud.

Let's consider an example of a fire and explosion in a batch rotating vacuum dryer used for drying a pharmaceutical powder. An operator had tested dryer samples on a number of occasions without any problems. After the last sampling, he closed the manhole cover, put the dryer under vacuum, and started rotation of the dryer. A few minutes later an explosion and flash fire occurred, which self-extinguished. Investigations revealed that after the last sampling, the dryer manhole cover had not been securely fastened. This allowed the vacuum within the dryer to draw air into the rotating dryer and create a flammable atmosphere. The ignition source was probably an electrostatic discharge (the Teflon coating on the internal lining of the dryer could have built up a charge). No nitrogen inerting had been used. A good industry practice to prevent this type of explosion is to provide nitrogen purging before charging or sampling of the dryer. The purging/inerting can be automated by applying a set point pressure in the dryer such that if the pressure rises above, say 4 psia, the rotation is stopped, an alarm sounds, and a nitrogen purge starts automatically.

Another common explosion scenario with dryers stems from the self-heating of products and intermediate materials being processed. This can lead to exothermic conditions, thus supplying an ignition source. Powders can accumulate on the walls of dryers and remain on surfaces for sufficiently periods of time so as to allow oxidative self-heating to bring the powdery material up to its ignition temperature. The result can be either a fire or a dust explosion inside the dryer or downstream in a dust collector.

2.3.4 Case Study of an Aluminum Dust Explosion

The following is a case study of a catastrophic fire incident that occurred at the Hayes Lemmerz International, Inc. facility in Huntington, Indiana on October 29, 2003. The U.S. Chemical Safety and Hazard Investigation Board (CSB) documented this incident in an investigation (Report No. 2004-01-I-IN). The following paraphrases their discussion of the incident. This incident resulted in one fatality and six serious injuries.

This facility manufactures cast aluminum alloy wheels. The dust explosion originated in a scrap re-melting system. The resultant explosion completely destroyed the dust collection equipment outside the building and damaged equipment inside the building. The explosion was so forceful that it lifted a portion of the building roof above one furnace and ignited a fire that burned for several hours. The basic manufacturing process is shown by the block diagram in figure 2.10. Figure 2.11 shows a diagram of the Furnace No. 5 chip feeder where the incident occurred.

The explosive properties of aluminum dust are well documented in the literature, including the National Fire Protection Association (NFPA) fire prevention standards. In NFPA 68, *Guide for Venting Deflagrations* (2002), aluminum is ranked among the most explosive of metal dusts. The CSB

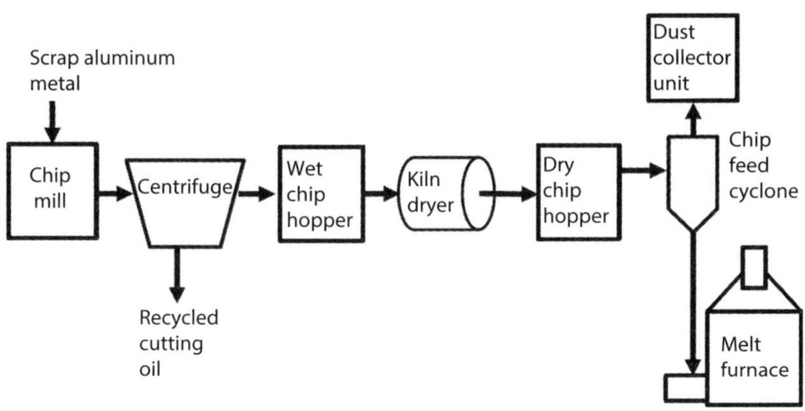

Figure 2.10 Process flow sheet showing the manufacturing process.

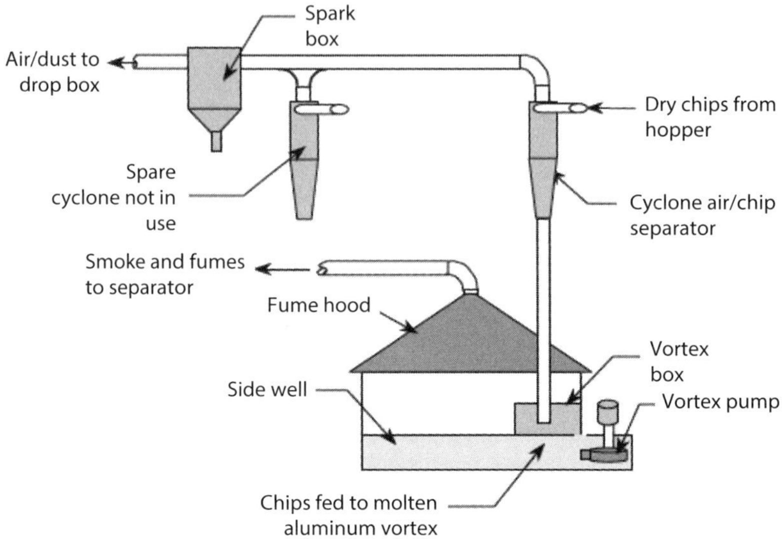

Figure 2.11 Details of Furnace No. 5 Chip Feeder.

reports that nearly one fourth of all dust explosions in the United States in the last 25 years involved metal dusts, and that aluminum accounted for the majority of these events. Metals account for about 19 % of dust explosions worldwide. This data indicates that metal dusts are particularly hazardous and all appropriate precautions need to be taken to prevent dust explosions.

Dust explosivity is typically expressed using the deflagration constant, K_{ST}.

$$K_{ST} = (dP/dt)_{max} V^{1/3}$$

where $(dP/dt)_{max}$ is the maximum measured rate of pressure rise and V is the container volume.

The K_{ST} constant is determined experimentally by measuring how fast the pressure rises when dust of a known concentration is ignited in a container of a specific volume (20 liters). The higher the K_{ST}, the more severe a dust explosion can be.

There are three "dust hazard classes" which are used to indicate relative explosiveness. For comparison, table 2.2 shows K_{ST} values for a few known explosive dusts. Pure aluminum has a high K_{ST} and is rated as a Class ST-328 dust. Actual explosivity values vary significantly with the size and shape of dust particles, the concentration of dust in the air, and the degree of surface oxidation (which, for metals, reduces flammability.

Table 2.2 Comparative Explosivities of Dusts (Source: Eckhoff, 1997).

Dust Type	Avg. Particle Size (microns)	K_{ST} (bar-meter/sec)	Dust Hazard Class
Pure aluminum powder	29	415	ST-3
Coal dust	<10–38	55–108	ST-1
Wheat flour	57	87	ST-1
Wood dust	43	102	ST-2

2.3.4.1 *The Incident*

At about 2:30 pm on October 29, 2003, the facility's maintenance personnel shut down the chip feed system because of a smoldering fire located near Furnace No. 5 inside the duct connecting the fume hood to the fume separator. The glowing red duct alerted employees to the fire. According to plant personnel interviewed by CSB, fires in this duct were not uncommon. On this particular day, employees followed the usual practice for dealing with these fires by shutting down the dry chip feed and the fume hood draft fan (but not the dust collector draft fan), and allowing the fire to burn itself out. Employees judged that the fire was out when the duct had cooled, and they found burnt material that had broken free in the fume separator solids drum. They then opened the fume duct and cleaned out the remaining residue within reach. The October 29 duct fire had been out for at least 2 hours before maintenance personnel restarted the dry chip feed.

At about 8:20 pm, three maintenance personnel restarted the dry chip feed system. Employees reported that after the restart, chips fed steadily to the Furnace No. 5 sidewell for about 10 minutes. At that point one of the workers — who was standing on the north side of the furnace, about 10 feet from his coworkers — noticed chips falling out of the spark box in the dust collector duct. Maintenance personnel were familiar with this phenomenon, which typically indicated that a crust had formed in the vortex, impeding chip feed and that chips were overflowing into the dust duct.

The mechanic who noticed the chips falling from the spark box turned to tell his coworkers to stop the dry chip feed. As he turned, a fireball erupted from beneath the furnace fume hood. The fireball ignited his clothing and totally engulfed another mechanic standing near the sidewell vortex pump. The flames also singed a third mechanic, who was standing near the system control panel.

The fireball expanded, rose upward, and blew open the roof of the building. A contractor was on the roof, having just finished testing emissions at Furnace No. 4 and was moving his equipment to Furnace No. 5, when the explosion occurred. He heard a boom, was knocked down as the roof panels beneath his feet heaved upward, and saw flames erupt through the roof. The contractor ran east approximately 200 feet, descended at a ladder and attempted to call 911. He was not seriously injured.

Outside the building, another stack emission-testing contractor was working in a mobile laboratory trailer parked near the drop box. He said he heard a "boom," and was knocked down as the trailer lurched backward. The contractor looked out the trailer door to see the dust collector totally involved in a bright, hot fire. Something was lodged against the rear door of the trailer, making that exit impossible to use. He was forced to exit the side trailer door, which was closer to the fire than the rear door, but he also escaped serious injury.

Within minutes of the explosion, employees used the intercom system to report the emergency. The plant emergency alarm sounded, and employees evacuated the plant. Coworkers assisted burn victims until emergency medical technicians (emergency responders) arrived. A headcount accounted for all employees and contractors.

The fireball ignited roofing material and insulation, along with a small accumulation of combustible material stored behind the kiln dryer control panel. North-west of the furnace sidewell, plastic pallets containing rejected aluminum wheels waiting to be re-melted were also ignited, and some of those wheels eventually melted.

2.3.4.2 Emergency Response

The Huntington fire chief resided close to the industrial park, and heard the explosion. Within minutes, he was at the nearby firehouse. Within 5 to 10 minutes of the explosion, he was on the scene. Because of prior visits to the plant, he knew that the plant worked with molten aluminum, and understood the proper techniques and materials for fighting aluminum fires. The fire was contained and extinguished in about 2 hours using Class D20 fire extinguishers. The fire chief reported that he inspected the facility for any signs of a natural gas fire and found the gas piping intact. Class D extinguishers are rated for metal fires and typically use sodium chloride or some other noncombustible salt as a smothering agent.

Fire department emergency responders attended the burn victims. There were no injuries to emergency responders in this event, and no fire equipment was damaged or lost.

2.3.4.3 *Injuries and Physical Damage to Facility*

One stack-testing contractor was on the roof above Furnace No. 5; the other was in a trailer just southeast of the drop box. The employee who was closest to the furnace sidewell was totally engulfed in flames and died of thermal burn injuries the day after the incident. A second employee received serious burns over nearly half his body. He remained in critical condition for days and was hospitalized for weeks. The third injured employee received only minor localized burns and returned to work to assist investigators the day after the explosion. Four additional workers (three Hayes employees and one contractor) received treatment by emergency responders for minor injuries on the scene; none of these workers required hospitalization.

The explosion vent panels in the dust collector were blown open, and three of four maintenance doors were blown open or off. The ensuing fire completely consumed the aluminum dust and combustible filter cartridge materials inside the dust collector, destroying it. The drop box was split open at the seams into four large pieces; one piece was thrown 20 feet to the south and lodged against the back of the contractor's mobile lab trailer. Another large fragment apparently hit the side of the foundry building, slightly deforming a vertical steel wall support beam near Furnace No. 5.

Overpressure caused the fume separator exhaust fan housing to bulge and disconnected the supply duct. The fume separator housing was undamaged, though its paint burned off. Also the facility air compressor room adjacent to the furnace area was damaged when the wall panel was blown out, and a water line ruptured causing water damage to some compressors. Exterior vent panels near the compressor room were blown out.

Inside the building, the large, approximately 200 lb sidewell doors on Furnace No. 4, which were closed before the explosion, were blown off by the pressure wave. One of the doors was found on the ground about 50 feet from the furnace. Overpressure also caused the dust collector duct to disconnect from the Furnace 4 cyclone. The fire damaged the chip system control console. Internal overpressure bulged the access doors on the side of the dry chip hopper and on the wet chip hopper inlet target box.

Much of the roof above Furnace No. 5 and the chip handling system was blown off. The west wall was deflected inward from the impact of drop box fragments. The explosion damaged the south wall adjacent to the compressor room, caused minor damage to the building walls, and caused significant roof damage (several trusses needed replacement.) A structural engineer verified the building was safe prior to reentry by inspectors and employees. The total value of the property losses resulting from the incident were not reported by the CSB.

2.3.4.4 *Root Causes of Incident*

The Chemical Safety and Hazard Investigation Board identified the following factors as the root causes for the incident:

- The facility did not perform a review to address why the chip system was releasing excess dust. The hazards of aluminum dust were neither identified nor addressed.
- The facility did not ensure the proper design of the dust collector system.
- The facility did not ensure that the dust collector design and installation followed the guidance in National Fire Protection Association (NFPA) 651.
- The facility had no formal, documented program to investigate and implement corrective action for incidents involving fires in the foundry area, especially those fueled by aluminum dust.
- Employees did not wear flame-retardant clothing when performing routine work near the melt furnace.
- Housekeeping and maintenance in the chip-processing and dust collector areas were inadequate.
- Facility personnel received no formal training for operating and maintaining the chip-processing and dust collection systems.
- The facility had not bothered to develop and implement a means of handling and processing aluminum chips that minimizes the risk of dust explosions, nor did it have a program to provide regular training for all facility employees on the fire and explosion hazards of aluminum dust.

2.4 Other Examples of Catastrophic Incidents

The following are few more examples of major dust explosions.
On the 15th March, 1987, in Harbin, China, an explosion of flax dust triggered a chain of explosions that destroyed the whole 13,000 m^2 linen plant, resulting in 58 deaths and 177 injuries. The strength and time of each explosion was detected at the nearby earthquake monitoring station.

On the 2nd March, 1982, in the UK, an articulated lorry carrying 19 metric tons of powdered resin crashed into a roadside cottage spilling part of its load. A dust cloud was thrown up, which ignited and exploded, setting fire to the entire load. Noxious fumes from the fire affected the driver, occupants of the cottage, workmen at the nearby building site and some fishermen in a boat about 1.5 kilometers out to sea.

In the US on 23rd December, 1977, at Westwego, Louisiana, a series of explosions destroyed 37 grain silos completely and damaged many others, causing 35 deaths and $100,000,000 worth of damage. The facility consisted of 73 silos with a capacity of 150,000 m3. It was one of 5 grain storage explosions that occurred over an 8 day period, and was attributed to unusually dry weather conditions.

On the 6th of February, 1979, in Bremen, Germany, the greater part of the 40,000 m^2 Roland Mill complex was destroyed by a series of explosions. The complex included a seven story flour store, six story mill, other silos and administration building. A pressure wave struck a loaded truck, throwing it against a wall spreading its load into the air and believed to have caused an open air explosion. No traces were found of seven of the fourteen killed in the fire, believed to be cremated in parts of the fire where all traces of combustibles were consumed; an estimated 1,000°C for several hours.

On 13 April 1965 a devastating explosion took place in the factory of General Plastics (New Zealand) Ltd., Masterton. The explosion was violent and spectacular. The blast burst floors upwards, blew the roof out of the ornamenting room, practically demolished a locker room, hurled 300 kg machines onto the roof and threw other machinery through a wall. The smoke from the resulting fire could be seen 40 km away. Four people died as a result of this explosion and 6 people were injured. The explosion was shown to be caused by a short circuit of an underfloor electrical socket of a button finishing machine. It acted as a source of ignition for plastic dust which had accumulated under the floor of the button ornamenting room and initiated a primary dust explosion in the floor cavity. The burning dust and flames from the initial explosion were driven under the floor of a locker room — causing a further explosion. The force of these initial explosions burst the floor upwards, taking the unburnt dust, gases and flames into both the ornamenting room and the locker room. These acted as a source of ignition for the secondary explosions that took place in these two areas, causing large-scale damage to the factory. Although the factory workroom was kept relatively free of accumulated plastic dust, over the years dust had filtered through the floor boards and collected in the floor cavity. This underlines the need to stop the accumulation of hazardous dusts. In this case the provision of solid floors and improved housekeeping would have prevented dust build-up.

2.5 Ignition Sensitivity

Dust explosions have been responsible for some of the most horrific industrial incidents. According to Bowen[5] there is more explosive energy in the dust from grains such as wheat, barley and corn, than in an equal amount of TNT. While fires are far more common, explosions are far more costly in terms of loss of life, injury and property damage.

A dust explosion is the very rapid combustion of a dust cloud that generates a flame and a pressure front. The flame front frequently causes loss of life, while the pressure front will often cause extensive damage to buildings and equipment. With some dusts, there is sometimes only a flame front and the pressure front is only minimal; with others, the flame spreads with the effect of an explosion.

An explosion occurs as the flame generates heat and combustion products, and expansion from both these sources causes an immediate pressure rise which must then move out as a pressure wave and impact on the surroundings. Pressures of up to 700 to 800 kPa can be generated, and if it occurs in a confined space such as a building or piece of equipment, the effects can be devastating (most buildings can only withstand 2 kPa at most).

Any combustible solid material which can be dispersed in air as a dust cloud is capable of causing a dust explosion. Explosions on record have originated from dusts from the following sources:

- Agriculture - grain dust, flour, sugar, milk powder, wool, paper, and wood
- Metals - aluminum, magnesium, zinc
- Mining - coal, combustible sulfide ores
- Chemical industry - sulfur, most plastics, pesticides, pharmaceutical's including aspirin and vitamin C

The Bureau of Mines of the U.S. Department of the Interior developed an arbitrary scale based on tests using small amounts of dusts, as a guide to the degree of hazard of each type of dust. This scale is reported in table 2.3.

---

[5] Bowen JE, "An Overview: Grain Dust Explosions." Fire Engineering, 136(1983):22,23,25,27

Table 2.3 Dust Explosion Hazard Scale[6].

Explosion Ranking	Ignition Sensitivity	Explosion Severity
Weak	< 0.2	< 0.5
Moderate	0.2–1.0	0.5–1.0
Strong	1.0–5.0	1.0–2.0
Severe	> 5.0	> 2.0

The higher the value on each of the scales (ignition sensitivity and explosion sensitivity), the greater the hazard that each dust represents. Values for different dusts are reported in table 2.4.

A similar dust hazard rating system has been devised by the New Zealand Department of Labor[7]. In this scale the degree of hazard depends mainly on the type of dust and the processing methods used. The index of explosibility is based on a scale of 0–100. A rating of 0 indicates no explosion hazard exists, while a result of 100 indicates the most severe explosion hazard. As in table 2.3, the scale is further subdivided into the categories: weak, moderate, strong or severe. These ratings are correlated with the index of explosibility as follows:

- Weak: 0–20
- Moderate: 20–50
- Strong: 50–80
- Severe: >80

Dusts are tesetd in a routine Godbert explosibility apparatus, which is standardized to simulate moderate conditions as in coal dust explosion. Explosive dusts are further classified by the New Zealand Department of Labor into different classes, depending on their explosibility ratings. Factory inspectors may require different precautions for dusts which fall into the different classes according to the following index:

- 80 Class 2 dust
- 20–80 Class 1 dust
- < 20 Fire hazard

[6] Fire Protection Handbook, Ed., McKinnon, G. P., 14th ed., 1976, NFPA, Boston, Mass
[7] New Zealand Dept. of Labor, Wellington, NZ, www.osh.dol.govt.nz

Table 2.4 Explosion Potential for Different Dusts (Source: McKinnon, 1976).

Dust	Ignition Sensitivity	Explosion Severity
Aluminum	7.3	>10.2
Aspirin	2.4	>4.3
Coal	2.2	1.8
Coffee Bean	0.1	0.1
Cotton	<0.1	<0.1
Egg White	<0.1	0.2
Flour	2.1	1.8
Grain Dust	2.8	3.3
Magnesium	3.0	7.4
Milk, Powdered	1.6	0.9
Polyethylene	24.0	2.2
Polystyrene	6.0	2
Rubber	4.6	1.6
Sugar	4.0	2.4
Sulfur	20.2	1.9
Vitamin C	1.0	2.2

Table 2.5 provides a tabulation of dust hazard ratings according to the New Zealand ranking system.

A comprehensive database of combustion and explosion characteristics of dusts has been created by the Berufsgenossenschaftliches Institute für Arbeitschutz, an institute for research and testing of the German Berufsgenossenschaften (BG). Important combustion and explosion characteristics of more than 4,000 dust samples from virtually all industry sectors have been determined and are summarized in a database located on the web site at: http://www.hvbg.de/e/bia/gestis/expl/index.html.

Dust clouds may be ignited by the effects of mechanical friction such as overheated bearings, motors overheating from air cooling vents being clogged with dust, particles of steel or stone caught up in grinding machinery producing sparks, overheated dust coated light bulbs, static electricity, electrical arcing, welding sparks and naked flames. As a general rule of

Table 2.5 New Zealand Dust Hazard Ratings for Different Materials.

Dust	Index	Hazard
Aluminum/resin mix	50–80	Strong
Aqualon fiber	>80	Severe
Bakelite	20–50	Moderate
Barley	20–80	Moderate to Strong
Barley/maize/lucerne mix	50–80	Strong to Severe
Biscuit	>80	Severe
Blood	20–80	Moderate to Strong
Blood and bone mix	0–20	Weak
Bread improver	20–50	Moderate
Bronzing powder	20–50	Moderate
Buttermilk powder	20–50	Moderate
Button (plastic)	> 80	Severe
Cane	20–50	Moderate
Caramel	20–50	Moderate
Carbon black	50–80	Strong
Cardboard	50–80	Strong
Cardboard/sisal mix	50–80	Strong
Casein	50–80	Strong
Cheesecake mix	> 80	Severe
Chicken flavoring	50–80	Moderate
Chocolate	20–50	Strong
Cinnamon	50–80	Strong
Clover	50–80	Strong
Cloves	50–80	Strong
Coal	20 – > 80	Moderate to Severe
Cocoa	50–80	Strong

(*Continued*)

Table 2.5 (*Cont.*)

Dust	Index	Hazard
Coconut shells	50–80	Strong
Coffee kernels	20–50	Moderate
Coriander	20–50	Moderate
Cornflour	20 – >80	Moderate to Severe
Cotton waste	50–80	Strong
Customwood	50–80	Moderate to Strong
Dog biscuit mix	50–80	Strong
Drug (mixture of various drugs)	50–80	Strong
Effluent (freezing works)	80	Severe
Epoxy powder/marble mix	20–50	Moderate
Fiberglass	>80	Severe
Fiberglass/plywood mix	50–80	Strong
Fiberglass/resin mix	20–50	Moderate
Fiberglass/talc mix	50–80	Strong
Fiberglass/wood mix	50–80	Strong
Fiberglass/wood/MDI foam mix	20–50	Moderate
Flax	50–80	Strong
Flour	0–20	Weak
Flour/sugar mix	20–50	Moderate
Food (mixture of various foods)	50–80	Strong
Garlic	20–50	Moderate
Grain	20 – >80	Moderate to Severe
Hardboard	20–50	Moderate
Icing sugar	50–80	Strong
Icing sugar/corn-flour/sugar mix	20–50	Moderate
Jute	50–80	Strong

Dust	Index	Hazard
Lactalbumen	20–50	Moderate
Lactose	50–80	Strong
Leather	>80	Severe
Leather (chrome tanned)	50–80	Strong
Leather/plastic/wood mix	50–80	Strong
Linen	50–80	Strong
Lucerne	20–80	Moderate to Strong
Maize	20–80	Moderate to Strong
Maize starch	50–80	Strong
Malt	20–80	Moderate to Strong
Malted barley	20–80	Moderate to Strong
Malt culms	20–80	Moderate to Strong
Meat meal	50–80	Strong
Milk powder	20–80	Moderate to Strong
Molasses	50–80	Strong
Molding starch	50–80	Strong
Oats	20–80	Moderate to Strong
Olive flour	20–80	Moderate to Strong
Paper	50 – > 80	Strong to Severe
Paper/wood/resin mix	50–80	Strong
Particleboard	20–80	Moderate to Strong
Pea-flour	20–80	Moderate to Strong
Pepper (black)	20–50	Moderate
Pepper (white)	0–80	Weak to Strong
Pinus radiata	50–80	Strong
Pinus radiata/particle board mix	50–80	Strong
Plastic powder	20 – >80	Moderate to Severe

(*Continued*)

Table 2.5 (*Cont.*)

Dust	Index	Hazard
Polyester powder	0 – >80	Weak to Severe
Polyethylene (polythene)	0 – >80	Weak to Severe
Polyurethane foam	0–80	Weak to Strong
Resin	>80	Severe
Resin/marble dust mix	0–20	Weak
Resin/sand mix	> 80	Severe
Rubber	20 – > 80	Moderate to Severe
Rubber/sulfur mix	>80	Severe
Rubber/leather/plastic, wood mix	>80	Severe
Rye	50–80	Strong
Saccharin	50–80	Strong
Sausage Meal	50–80	Strong
Skim milk	50–80	Strong
Sodium caseinate	50–80	Strong
Soyabean flour	20–50	Moderate
Sphagnum moss (dried)	20–50	Moderate
Spice mix (tumeric,pimento, chilies, cinnamon, ginger mix)	50–80	Strong
Stock feed	50–80	Strong
Sulfur (100%)	>80	Severe
Sulfur (50%)/inert filler	(50%)>80	Severe
Tannaphen (bark extract)	50–80	Strong
Toilet soap	>80	Severe
Wheat	20–80	Moderate to Strong
Wheat/barley mix	50–80	Strong
Wheat flour/gluten mix	0–20	Weak
Whey protein (soluble)	50–80	Strong

Dust	Index	Hazard
Whipping fat (70% coconut fat)	50–80	Strong
Whiting/emery/wax mix	0–20	Weak
Wholemeal (wheat)	20–50	Moderate
Wood	20–80	Moderate to Strong
Wood/formica mix	> 80	Severe
Wool	50–80	Weak to Strong

thumb, dusts require 20 to 50 times more energy from an ignition source compared with a flammable vapor, or they need direct contact with surface temperatures ranging from 300 to 600°C.

One should bear in mind that the finer the dust the greater the hazard. Not only can it be more easily blown into the air, it will stay suspended in air much longer. Fine dust has a greater surface area per unit volume so that it can burn all the more rapidly, increasing the intensity of the flame front and the violence of the explosion.

Dusts like flammable vapors, have lower and upper explosive limits. The lower limit is the concentration of dust in air to just sustain the flame front. The lower flammability limit ranges from about 10 to 40 g/m^3 depending on the type of dust. At these concentrations the dust is noticeably visible to the naked eye as a fog or cloud. The upper limit is usually difficult to measure since there appears to be no clear cut-off point. Instead it may, or may not, ignite at a given concentration. If it does ignite, it tends to leave behind increasing amounts of charred residue.

Concentrations of dust which are potentially explosive are intolerable for people to remain in and are not likely to be found in the open, however they can be found around machinery used for crushing, grinding, sanding, milling, filtering, blending, shredding, spray drying, or conveying bulk quantities of solid materials.

Recommended References

Abbot, J. *Prevention of Fires and Explosions in Dryers*, Institute Chem. Engrs, 1990.

ASTM E1226-05, "Standard Test Method for Pressure and Rate of Pressure Rise for Combustible Dusts," American Society for Testing and Materials, 2005.

ASTM E1491-06, "Standard Test Method for Minimum Autoignition Temperature of Dust Clouds," American Society for Testing and Materials, 2006.

ASTM E1515-07, "Standard Test Method for Minimum Explosible Concentrations of Combustible Dusts," American Society for Testing and Materials, 2007.

ASTM E 2019-03 "Standard Test Method for Minimum Ignition Energy of a Dust Cloud in Air," American Society for Testing and Materials, 2003.

ASTM E 2021-01 "Standard Test for Hot Surface Ignition Temperature of Dust Layers," American Society for Testing and Materials, 2001.

Babrauskas, V., *Ignition Handbook, Fire Science Publishers, 2003.*

Britton, L., *Avoiding Static Ignition Hazards in Chemical Operations, AIChE CCPS, 1999.*

CCPS, *Guidelines for Safe Handling of Powders and Bulk Solids, AIChE Center for Chemical Process Safety, 2005.*

CSB, 2005. *Investigation Report. Combustible Dust Fire and Explosions, CTA Acoustics, Inc., Chemical Safety Board, February 2005.*

Eckhoff, R., *Dust Explosions in the Process Industries, 3rd Edition, Gulf Professional Publishing, 2003.*

Gray, B., "Spontaneous Combustion and Self-Heating," *SFPE Handbook of Fire Protection Engineering, 3rd Edition, pp. 2-211-228, SFPE, NFPA, 2002.*

Glor, M., "Electrostatic Ignition Hazards in the Process Industries," J. of Electrostatics, v 63, pp. 447–453, 2005.

Glor, M. and Schwenzfeuer, K., "Direct Ignition Tests with Brush Discharges," J. of Electrostatics, v 63, pp. 463–468, 2005.

Gummer, J. and Lunn, G., "Ignitions of Explosive Dust Clouds by Smouldering and Flaming Agglomerates," J. of Loss Prevention in the Process Industries, v. 16, pp 27–32, 2003.

Jaeger, N., "Safety strategy against potential hazards due to the handling of powders in a blending unit," J. Loss Prevention in the Process Industries, v. 14, pp 139–151, 2001.

Lunn, G., Holbrow, P., Andrews, S., and Gummer, J., "Dust explosions in totally enclosed interconnected vessel systems," J. Loss Prevention in the Process Industries, v. 9, pp. 45–58, 1996.

NFPA 68, "Explosion Protection by Deflagration Venting," National Fire Protection Association, 2007.

NFPA 69, "Standard on Explosion Prevention Systems," 2008.

NFPA 70, "National Electrical Code," National Fire Protection Association, 2008.

NFPA 77, "Recommended Practice on Static Electricity," 2007

NFPA 499, "Recommended Practice for the Classification of Combustible Dusts and of Hazardous (Classified) Locations for Electrical Installations in Chemical Process Areas," National Fire Protection Association, 2008.

NFPA 654, "Standard for the Prevention of Fire and Dust Explosions from the Manufacturing, Processing, and Handling of Combustible Particulate Solids," National Fire Protection Association, 2006.

Scherpa, T., "Secondary Dust Cloud Formation from an Initiating Blast Wave," WPI M.S. Thesis, 2002.

Skjold, T., "Review of the DESC project," Journal of Loss Prevention in the Process Industries v. 20: 291–302, 2007.

UL 913, "Standard for Safety, Intrinsically Safe Apparatus and Associated Apparatus for Use in Class I, II, and III, Division 1, Hazardous (Classified) Locations," Underwriters Laboratories, 1997.

UL 1203, *Explosion-proof and Dust-Ignition-proof Electrical Equipment for Hazardous (Classified) Locations*, Underwriters Laboratories, 1994.

Zalosh, R., Grossel, S., Kahn, R., and Sliva, D., "Safely Handle Powdered Solids," Chemical Engineering Progress, v. 101, pp. 23–30, 2005.

3

Factors Influencing Dust Explosibility

3.1 Introduction

There are a number of factors that influence dust explosibility, including but not limited to:

- particle size and particle size distribution
- dust concentration
- oxidant concentration
- ignition temperature
- turbulence of the dust cloud
- maximum rate of pressure rise
- admixed inert dust concentration
- presence of flammable gases

Each of these is touched upon in the discussions presented in this chapter.

3.2 Particle Size and Dust Concentration

The larger the surface area per unit mass of a dust particle, the greater the hazard it poses. Small particle sizes will provide large surface areas.

In some cases, there is a likelihood of very small particles agglomerating into lumps. If this happens, the explosibility of the dust decreases and when the particle size increases beyond about 500 microns. Explosibility of dusts do not vary linearly with particle surface area although it depends on it. This dependence is influenced by the actual speed of combustion of volatiles and the concentration of dusts.

Dust clouds explode only if the dust concentration is within certain limits. These generally are 50–100 g/m^3 (widely considered to be the lowest concentration) to an upper limit of 2–3 kg/m^3.

There is a certain stoichiometric concentration of volatiles in air of the solid phase fuel must be generated for a flame to propagate rapidly through the mixture before more fuel volatiles are produced. This indicates that the lower concentration limit is determined by the minimum quantity of fuel particles that must exist in order to sustain combustion. A near parabolic relationship exists between dust concentration and ignition energy. The latter is high at high dust loadings and decreases to a minimum value with decreasing concentrations. A further decrease in dust concentrations result in an increase in the ignition energy.

The upper concentration limits are dictated by the minimum amount of oxygen needed for explosion. As is the case with flammability characteristics, data on the explosibility of the same dust differs from test to test.

3.3 Particle Volatility

Anezaki, et.al.[1] have studied the phenomenon in the dust explosion is flame propagation and have found that the properties of the dust have a significant impact on flame propagation during dust explosions. These investigators report that the effect of particle volatility is critical and that particle materials affect the generation of luminous flame, which causes the acceleration of the flame propagation. They note that when a certain

---

[1] T. Anezaki and R. Dobashi, Effects Of Particle Materials On Flame Propagation During Dust Explosions Proceedings of the 5th International Seminar on Fire and Explosion Hazards, Edinburgh, UK, 23–27 April 2007

amount of energy (static electricity, etc.) is applied to combustible dusts which are dispersed in the air, ignition might occur and a flame starts to propagate in the dust clouds at a high speed.

Gas-phase flame is formed in most dust explosions. In such cases, combustible gases are generated by the vaporization or degradation of dust particles in the combustion zone. During a flame propagation, the heat from the flame vaporizes or degrades the unburned particles in front of the flame and generates combustible gases, which form the gas-phase flame, and then the flame continues to propagate into the next unburned particles. Therefore, the volatilizing characteristic (volatility) of the particle material should affect the flame propagation behavior strongly.

Flame structure varies considerably depending on the particle materials. Anezaki reports that the flame front formed in sprayed decane droplets of higher volatility is smooth in shape and propagates continuously. In contrast, the flame propagating through dust clouds of stearic acid, of lower volatility, forms a complicated structure. The flame zone of the latter case consists of blue spot flames at the leading zone and luminous flames behind them. The flame zone appears as discrete structures.

In their experiments the dust explosion of highly volatile solid particles such as myristic acid and behenic acid, two kinds of flames are observed during propagation. There are many spherical flames at the forefront of the combustion zone, which are isolated from one another and emit faint blue lights. These investigators refer to these as 'blue spot flames'. So-called luminous flames are formed behind the blue spot flames, which emit strong lights, i.e. radiations from combustion products like soots.

Focusing on the behavior of luminous flames, it was found that there are two different patterns in flame propagation. In the first pattern, a luminous flame grows at an accelerated pace soon after its generation and gets close to the blue spot flames. In the second pattern, the initial luminous flames are quenched before growing up to a big block.

The study shows that that volatile emissions from unburned particles maintain flame propagation. Particles of low volatility cannot volatilize significantly before they are burned. As a result, the combustion zone becomes heterogeneous where solids, liquids, and gases are mixed. In such a heterogeneous region, combustible gas density is locally high in the neighborhood of each particle where soots are likely produced due to imperfect combustion. One would expect then that the flame might propagate more rapidly in dust mixtures comprised of more volatile particles. The generation of intense luminous flames accelerates the flame propagation even if the particle volatility is lower.

3.4 Heats of Combustion

As explained early on in this volume, a dust explosion is initiated by the rapid combustion of flammable particulate matter suspended in air. Any solid material that can burn in air will do so with a violence and speed that increases with the degree of sub-division of the material (see Eckhoff, 2003[2]). The more finely divided the particles (i.e., the higher the degree of sub-division, or in other words smaller the particle size) the more rapid and explosive the burning. There is however, a limiting stage when particles too fine in size tends to lump together.

If the ignited dust cloud is unconfined, the incident results only in a flash fire. But if the ignited dust cloud is confined, even partially, the heat of combustion may result in the rapid development of pressure, with flame propagation across the dust cloud and the evolution of large quantities of heat and combustion reaction products.

In addition to particle size, the severity of the resultant explosion depends on the rate of energy release due to combustion relative to the degree of confinement and heat losses. We should bear in mind that combustion itself is a series of complex chemical reactions. The oxygen required for combustion is mostly supplied by air. The condition necessary for any dust explosion is a simultaneous presence of dust cloud of appropriate concentration in air that will support combustion throughout the process and a suitable ignition source of sufficient energy.

With dusts that are comprised of volatile substances, the explosion may occur in three steps which may follow each other in very quick succession:

1. Devolatization (where volatile materials are released from the particles or the particles themselves vaporize
1. Gas phase mixing of fuel (released by dusts) and oxidant (usually air)
1. Gas phase combustion

Many combustible dusts if dispersed as a cloud in air and ignited, will allow a flame to propagate through the cloud in a manner similar to the propagation of flames in premixed fuel – oxidant gases (see discussions by

[2] R.K. Eckhoff, Dust Explosions in the Process Industries, 3rd ed., Gulf Professional Publishing, USA, 2003.

Proust, 2005[3]). Examples of such dusts include foodstuffs like sugar flour, cocoa, synthetic materials such as plastics, chemicals and pharmaceuticals, metals such as aluminum and magnesium and fuels such as coal and wood. In the most general sense, a dust explosion involves oxide formation; i.e.

$$\text{fuel} + \text{oxygen} \rightarrow \text{oxide} + \text{heat}$$

Metal dusts can also react with nitrogen or carbon dioxide to generate the necessary heat for an explosion.

The interdependence of the various parameters which influence the explosion pressure is described by the equation of state for ideal gases (i.e., the Ideal Gas Law):

$$P = nRT/V$$

where P is pressure, V is volume, R the universal gas constant, n the number of moles of gas and T is the temperature.

All things being equal, the increase of T due to the heat developed in a burning dust cloud essentially has the deciding influence on the explosion pressure. It is logical that the higher the heat of combustion of a given dust per mole of oxygen (O_2) consumed, the greater is the likely severity of an explosion. The heat of combustion is therefore a thermodynamic property that we need to pay attention to when assessing dust explosion potential.

The reader should recall from basic chemistry that the heat of combustion ($\mathbf{\Delta H^\circ_c}$) is the energy released as heat when a compound undergoes complete combustion with oxygen under standard conditions. The chemical reaction is typically a hydrocarbon reacting with oxygen to form carbon dioxide, water and heat. It may be expressed with the quantities:

- energy/mole of fuel (kJ/mol)
- energy/mass of fuel
- energy/volume of fuel

Heats of combustion are conventionally measured with a bomb calorimeter. It may also be calculated as the difference between the heat of formation ($\mathbf{\Delta H^\circ_f}$) of the products and reactants.

See table 3.1 for some typical values of heats of combustion. One can see from the table that metals generally have high heats of combustion

[3] C. Proust, A few fundamental aspects about ignition and flame propagation in dust clouds, J. Loss Prevent. Process Ind. 19 (2005) 104–120.

Table 3.1 Heats of combustion of dusts.

Material	Oxidation Products	Heat of Combustion (kJ/mol O_2)
Calcium	CaO	1,270
Magnesium	MgO	1,240
Aluminum	Al_2O_2	1,100
Silicon	SiO_2	830
Chromium	Cr_2O_2	750
Zinc	ZnO	700
Iron	Fe_2O_3	530
Copper	CuO	300
Sucrose	$CO_2 + H_2O$	470
Starch	$CO_2 + H_2O$	470
Polyethylene	$CO_2 + H_2O$	390
Carbon	CO_2	400
Coal	$CO_2 + H_2O$	400
Sulfur	SO_2	300

and hence may be considered dangerous from the standpoint of dust explosibility.

3.5 Explosive Concentrations and Ignition Energy

There exists a range of concentrations of the dust and air within which a mixture can explode; i.e., mixtures above or below this concentration range will not explode. The lowest concentration of dust capable of exploding is referred to as the lower explosive limit and the concentration above which an explosion will not take place as the upper explosive limit.

The lower explosive limits of many materials have been measured and are reported in the literature. Values vary from 10 grams per cubic meter (g/m^3) to about 500 g/m^3. For practical purposes it may be assumed that 30 g/m^3 is the lower explosive limit for most flammable dusts. While this may seem to be a particularly low concentration, in appearance a cloud of dust of such a concentration would resemble a dense fog. The upper explosive limits are not well defined and have poor repeatability under laboratory test conditions. Since the upper explosive limit is of little practical importance, data for this parameter is rarely reported. The most violent explosions are produced when the proportion of oxygen present is not far removed from that which will result in complete combustion. The range of the explosive concentrations of a dust cloud is not simply a function of the chemical composition of the dust; the limits vary with the size and shape of the particles in the dust cloud.

Although mixtures of dust and air within the flammable range are capable of explosion, they will not explode unless they are ignited; i.e., there must be a suitable source of ignition present. Once a source of ignition is presented to the flammable mixture, flame will propagate throughout the cloud. The mode of ignition of a dust cloud is typically a hot surface, an electrical spark or a mechanically generated frictional spark. The minimum condition necessary to initiate a dust explosion with certain modes of ignition have been measured; some comparative data reported in the literature are reported in table 3.2.

Table 3.2 Critical Temperatures and Dust Concentrations of Common Dusts[4].

Substance	**Ignition Temperature of Dust Cloud (°C)**	**Minimum Explosive Concentration (oz/ft^3)**	**Relative Explosion Hazard**
Alfalfa	460		
Aluminum	650	0.045	Severe
Al-Mg alloy		0.02	Severe
Cereal grass	550		
Chromium		0.23	Strong
Coal	610	0.055	Strong
Copper	900		Fire
Corn	400		
Epoxy Resin	530	0.02	Severe
Flax shive	430		
Grain dust, winter wheat, corn, oats	430		
Iron	420	0.1	Strong
Magnesium	520	0.02	Severe

(*Continued*)

[4] http://www.dustexplosion.info/dust%20explosions%20-%20the%20basics.htm, and http://www.explosiontesting.co.uk/ignition_temp_9.html

Table 3.2 (*Cont.*)

Substance	Ignition Temperature of Dust Cloud (°C)	Minimum Explosive Concentration (oz/ft³)	Relative Explosion Hazard
Rice	440		
Silicon		0.11	Strong
Soy flour	540		
Tin	630	0.19	Moderate
Titanium	460	0.045	Severe
Uranium (finely divided metal dust)	20	0.06	Severe
Wheat flour	380		
Wheat straw	470		
Zinc	600	0.48	Moderate

[1] *ft³ = 0.02832 m³= 28.32 dm³ = 0.03704 yd³ = 6.229 Imp. gal (UK) = 7.481 gal (US) = 1,728 cu.in. ; 1 oz (ounce) = 28.35 g = 437.5 grains = 0.0625 lb = 0.0000279 long ton (UK) = 0.00003125 long ton (US) = 0.000558 long hundredweight (UK) = 0.000625 long hundredweight (US) = 0.004464 stone = 16 dram*

In table 3.2, the term minimum ignition temperature (MIT) is used. The MIT is defined by convention to be the lowest temperature of a hot surface that will cause a dust cloud, rather than a dust layer, to ignite and propagate flame. The test follows BS EN 50281-2-1:1999 (methods of determining minimum ignition temperatures). Roughly 0.1 g of combustible dust is placed in a dust holder at the top of a temperature controlled furnace with an open bottom. The dust is dispersed by compressed air downwards past the hot surface of the furnace to see if ignition occurs and flames are produced. If the dust does not ignite, the furnace temperature is increased and the test repeated until ignition of the dust occurs. Once ignition has been established, the mass of the dust sample and injection pressure are varied to find the most vigorous explosive flame discharge. The temperature of the furnace is then reduced incrementally until flame propagation is no longer observed. At this temperature, the dust mass and injection pressure are varied to confirm that no ignition is found over ten consecutive tests. The minimum ignition temperature (MIT) is the lowest temperature of the furnace at which flame is observed minus 20°C for furnace temperatures over 300°C or minus 10°C for furnace temperatures under 300°C.

For plant processing equipment such as driers, testing the minimum ignition temperature is important to prevent a dust explosion occurring through contact with a hot surface. The internal temperature is generally limited to two thirds of the MIT when measured in degrees Centigrade. For example, a dust with a MIT of 450°C would require a maximum operating temperature in the drying process of 300°C, giving a safety margin of 150°C.

ASTM E2021 - 09 is the standard test method for hot-surface ignition temperature of dust layers. The test method is applicable to dusts and powders, and provides a procedure for performing laboratory tests to evaluate hot-surface ignition temperatures of dust layers. The test data can be of value in determining safe operating conditions in industrial plants, mines, manufacturing processes, and locations of material usage and storage. Due to the variation of ignition temperature with layer thickness, the test data at one thickness may not be applicable to all industrial situations. Tests at various layer thicknesses may provide a means for extrapolation to thicker layers.

3.6 Classification of Dusts

It is expected that all explosible dusts are combustible; however not all combustible dusts are easily explosible[5]. For example, anthracite and graphite are not easily explosible, although they have high heats of combustion. If the composition of the dust is known, one may check whether it is explosible by consulting the list of experimentally tested dusts, published by HM Factory Inspectorate of the Department of Employment, UK (see Barton, 2002[6]). According to this classification, dusts, which propagated a flame when ignited have been classified under Group A. The dusts, which did not propagate a flame have been classified under Group B. This classification is applicable to dusts which are at or near the atmospheric temperature (25°C) at the time of ignition. At higher temperature some of the Group B dusts can become explosible.

Dusts that are ignitable but not explosible can become explosive if admixed with fuel dust; for example the ignitable but non-explosible fly ash becomes explosible when spiked with pulverized coal or petroleum coke (see discussions by Amyotte, et.al., 2005[7]). This tends to happen due to increased volatile matter provided by fuel dusts.

---

[5] G. Vijayaraghavan, Impact assessment, modeling, and control of dust explosions in chemical process industries, MTech Thesis, Department of Chemical Engineering, Coimbatore Institute of Technology, 2004

[6] J. Barton (Ed.), Dust Explosion Prevention and Protection: A Practical Guide, Institution of Chemical Engineers (IChemE), Warwickshire, 2002.

[7] P.R. Amyotte, A. Basu, F.I. Khan, Dust explosion hazard of pulverized fuel carry-over, J. Hazard. Mater. 122 (2005) 23–30.

Another measure of the ignitability of a dust layer and burning intensity of a dust layer is the Combustion Class (see ISSA, 1998[8] and Gummer et.al., 2003[9]). This classification is based on the behavior of a defined heap when subjected to a gas flame or hot platinum wire. This is a six-tier classification system defined as follows:

i. CC1: No ignition; no self-sustained combustion
ii. CC2: Short ignition and quick extinguishing; local combustion of short duration
iii. CC3: Local burning or glowing without spreading; local sustained combustion but no propagation
iv. CC4: Spreading of a glowing fire; propagation smoldering combustion
v. CC5: Spreading of an open fire; propagating open flame
vi. CC6: Explosible burning; explosive combustion

There is also a 3rd categorization of dusts based on the so-called 'K_{St} value'. This is a term which quantifies the maximum rate of pressure rise in $1m^3$ vessel when a dust is ignited. It may be thought of as a parameter that defines the degree of 'dust explosion violence'. Detailed discussions are given by Eckhoff, 2004[10], Bartknecht, 1978[11] and Bartknecht, 1981[12].

For many dusts for vessel volumes > $0.04m^3$, the K_{St} constant follows the following relationship:

$$(dP/dt)_{max} V^{1/3} = \text{constant} = K_{St}$$

The K_{St} value has units of bar m/s, being numerically identified with the $(dP/dt)_{max}$ (bar/s) in the $1m^3$ standard International Standards Organization (ISO) test and is denoted as a 'specific dust constant'.

The following is the general ranking:

- K_{St} = Group St_0: Non-explosible
- $0 < K_{St} < 200$ = Group St_1: Weak explosibility

---

[8] ISSA, Determination of the Combustion and Explosion Characteristics of Dusts, International Social Security Agency, Mannheim, 1998.

[9] J. Gummer, G.A. Lunn, Ignitions of explosive dust clouds by smoldering and flaming agglomerates, J. Loss Prevent. Process Ind. 16 (2003) 27–32.

[10] R.K. Eckhoff, Partial inerting—an additional degree of freedom in dust explosion protection, J. Loss Prevent. Process Ind. 17 (2004) 187–193.

[11] W. Bartknecht, Brenngas-und Staubexplosionen, Forschungsbericht F45. Koblenz Bundesinstitut fur Arbeitsschutz, Federal Republic of Germany, 1978.

[12] W. Bartknecht, Explosionen, Ablauf und Schutzmassnahmen, Springer- Verlag, Berlin, 1981.

- $200 < K_{St} < 300$ = Group St_2: Strong explosibility
- $300 < K_{St}$ = Group St_3: Very strong explosibility

In the U.S., the preferred index of explosibility is that devised by the U.S. Bureau of Mines which is based on ranking dusts relative to Pittsburgh Coal. The index of explosibility IE is defined as the product of the explosion severity ES and the ignition sensitivity IS:

$$IE = IS \times ES$$

The ignition sensitivity is defined as:

$$IS = [(MIT \times MIE \times MEC)_{Pc}] \div [(MIT \times MIE \times MEC)_{sample}]$$

The explosion severity parameter is defined as:

$$ES = [(MEP \times MRPR)_{Pc}] \div [(MEP \times MRPR)_{sample}]$$

Where:

MEC = minimum explosive concentration
MEP = the maximum explosion pressure
MIE = minimum ignition energy
MIT = minimum ignition temperature
MRPR = the maximum rate of pressure rise

The subscripts 'Pc' and 'sample' denote Pittsburgh Coal and sample. This index of explosibility is a relative relationship, and is to this extent less dependent on the apparatus used, but its determination requires the conduct of the full range of tests. See discussions by Lees on the index (1996[13], 2005[14]).

3.7 Oxidant Concentration

The reader may recall that one of the sides of the 'dust explosion pentagon' is the oxidant, which usually is oxygen in air. Oxygen influences the dust explosion process significantly. An oxygen concentration greater than

[13] F.P. Lees, Loss Prevention in the Process Industries—Hazard Identification, Assessment and Control, vol. 2, Butterworth-Heinemann, London, 1996. F.P. Lees, Loss Prevention in the Process Industries—Hazard Identification, Assessment and Control, vol. 2, Butterworth-Heinemann, London, 1996.

[14] F.P. Lees, Lees' Loss Prevention in the Process Industries (Partially updated by S. Mannan), vols. 1–3, Elsevier/Butterworth-Heinemann, Oxford, 2005.

21% tends to increase the burning velocity of the fuel. But for a concentration less than 21% the burning velocity is reduced. The reason for this is that oxygen is consumed by the fuel in the combustion process, thereby decreasing the oxygen concentration. Consequently, the rate of combustion of the dust comes down. Eventually, the combustion may die down, or, if an explosion does occur, it may be less severe. Fires are sustained only when the oxygen concentration in air is greater than about 10% by volume.

3.8 Turbulence

Turbulent mixing plays a major role whenever there is mass and heat exchange accompanying complex chemical reactions, as is the case with combustion. A highly turbulent dust cloud will result in widely dispersed particulate matter. When such a cloud catches fire, the turbulence will cause a mill-like effect described as the mixing the hot burnt/burning parts of the cloud with the cold unburned portions, generating a three-dimensional laminate of alternating hot burnt/burning and cold unburned zones. Flames will tend to propagate rapidly through a dust cloud if the latter has a high degree of turbulence, resulting in a violent explosion. Turbulence affects the rate of pressure rise much more than the peak pressure as reported by Torrent[15].

When a less turbulent dust cloud is ignited, it releases an initial large amount of heat, which is locally concentrated due to its low rate of heat dissipation. Further propagation of a flame produced in the dust cloud is due entirely to the degree of dust dispersion. A more evenly dispersed dust burns more readily.

There are two types of turbulence described in the general literature. One type is generated by the dust production operations such as by air jet mill, mixer, bag filter, pneumatic transport pipe and bucket elevator. This type of turbulence is referred to as initial turbulence. The second kind of turbulence is generated during the combustion process after the dust cloud has ignited.

It is an expansion-induced flow of unburned dust cloud ahead of the propagating flame. The speed of the flow and the geometric constrictions present at the operation site establish the degree of turbulence generated. Vent openings and obstructions caused by such equipment like buckets in a bucket elevator, enhance the turbulence generation process. In general, the turbulence generated by the flame front is much greater than the initial turbulence.

[15] J. Gracia-Torrent, E. Conde-Lazaro, C.Wilen, Biomass dust explosibility at elevated initial pressures, Fuel 77 (1998) 1093–1097.

Given that the rates of combustion and other chemical reactions associated with dust explosions are characterized through a set of fundamental properties such as burning velocity, turbulence has been regarded by some investigators as 'the single most important factor in dust explosions[16].

3.9 Maximum Rate of Pressure Rise

The rate of pressure rise, when a dust is ignited, is both a measure of the 'explosibility' of a dust and a key property on which the design of several explosion detection systems and vents are based.

The absolute pressure as a function of time, P(t), in a constant volume, spherical explosion, is related to the fractional volume, V(t), occupied by the fireball during the time of propagation, t, as follows:

$$\{P(t) - P_0\}/\{P_{max} - P_0\} = k\,V(t)/V_0$$

Where P_0 is the initial absolute pressure, V_0 the chamber volume, and k is a correction factor related to the difference in compressibility between burned and unburned gases. For spherical propagation from a point source:

$$V(t)/V_0 = [r(t)/r_0]^3 = [S_b t/r_0]^3$$

Where r(t) is the fireball radius, r_0 the chamber radius, and S_b is the flame speed given by the following expression:

$$S_b = dr(t)/dt = (\rho_u/\rho_b)\,S_u$$

Where ρ_u/ρ_b is the density ratio of unburned to burned gases (at constant pressure), and S_u is the burning velocity, i.e., the rate of flame propagation relative to the unburned gas ahead of it. The flame speed, S_b, is relative to a fixed reference point. Both S_b and S_u are for turbulent non-laminar conditions for dust explosions. For spherical propagation in a spherical chamber, the maximum pressure is reached just as the flame contacts the wall. At that instant, k = 1.

Vijayaraghavan, 2004 has prepared an eloquent derivation combining the above relationships to provide an expression for the size normalized maximum rate of pressure rise, stated as follows:

$$K_{St} = [dP(t)/dt]_{max}\,V_0^{1/3} = 4.84\,[(P_{max}/P) - 1)]\,P_{max} S_u$$

[16] F. Tamanini, The role of turbulence in dust explosions, J. Loss Prevent. Process Ind. 11 (1998) 1–10.

Note that the subscript "St" is by convention and refers to the German word for dust.

Various investigators (see Bartknecht[17] and Wiemann[18]) have reported the effect of initial pressure on the P_{max} and K_{St} values in which they found from experiments that P_{max} increases linearly with increase in initial pressure, over the range of 1 to 4 bars. They also report that K_{St} increases with initial pressure.

The relevance of the K_{St} concept is that it serves as a key parameter for the design of explosion vents. See NFPA Standard 68[19] for a detailed discussion of explosion vents and the application of the K_{St} parameter.

3.10 Presence of Volatile and Flammable Gases

The presence of volatile organic compounds (VOCs), in particular, those which are considered flammable, can greatly enhance the explosibility of dusts. The net result of having such constituents present is a reduction in the minimum explosive concentration, minimum ignition temperature, and the minimum ignition energy, while at the same time there will be an increase in the maximum rate of pressure rise. The minimum ignition energy of the dust–flammable gas mixtures will always tend to be lower than that of the dust alone.

Flammable gases can create an explosive dust–gas mixture at dust concentrations which are well below the normal lower explosive limit for the dust and at a gas concentration below the normal lower explosive limit for the gas. These constituents may also make explosive a dust of such large particle size which would otherwise have been considered non-explosive. As an example, for a dust of class St 0, the presence of flammable VOCs can change the class to St 1, 1/2, 2 and 3 at methane concentrations of 1%, 3%, 5% and 7%, respectively, and to classes 1, 2/3 and 3 at propane concentrations of 0.9%, 2.7% and 4.5%, respectively (see Lees, 2005, 1996).

Hybrid mixtures of polyurethane–cyclopentane and plastic dust–cyclopentane were found to be two times more sensitive to dust explosion than the dust without the cyclopentane gas (see Nifuku *et.al*[20]).

---

[17] W. Bartknecht, Dust Explosions: Course, Prevention, Protection, Springer, Berlin, 1989.

[18] W. Wiemann, Influence of temperature and pressure on the explosion characteristics of dust/air and dust/air/inert gas mixtures, in: Proceedings of the Industrial Dust Explosions, STP 958, American Society for Testing and Materials, West Conshohocken, PA, 1987, pp. 33–44.

[19] NFPA 68, Guide for Venting of Deflagrations, National Fire Protection Association, 1998.

[20] M. Nifuku, J. Gatineau, C. Barre, I. Sochet, H. Katoh, Explosibility assessment of dusts produced in the recycling process of electrical appliances, J. Phys. IV France 12 (2002) 133–140.

As a matter of practicality, one should always pay close attention to the so-called *flammable range* (Explosive Range) of gases or vapors that will burn (or explode) if an ignition source is introduced. Recall that below the explosive or flammable range the mixture is too lean to burn and above the upper explosive or flammable limit the mixture is too rich to burn. The limits are commonly called the "Lower Explosive or Flammable Limit" (LEL/LFL) and the "Upper Explosive or Flammable Limit" (UEL/UFL). The lower and upper explosion concentration limits for some common gases are reported in table 3.3.

Table 3.3 Flammability Limits of VOCs.

VOC	"Lower Explosive or Flammable Limit" (Vol. %)	"Upper Explosive or Flammable Limit" (Vol. %)
Acetaldehyde	4	60
Acetic acid	4	19.9
Acetone	2.6	12.8
Acetyl chloride	7.3	19
Acetylene	2.5	81
Acrolein	2.8	31
Acrylonitrile	3	17
Allyl chloride	2.9	11.1
Ammonia	15	28
Arsine	5.1	78
Benzene	1.35	6.65
1,3-Butadiene	2	12
n-Butane	1.86	8.41
iso-Butane	1.8	8.44
iso-Butene	1.8	9
Butyl alcohol, Butanol	1	11
Butylene	1.98	9.65
Carbon Disulfide	1.3	50

(Continued)

Table 3.3 (*Cont.*)

VOC	"Lower Explosive or Flammable Limit" (Vol. %)	"Upper Explosive or Flammable Limit" (Vol. %)
Carbon Monoxide	12	75
Cyanogen	6	42.6
Cyclobutane	1.8	11.1
Cyclohexane	1.3	8
Cyclohexanol	1	9
Cyclopropane	2.4	10.4
Dekane	0.8	5.4
Diborane	0.8	88
1,1-Dichloroethane	6	11
Diethyl Ether	1.9	36
Diesel fuel	0.6	7.5
Diethylamine	2	13
Diethyl ether	1.9	48
Disobutyl ketone	1	6
Ethane	3	12.4
Ethylene	2.75	28.6
Ethyl Alcohol, Ethanol	3.3	19
Ethyl acetate	2	12
Ethylamine	3.5	14
Ethylbenzene	1	7.1
Ethyl Chloride	3.8	15.4
Etylene glycol	3	22
Ethylene oxide	3	100
Fuel Oil No.1	0.7	5
Furan	2	14

VOC	"Lower Explosive or Flammable Limit" (Vol. %)	"Upper Explosive or Flammable Limit" (Vol. %)
Gasoline	1.4	7.6
Glycerol	3	19
Heptane	1	6.7
Hexane	1.1	7.5
Hydrogen	4	75
Hydrogen sulfide	4.3	46
Isobutane	1.8	9.6
Isobutyl alcohol	2	11
Isophorone	1	4
Isopropyl Alcohol, Isopropanol	2	12
Kerosene Jet A-1	0.7	5
Methane	5	15
Methyl Acetate	3	16
Methyl Alcohol, Methanol	6.7	36
Methyl Chloride	10.7	17.4
Methyl Ethyl Ketone	1.8	10
Mineral spirits	0.7	6.5
Naphthalene	0.9	5.9
n-Heptane	1	6
n-Hexane	1.25	7
n-Pentene	1.65	7.7
Naphtalene	0.9	5.9
Neopentane	1.38	7.22
Neohexane	1.19	7.58
Nitrobenzene	2	9

(Continued)

Table 3.3 (*Cont.*)

VOC	"Lower Explosive or Flammable Limit" (Vol. %)	"Upper Explosive or Flammable Limit" (Vol. %)
Nitromethane	7.3	22.2
n-Octane	1	7
iso-Octane	0.79	5.94
n-Pentane	1.4	7.8
iso-Pentane	1.32	9.16
Propane	2.1	10.1
Propyl acetate	2	8
Propylene	2	11.1
Propylene oxide	2.3	36
Pyridine	2	12
Silane	1.5	98
Styrene	1.1	6.1
Toluene	1.27	6.75
Triptane	1.08	6.69
Turpentine	0.8	
Vinyl acetate	2.6	13.4
Vinyl chloride	3.6	33
p-Xylene	1	6

3.11 Limiting Oxygen Concentration

The Limiting Oxygen Concentration or LOC, is an important parameter that we will turn attention to again in a later chapter. The LOC is also known as the Minimum oxygen concentration, (MOC), and is defined as the limiting concentration of oxygen below which combustion is not possible, independent of the concentration of fuel. It is expressed in units of volume percent of oxygen. The LOC varies with pressure and temperature. It is also dependent on the type of inert (non-flammable) gas.

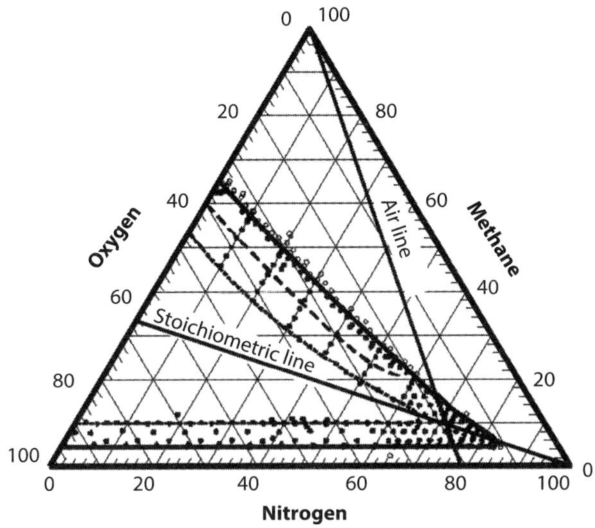

Figure 3.1 Example of a flammability diagram.

Figure 3.1 shows a flammability diagram. Engineers use these types of diagrams to determine the combustibility of mixtures. The illustrative diagram shows the regimes of flammability in mixtures of fuel, oxygen and an inert gas, in this case nitrogen. Mixtures of the three gasses are usually depicted in a triangular diagram, also known as a Ternary Plot.

A number of such charts have been developed and reported in the literature for various combinations of flammable gases. A good source for this information is Perry's Chemical Engineer's Handbook (Edition 5). See also Zabetakis[21].

The example in figure 3.1 shows the possible mixtures of methane, oxygen and nitrogen. Air is a mixture of about 21 volume percent oxygen, and 79 volume percent inerts (nitrogen). Any mixture of methane and air will therefore lie on the straight line between pure methane and pure air – see the line designated as 'air-line' in the figure. The upper and lower flammability limits of methane in air are located on this line, as shown.

The stoichiometric combustion of methane is: $CH_4 + 2O_2 \rightarrow CO_2 + 2H_2O$. The stoichiometric concentration of methane in oxygen is therefore 1/(1+2), which is 33 percent. Any stoichiometric mixture of methane and oxygen will lie on the straight line between pure nitrogen (and zero percent

[21] Zabetakis, Michael G. (1965), Flammability characteristics of combustible gases and vapors (Bulletin 627), Bureau of Mines, Wash., D.C. p. 129.

methane) and 33 percent methane (and 67 percent oxygen) – see the line designated as 'Stoichiometric Line' in the figure. The upper and lower flammability limits of methane in oxygen are located on the methane axis, as shown.

The actual envelope defining the flammability zone can only be determined based on experiments. The envelope will pass through the upper and lower flammability limits of methane in oxygen and in air, as shown. The nose of the envelope defines the limiting oxygen concentration (LOC)).

The LOC for various common gases and solids are reported in table 3.4.

3.12 Important Definitions and Concepts

Combustible Dust Concentrations – Dusts are combustible within certain concentrations. General ranges are 20–60 g/m^3 air as the lower combustibility limit to 2–6 kg/m^3 air as the upper combustibility limit. These ranges

Table 3.4 Limiting oxygen concentration for selected gasses and solids/dusts (volume % O_2).

Material	Nitrogen/ Air	Carbon dioxide/ Air
Gases/Vapors		
Hydrogen	5	5.2
Methane	12	14.5
Ethane	11	13.5
Propane	11.5	14.5
n-Butane	12	14.5
Isobutane	12	15
Solids/Dusts		
Polyethylene (High Density)	16	
Polypropylene	16	
PMMA (Poly(methyl methacrylate)	15.9	
PVC	16.9	
Polyetheylene (Low Density)	15.9	

Material	Nitrogen/ Air	Carbon dioxide/ Air
Fir wood (e.g., Douglas fir, with the scientific name Pseudotsuga menziesii, also known as Oregon pine or Douglas spruce, is an evergreen conifer species native to western North America. The common name is misleading since it is not a true fir, i.e., not a member of the genus Abies. For this reason the name is often written as Douglas-fir (a name also used for the genus Pseudotsuga as a whole).	17	
Corrugated board	15	
Cardboard palletized	15	
Paper	14.1	

vary considerably. Highly combustible dusts can form flammable mixes at concentrations less than 15 g/m^3.

Connecting Components – As in gas explosion protected areas, electrical equipment is connected to the outer power circuit using clamps. The equipment can also be connected using a cable that is continually connected. Devices with attached cables are an exception when only one end of the cable is permanently connected. These devices must be labeled with the mark X. The operator must be given instructions for the unattached end of the cable.

Dustproof Enclosure – An enclosure which prevents visible dust infiltration. This means that a non-risk area is created inside the enclosure. Pressure-resistant enclosures are not in themselves dustproof and must be separately checked and certified according to their condition.

Dust-protected Enclosure – Refers to an enclosure which does not entirely prevent dust infiltration, but which does not allow sufficient dust to enter to cause difficulties with the safe operation of the equipment. Dust must not be allowed to accumulate where it could cause risk of explosion. The material used for the enclosure is especially important. It must be subjected to material tests. The enclosure must provide the necessary protection from dust in spite of the deterioration of the material and usual mechanical wear and tear. Suitable materials include metals (such as coated steel plates, high-grade steel, light metal); glass (for enclosure parts, e.g., viewing panes); molded plastic. Metals used for this purpose may have to be subjected to an impact test at low temperatures, as some metals (light metals) have less favorable mechanical properties at low temperatures than at higher ones. In addition, light metal may contain a maximum of 6% magnesium, as it otherwise tends to throw off sparks upon impact with materials such as rusty iron. Glass must withstand a thermal shock without cracks or without such extensive damage that it breaks during a subsequent impact test.

Electrostatics – An electrostatic discharge is an effective ignition source. When molded plastic is used for enclosures, the outer surface must be prevented from becoming charged. Otherwise, one of the following types of discharge can occur:

- Spark discharge - These discharges take place between grounded and ungrounded components and are sufficient to ignite all gases and vapors, and almost all dusts.
- Brush discharge - This is a special form of the corona discharge. Pipes, elbows, screws, and tools may serve as electrodes with the maximum field strength. This type of discharge poses no danger to most dusts, but caution is warranted with regard to gases and vapors.
- Propagating brush discharge - This is a discharge of a chargeable material in a thin layer (< 8 mm) atop a sufficiently conductive lower layer.

Explosion Suppression – This process is generally used in containers and production facilities for which an explosive pressure exceeding the explosion resistance of the container/facility is predicted. The explosion is suppressed in its initial stages, before a hazardous rise in pressure can occur. To accomplish this, an extinguishing agent is used in the protected area within fractions of a second of the explosion being detected. For the suppression of an explosion (use of extinguishing agent) it is mandatory that the explosion be detected promptly. In the case of explosions that begin slowly, the initial pressure build-up is not adequate for prompt detection. Additional measures such as optical fire detectors or supplementary pressure detectors may be necessary.

Explosion Barriers – Refers to the prevention of explosion spread, isolation of devices/facilities. Isolation as an explosion protection measure allows the explosion to reach full force, but prevents it from spreading to other, unprotected parts of the facility. This is accomplished by mechanical barriers which immediately block connecting pathways, or by a barrier consisting of chemical extinguishing agent(s).

Explosion-Resistant Construction – This type of construction limits an explosion to the inside of compression-proof or blast-proof containers – which, however, also means that connected equipment such as tubes/pipes and decoupling mechanisms must fulfill the same requirements. Explosion-resistant containers or equipment are those that can withstand many times the predicted explosive pressure without being permanently deformed.

Explosion Venting – Defined venting by means of bursting discs, pressure-relief flaps, etc. This measure is intended to prevent the build-up of excessively high explosive pressure in the inside of containers by prompt release through certain openings. This measure addresses only the effects of the explosion, and can be implemented without additional control mechanisms. As soon as the static response pressure is reached or exceeded, an outflow process from the protected apparatus into the surrounding area begins. Apart from the flame and pressure wave, this outflow from the venting openings, which is a part of explosion venting, also contains combusted and non-combusted substances. It must always be established whether the effects of the explosion in the location in question can be managed.

Ignition Sources – These are the causes that can trigger a dust explosion; the most common ones are flames and direct heat, hot work, incandescent material, hot surfaces, electrostatic sparks, electrical sparks, friction sparks, impact sparks, self-heating, static electricity, lightning and shock waves. These ignition sources differ in terms of temperature, energy and power; the dusts can be ignited by low energy as well as high energy ignition sources. Of these, the ignition sources which can occur inside the plant are of particular importance, and include incandescent material, hot surfaces, sparks, self-heating and static electricity.

- Flames and direct heat - This trigger can be eliminated by using indirect heating methods like circulating hot water or steam through pipes and using hot water/steam baths.
- Self-heating - Self-heating or spontaneous combustion may occur due to exothermic reactions. A variety of reactions can give rise to self-heating. These include oxidation reactions as well as reactions of certain dusts with water or wood. In most cases the reaction rate accelerates with temperature, but there are also autocatalytic reactions which may accelerate due to production of a catalyst or removal of an inhibitor. Induction times may be long and the self-heating may be slow to start but may then proceed undetected for a long period.

Contaminants such as oil and products of thermal degradation can also contribute to self-heating. Dust should be screened to determine whether it is prone to self-heating. The dust temperature during the process and in storage should be controlled. One aspect of this is control of hot surfaces, which may arise in normal operation. The unintended accumulation of dust deposits, which could undergo self-heating, should be avoided. Situations in which there is a large mass of dust stored at a high initial temperature (to keep the dust dry) are hazardous. Dust in a pile has a high surface area and sufficient air circulation, both of which favor self-heating. The risk of accident is further enhanced during the discharge of hot dust from a drier into a hopper. It may be necessary to cool the dust prior to storage. Another measure which is sometimes used is to recirculate the hot dust through a cooling system prior to its further use.

- Hot work - Excessive heat generated during operations such as welding and cutting is another obvious trigger more so when a dust of low ignition threshold (100–200°C) is present nearby. Accidents often occur because this hazard is not appreciated and the dust is not cleaned out of the equipment before hot work is started.
- Incandescent material - Smoldering particles or other incandescent material can trigger a dust explosion inside dust handling equipment. The explosion may then travel through the ducts and connected vessels. Direct firing systems are potential sources of incandescent particles. In direct-fired driers the air inlet should be protected by a fine screen to prevent ingress of such incandescent material.
- Hot surfaces - Equipment with a hot surface such as a steam pipe or electric lamp, or overheated moving equipment such as distressed bearing, falls under this category of triggers. The surface temperature that can cause ignition of a dust layer is frequently in the range of 100–200°C. The ignition temperature moves closer to the lower limit of this range as the thickness of the layer increases. Findings from investigations of dust related accidents often reveal that ignitions occur at unexpectedly low temperatures. A dust may contribute to its own ignition; dusts being poor conductors of heat, a layer of dust on the equipment may prevent heat loss to the atmosphere and thus raise the temperature below the

surface of the dust heap to the point of ignition. A smoldering or burning layer can act either directly as an ignition source for a dust cloud or by means of agglomerations or 'nests' of burning material that break away from deposits and ignite a dust cloud in another part of the plant. When dusts accumulate on hot surfaces, they may go through different and complex stages before combusting. Some dusts burn directly in solid phase with a flame or by smoldering, others melt and burn as liquids. Some dusts can give off large amounts of flammable gases. The size of the flames produced by different dusts also vary. Hot-surface ignition is a particular problem with driers of various types.

- Electrostatic sparks - Electrostatic discharge from electrical equipment may cause a spark which in turn may ignite a dust cloud. Protection against such discharges is based on hazardous area classification and the associated safeguarding. Electrical equipment is designed so that incendive capacity or inductive discharges cannot occur.
- Electrical sparks -Electrical sparks occur in the normal operation of switches and relays and in malfunctioning electrical equipment. To protect against electrical sparks, hazardous areas ought to be classified and safeguarded. In particular, flameproof equipment must be used, and should exclude dusts. A distinction may be drawn between equipment which is dust tight and excludes dust entirely and equipment which is dustproof and lets in only an insignificant amount of dust.
- Friction sparks and hot spots - Frictional sparks can occur wherever there is rubbing of one solid with another or during grinding. Foreign materials such as tramp iron can also cause sparks. The dust itself may block the equipment and cause overloading, leading to spark generation. To prevent frictional sparks, dust flow should be controlled and machine overload trips should be installed. The removal of foreign objects should be effected by magnetic or pneumatic separation, especially when the material is to pass through a mill. Friction-induced heating can also raise dust temperature. Pulling the dust through drag conveyer heats it up a bit, so do mixing operations. But more serious friction-induced heating can occur when hot spots are formed in localized areas of blenders due to the blenders' shearing action.

- Impact sparks - Hand tools may create an incendive impact spark, although there is little evidence from incidents of single impact ignition. The incendive potential of an impact such as that of a metal tool on a metal surface arises from the heating of that surface. The ignition source is not the spark itself but the heated surface, and the heat is transferred from the metal surface to the dust.
- Static electricity - Static electricity may turn to sparks when an object moves rapidly into or out of its field. It is more strongly influenced by the process than by the material. For sieving and pouring the charges are low, but for size reduction they are much higher.
- Lightening, shock waves - Lightening can initiate dust explosions.

Ignition Temperature – Defined as the lowest temperature of a heated wall that will ignite a dust/air mixture upon brief contact. The shape of the vessel in which the ignition temperature is measured is important. Ignition on differently shaped surfaces is, in practice, only possible at much higher temperatures. In the case of dust involving food products and animal feeds, the value is typically between 410 and 500°C, depending on the material.

K_{St} – This is a classifying parameter that defines the volatility of the combustion. The value reflects the maximum speed of pressure build-up during the explosion of a dust/air mixture in a container of size $1m^3$. The K_{St} value is used as a basis for calculating pressure discharge surfaces.

Maximum Explosive Pressure – In simple closed containers, flammable dusts can cause an explosive pressure between 6 and 10 bar. With light metal dusts, explosive pressures of up to 20 bar are possible.

Minimum Ignition Energy – Defined as the minimum energy of an electrical spark, under defined conditions, that is capable of igniting the explosive dust/air mix. Not every spark will cause ignition. The decisive factor for ignition to occur is whether there is sufficient energy to cause combustion of the entire mix.

Moisture – Moisture or relative humidity of a dust mixture is an important factor for potential ignitions and explosions. Although there are no engineering or empirical formulae, it is understood that a moister dust requires a higher ignition energy and is less likely to ignite.

Particle Size – Chapter 1 provides several definitions of particle size. Dust particles larger than 400 μm are not combustible. Dust particles that measure less than 400 μm and typically in the league of 20 μm are considered combustible. Handling, processing, transporting granular materials, pellets and large particles cause attrition, producing finer particle sizes.

Smoldering Temperature – Defined as the lowest temperature of a hot surface on which a 5mm dust deposit is ignited. The smoldering temperature describes the ignition characteristics of thin dust layers. If the layer is thicker, or if the ignition source is completely buried in dust, the thermal insulation provided by the dust layer increases, which changes the smoldering temperature, sometimes lowering it considerably, which could trigger an exothermal reaction. Experiments have shown that the smoldering temperature decreases nearly linearly as the thickness increases. Smoldering temperature can be considerably lower than the ignition temperature for an airborne mixture of the same dust. The estimated maximum permissible surface temperature for electrical equipment may be higher, depending on the dust's thermal conductivity. Unnoticed smolder spots can be present for long periods in thick layers of dust and can, if the dust is swirled up, become highly effective ignition sources.

Thermal Resistance – Molded plastic must certainly fulfill the most complex requirements. For electrical devices from Categories 1D and 2D, the temperature index "TI" must be known, according to EN 50281. This figure allows conclusions about the long-term mechanical properties of molded plastic to be drawn. The temperature index is identical to the 20,000-h point on the thermal resistance diagram, with a reduction of the bending strength (tensile strength) of < 50 %. This figure must be 20K higher than the temperature at the hottest area of the enclosure. In addition, the molded plastic must be proven to have sufficient thermal resistance for the intended application. Enclosures or enclosure parts made of molded plastic for electrical equipment from the categories 1D and 2D must be subjected to heat and cold resistance tests according to EN 50014 (artificial deterioration). The deterioration process caused by extreme temperatures must not cause the molded plastic to become brittle, and thus unable to provide protection according to IP regulations.

Recommended References

K.L. Cashdollar, M. Hertzberg, and I.A. Zlochower, Effect of Volatility on Dust Flammability Limits for Coals, Giolsonite, and Polyethylene, 22nd Symp. (Int.) on Combustion, The Combustion Institute, 1757 (1988).

K.L. Cashdollar, *Overview of dust explosibility characteristics*, J. Loss Prev. Process Ind., 13(3–5), 183–199 (2000).

R.K. Eckhoff, *Current status and expected future trends in dust explosion research*, J. Loss Prev. Process Ind., 18(4–6), 225–237 (2005).

J-L.Chen, R.Dobashi, and T.Hirano, *Mechanisms of Flame Propagation through Combustible Particle Clouds*, J.Loss Prev. Process Ind.,9(3),225 (1996).

W-J.Ju, R.Dobashi, and T.Hirano, *Dependence of flammability limits of a combustible particle cloud on particle diameter distribution*, J.Loss Prev.Process Ind., 11(3), p.177(1998).

W-J.Ju, R.Dobashi, and T.Hirano, *Reaction zone structures and propagation mechanisms of flame in stearic acid particle clouds*, J.Loss Prev.Process Ind., 11(6), p.423(1998).

R. Dobashi, J-H. Sun, W-J. Ju, and T. Hirano, Flame Propagation through Combustible Particle Clouds, Fire and Explosion Hazards - Proc. of the Third International Seminar, 569–578 (2001).

Center for Chemical Process Safety, *Guidelines for Safe Handling of Powders and Bulk Solids*, AIChE, 2005.

Febo, H., FM Global Personal communication to S.S. Grossel, 2001.

Thornberg, W., "Personal communication from Mr. Thornberg of IRI to S.S. Grossel, 2001.

Eckhoff, R., *Dust Explosions in the Process Industries*, Butterworth-Heineman, 2nd Edition, 1997.

Matsuda, T. and Yamaguma, M. "Tantalum Dust Deflagration in a Bag Filter," J. of Hazardous Materials, v A77, (2000) pp. 33–42.

Tyldesley, A., Personal communication from Alan Tyldesley, HSE, Bootle, UK to S. S. Grossel (February 19, 2004).

Drogaris, G. *Major Accident Reporting System:* Lessons *Learned from Accidents Notified.* Elsevier Science Publishers, 1993.

Pickup, R. D. "Dust Explosion Case Study: Bad Things Can Still Happen to Good Companies," Proc. Safety Prog., Vol. 20, pp. 169–172, September 2001.

NFPA 69, "Standard on Explosion Prevention Systems," National Fire Protection Association, 2002.

FM 7-76, "Prevention and Mitigation of Combustible Dust Explosions and Fires," Loss Prevention Data Sheet 7-76. FM Global, 2001.

CSB, "Dust Explosion (6 Killed, 38 Injured)," Chemical Safety Board Investigation Report No. 2003-07-1-NC, September 2004.

NFPA 654, "Standard for the Prevention of Fire and Dust Explosions from the Manufacturing, Processing, and Handling of Combustible Particulate Solids, National Fire Protection Association, 2000.

4

Explosion Prevention in Grain Dust Elevators

4.1 Introduction

Dust explosions have been the concern of several industry sectors such as the coal mining industry, facilities handling agricultural products, and those industries involved in the manufacture of finely divided organic materials such as plastics, pharmaceuticals, etc. In spite of industry familiarity with the problem and a rich scientific a industry library of good industry practices to prevent explosions, the problem still persists.

The first recorded incident of a dust explosion was in 1785, in a flour mill in Turin, Italy. A series of accidents during World War I led to a flurry of scientific activity culminating in the publication of the classic work of Price and Brown[1] along with numerous pamphlets and bulletins by the U.S. Department of Agriculture. This work identified grain dust as the specific

[1] D. J. Price and H. E. Brown, Dust Explosions: Theory and Nature of, Phenomena, Cause and Methods of Prevention, Published by National Fire Protection Association, Boston, Mass, 1922

ingredient common to all accidents, and recommended best practices were made in order to prevent the occurrence of these accidents. With minimal updating, many of these good practices can still be used, but the continuing accident record would indicate that they have not received widespread implementation. The revival of interest in agricultural dust explosions in the U.S. was a series of explosions which occurred in four U.S. grain elevators during the Christmas season of 1977:

- Behimer and Kissner in Wayne City, Illinois (one death, no injuries, and $1.5 million property damages)
- Continental Grain in Westwego, Louisiana (thirty-six deaths, ten injured, $30 million property damages)
- Sunshine Mills in Tupelo, Mississippi (four deaths, fifteen injured, $1 million property damages)
- Farmers Export in Galveston, Texas (eighteen deaths, twenty-two injured, $15 million property damages)

These explosions constituted an enormous human and financial loss. Many of the dead and injured were federal employees and the loss of two export elevators decreased the number of elevators by approximately 2.5 percent; the consequence of which was that considerable federal attention was immediately given to the problem of safety in grain handling facilities.

The FBI, BATF, OSHA, and state and local authorities conducted accident investigations, a symposium was held, a hazard alert was issued and a National Academy of Sciences panel was created. The private sector responded with establishing revised fire codes relating to grain elevators and a large research fund was established by the grain industry trade association focusing on prevention.

Despite all of this effort and attention, 29 major incidents occurred in 1979 resulting in two deaths and nineteen injuries. In 1980, forty-five accidents occurred with twenty deaths and fifty injuries. In 1984, there were at least fifteen explosions, resulting in eight deaths and twenty-four injuries. Reporting of incidents over the subsequent years has been spotty and reliable statistics have not been found.

Hajnal[2] reported on grain elevator dust explosions in South America. Between 2001 and 2002 there were three major dust explosions

[2] R. D. Hajnal, Grain Dust Explosions: A Report from South America, Oil Mill Gazetteer, Volume 111, August 2005.

documented: Toepfer, Puerto San Martin (Argentina), ACA San Lorenzo (Argentina) and Coinbra, Paranaguá (Brazil). In these incidents there was both great loss of life and materials. In October 2001, a severe explosion left three dead and seven injured in the Terminal of A.C. Toepfer in Puerto San Martín, Santa Fe province, Argentina. A month later, a similar disaster destroyed the port terminal of Coinbra, Louis Dreyfus' Brazilian grain subsidiary, in Paranaguá, Paraná State, Brazil. Fortunately, on this occasion, it was without casualties but it did cause complete material damage. In April 2002, the ACA Terminal (Asociación de Cooperativas Argentina) exploded in San Lorenzo, Santa Fe province, Argentina. The result was tragic: three people were killed, 19 injured, and there was total destruction of the main infrastructure, resulting in millions of dollars of lost materials. Other recent explosions, although less damaging, include the terminal of Productos Sudamericanos, in Punta Alvear, near Rosario, Argentina, on the Paraná river, in August 2000; the Louis Dreyfus Terminal in General Lagos, north of Rosario (where the world's largest oilseed crushing plant is located, with a production capacity of 12,000 metric tons per day) - one person was killed in this explosion. An explosion in a flour silo at Molino Argentino (a wheat flour mill) in 1995 in the Buenos Aires metropolitan area killed three but occurred without any material losses. One of the largest grain elevator dust explosions in South America was in 1990, at the Genaro Garcia Terminal in the port of Rosario, resulting in 10 deaths. The worst explosion was in 1985 at the silos of la Junta Nacionalde Granos, in Bahía Blanca, an ocean terminal, killing 22 and injuring more than 10 people.

Grain dust explosions have resulted in spectacular dollar losses and human tragedies in many parts of the world. Despite rather extensive safety precautions and standards, many of which were introduced during the World War I post era, these operations constitute to be high risk.

4.2 Causes

As explained in previous chapters, the requirements for a fire are fuel, oxidizer, and ignition as represented by the familiar "fire triangle." These also represent a portion of the elements required for a successful dust explosion, but in addition, two more requirements must be met. The fuel must be well mixed with the oxidizer, and the mixture must be confined. Hence, the concept of the fire explosion pentagon was devised.

The mixing requirement of the explosion pentagon allows two major classifications to be made with regard to dust explosions. The primary explosion is one in which the dust and the oxidizer are mixed before the introduction of the ignition source. This mixing usually takes place inside of equipment or a structure where for one reason or another agitation of the product occurs producing a dust cloud.

The primary explosion is almost always followed by secondary explosions which occur in open areas of the building or outside of the equipment. Here the dust is layered on horizontal and vertical surfaces and initially unmixed with the oxidizer. The dynamics of the explosion process create essentially a windstorm which mixes the two, again producing a dust cloud.

In the primary explosion, the confinement requirement is usually met by the enclosure which surrounds the equipment. Most often its main design function is to prevent the escape of suspended dust. For the secondary explosion, major portions of the building structure unto itself, such as tunnels, silos, head houses, galleries, provide the confinement. The oxidizer component of the explosion pentagon is provided by atmospheric air.

The fuel for the explosions in grain handling and agricultural products facilities is grain dust or the product. These are highly explosive materials. The explosive characteristics of these materials may be measured by using a standard dust explosion testing device known as a Hartmann bomb. In a comparison with TNT, agricultural dust in general is easier to ignite and results in a more severe explosion than TNT dust. It is little wonder that for decades large agricultural insurance companies have been distributing placards to the insured that read:

"Warning, Grain Dust Is Like High Explosives!"

The ignition source for a dust explosion is usually provided by a mechanical or electrical device. The mechanical and electrical devices both can produce either sparks or a hot surface. In some cases, ignition of the dust cloud can occur directly, in other cases, a fire with or without an open flame may first result which then itself acts as the ignition source for the dust cloud. The energy requirements for the ignition of a dust cloud are quite small, and it has been found that most actual ignition sources are capable of supplying considerable amounts of energy.

The removal of any one of these five components will prevent a dust explosion from occurring; however, it may be more effective or convenient to deal with certain ones.

Figure 4.1 shows the typical details of a grain elevator system. Explosions in grain elevators are not caused by spontaneous combustion. To cause an explosion, the dust must be mixed with air to form a dense cloud (layered

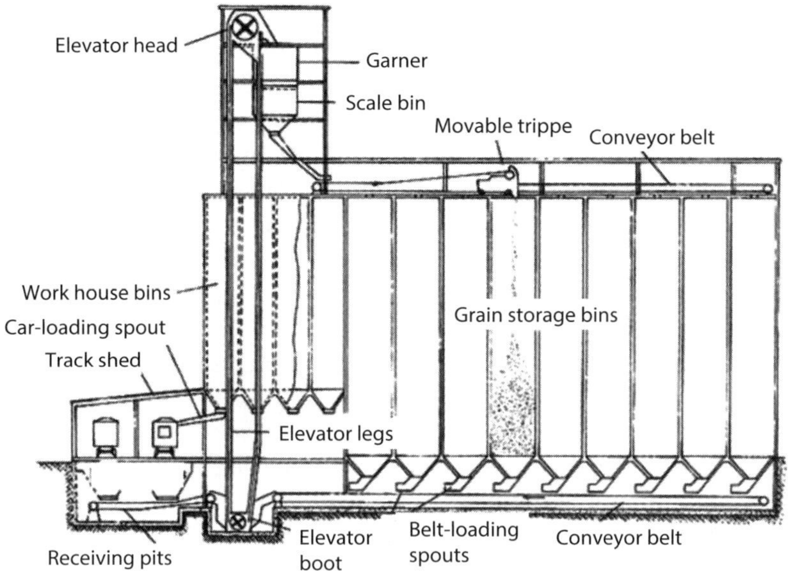

Figure 4.1 Typical design features of a grain handling process.

dust only smolders upon ignition.) The airborne dust concentrations must also be above the lower explosive limit (LEL), a condition so opaque that visibility is minimal and breathing is difficult. Concentrations above the LEL occur regularly inside most bucket elevators but seldom elsewhere in a facility. Contrary to popular belief, grain elevator explosions are not one big blast. They usually involve a primary explosion and a series of secondary ones. The primary explosion generates shock waves throughout the elevator, often raising into suspension layered dust on walls, rafters, equipment, and the floor. Accumulations of as little as one-hundredth of an inch will propagate the flame from an initial explosion. In other words, layered dust provides the fuel to turn a primary explosion—often itself quite minor—into a major one.

Dust explosions usually occur at grain transfer points – in bucket elevators or enclosed conveyors – where small dust particles become dislodged from kernels due to tumbling, agitation, and kernel impacts, as fast-flowing grain hits bucket elevator cups or changes direction in drag or belt conveyors. The turbulent grain movement causes high levels of suspended dust particles (two to 20 microns in diameter) in the airspace, often close to a hot leg boot section bearing or a spark from tramp metal in a dump pit or drag conveyor. According to national survey data, of 106 reported grain dust explosions in the U.S. since 1988, 51 were in grain elevators and 34 were in grain milling facilities (wheat, corn, oat, and rice mills)[3].

Many primary explosions originate in elevator legs. Stored grain typically contains 2 to 10 pounds of grain dust per ton (see Parnell[4]).

If a 12,000-bushel per hour leg handles wheat at 360 tons per hour, at the lower level of two pounds of dust per ton, 720 pounds per hour of grain dust is moving with the grain. If this leg is 130 feet high, the leg trunk casing volume is about 500 cubic feet. At the MEC level of 0.05 ounces per cubic foot, only 25 ounces, or 1.56 pounds, of free grain dust recirculating in the air inside the leg is needed to reach the MEC.

An NGFA report on grain dust levels in bucket elevators states that "concentrations in the bucket elevator almost always exceed the minimum limits and thus constitute an explosive condition" (see Buss[5]). Hence, when only 0.05 ounces of dust per cubic foot are needed to reach the MEC, as dust concentrations build inside a leg, they can quickly exceed the MEC, even in some aspirated or ventilated legs when excessively dusty grain, like sorghum, is being transferred.

Belt speeds for a 12,000-bushel per hour leg typically run between 600 and 800 feet per minute, or about 10 to 13 feet per second. The belt in a 130-foot leg makes one revolution in about 20 seconds. Part of the airborne dust tends to circulate continuously as the air is dragged along by the cups in the leg casing. Even though only a portion of the total dust is entrained in the air in the leg casing, much of the dust in non-ventilated legs remains concentrated in the air circulating in the leg housing during continuous operation, usually exceeding NGFA's MEC value of 0.05 ounces per cubic foot (see Buss, 1981).

4.3 Properties of Grain Dusts

Parnell, et.al.[6] reported on the properties of several grain dusts. Grain dust physical properties are important to understanding the causes of

---

[3] Schoeff, Robert W. 2006. Agricultural Dust Explosions in 2005. Department of Grain Science, Kansas State University, Manhattan, KS., March 20, 2006, 5 pp.

[4] Parnell, Calvin. 1998. Personal Conversation and E-mail. Texas A&M University, College Station, TX. June 15, 1998.

[5] Buss, Kenneth L. 1981. Dust Control for Grain Elevators, National Grain and Feed Association, Washington, DC, pp. 64–87.

[6] C. B. Parnell, Jr., D. Jones, R. Rutherford, and K. Goforth, Physical Properties of Five Grain Dust Types, Environmental Health Perspectives, Vol. 66, pp. 183–188, 1986

explosions. In addition, cyclone efficiency evaluations and design criteria require substantial data on dust physical properties. Cyclones are standard dust control devices used with grain handling operations.

As already noted, for grain dust explosions to occur, four ingredients must be present. These ingredients are fuel, confinement, ignition source, and oxygen. The fuel for a grain dust explosion is grain dust in suspension above the minimum explosive concentration (MEC). Containment is a requirement for an explosion to occur in that it allows a buildup of pressure resulting in rupture of the confinement. Containment is also necessary to achieve the MEC of grain dust; Palmer[7] reports this to be in the range of 50 g/m^3.

Parnell, et.al. note that the dispensability and combustion rates of dust are governed by chemical and physical properties of the dust involved. How easily and uniformly a dust is suspended into the air depends on its particle size distribution and density. The rate of combustion is highly dependent on the exposed surface area of dust that can readily react with oxygen. These physical properties are vital to defining dust explosibility and in defining dust explosibility and developing explosion hazard indication criteria.

The investigators note further that different laboratory techniques have been employed by various researchers in an effort to quantify dust characteristics, resulting in reporting particle size distributions (see table 4.1 for literature reported data). Plemons[8] reports that by far, the most explosive grain dust fraction is that associated with aerodynamic particle sizes of less than 100μm. The smaller fractions of grain dust are most explosive because the surface area per unit mass increases as the particle size decreases. However, larger fractions (250–500μm) in sufficient concentrations can also be explosive.

The surface areas of grain dust have been determined and reported on by Deshpande and Matthews[9] and Martin (1981). Martin reported that the surface area for grain dust varied from 0.6 to 0.9 m^2/g. Deshpande and Matthews found that the surface area for grain dust ranged from 0.6 to 1.96 m^2/g.

---

[7] Palmer, K. N. Dust Explosions and Fires. Chapman and Hall, London, 1973.

[8] Plemons, D. S., and Parnell, C. B. Developing an explosion index based on chemical and physical properties of grain dust. ASAE Paper No. 81-3068, American Society of Agricultural Engineers, St. Joseph, MI, 1981.

[9] Deshpande, U. A., and Matthews, J. C. Adsorption of CO and CH4 on grain dust: surface area measurements by adsorption of N2 and CO2. In: Proceedings of the International Symposium of Grain Dust, Kansas State University, Manhattan, KS, 1979.

Table 4.1 Literature reported particle sizes.

Dust	Measurement Method	Mass Mean Diameter, microns	Noted	Investigator
Rice	Coulter Counter	21.75	Mass mean diameter of whole grain dusts determined by Coulter technique with a 400 micron aperture.	Plemons[10]
Corn	Wet Sieving	19.57		
Soybean	Dry Sieving	25.17		
Wheat		32.97		
Sorghum		36.92		
Soybean	Coulter Counter	30.00	Mass mean diameters of whole soybean dusts determined by Coulter Counter and read from graphical results.	Martin[11]
	Wet Sieving			
	Dry Sieving			
	Capture Velocity Technique			
Corn	Coulter Counter	13.70	Mass mean diameters of whole corn and soybean dust determined by Coulter Counter technique	Wade[12]
Soybean		15.50		

[10] Plemons, D. S., and Parnell, C. B. Developing an explosion index based on chemical and physical properties of grain dust. ASAE Paper No. 81–3068, American Society of Agricultural Engineers, St. Joseph, MI, 1981.

[11] Martin, C. R. Characterization of grain dust properties. Trans. ASAE 24: 738–742 (1981).

[12] Wade, F. J., Hawk, A. L., and Watson, C. A. A survey of grain dust properties at large grain terminal. In: Proceedings of the International Symposium of Grain Dust, Kansas State University, Manhattan, KS, 1979.

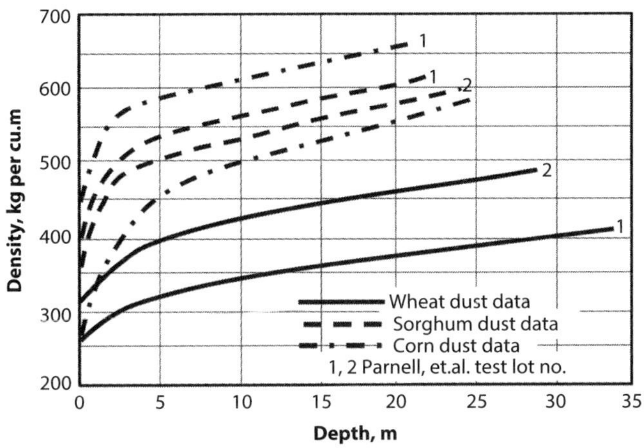

Figure 4.2 Literature reported vertical density distribution of grain dust in self-packed dust columns.

The bulk density and particle density affect the handling and conveying characteristics of particulate material. Chang and Martin[13] developed models to predict the density distribution and weight of grain dust in self-packed columns. They reported that the bulk density of self-packed dust increased linearly as the depth of the pile increased. These tests were performed on wheat, sorghum, and corn dust. Figure 4.2 shows the relationships between the vertical density distribution of grain dust in a self-packed grain dust column.

Plemons and Parnell (1981) reported that the particle densities for rice, corn, wheat, soybean, and sorghum dust which ranged from 1.41 to 1.90 g/cm^3 for wheat and soybean dust, respectively. Martin (1981) found the particle density of whole "grain" dust to be 1.49 g/cm^3.

Physical properties of grain dust play an important role in explaining dust explosibility and handling characteristics. Analysis of interaction between these properties aid in the development of an explosion hazard indicator and in the design and evaluation of dust handling/separation equipment. Results of the laboratory analysis of wheat, corn, rice, soybean, and sorghum dust reported by Parnell (1986) are as follows:

- bulk density, 0. 150–0.308 g/cm^3
- particle density, 1.43–1.69 g/cm^3;
- % < 100μm (by weight), 34.3–50.6%

[13] Chang, C. S., and Martin, C. R. Bulk density characteristics of grain dust. Trans. ASAE 27: 898–902 (1984).

- mass mean diameter (dust < 100μm), 10.7–14 μm
- ash content, 5.12–30.6%

4.4 Case Studies

4.4.1 Toepfer Puerto San Martín Explosion,Argentina, October 2001

The explosion happened during lunchtime in the spring, one hour after loading a ship, in a tunnel underneath five steel bins. Parallel to and beside these bins there was a horizontal flat silo, that was empty, where five workers were doing civil maintenance work. Between the steel bins and the horizontal silo there was a connecting tunnel that held a belt conveyor that collected both from these silos and from others, conveying to the shipping bucket elevators. The first or primary explosion started underneath the steel bins, in the tunnel. Standing dust on the floors and edges was stirred up by the shock wave caused by this primary explosion, and provided the fuel for a secondary explosion, which was much more violent than the first, expanding quickly through the connecting tunnel to the tunnel underneath the horizontal silo, where the workers were. The horizontal silo was empty but two workers inside the tunnel were killed instantly and a third, working on the floor of the flat silo, was killed as the concrete tunnel roof blew up. A chain reaction of ever-increasing intensity had been set in motion that culminated with a third explosion that impacted the reception area (about 300 meters away from the starting point), and other concrete and metallic structures.

The conveyor belt had stopped operating inside the tunnel where the explosion had started one hour before the explosion. Therefore, everything was still and quiet, and there was no dust-air mixture in suspension. An ignition source of sufficient energy, temperature, and duration to initiate the explosion has to be present. Without an electrical spark or an overheated bearing, and with no dust in suspension and nothing moving, an explosion would not have been expected.

Experts theorized that the Toepfer explosion was due to hexane gas rather than grain dust. Since rebuilding hexane monitors have been installed in the tunnels which detected gas leaking into the tunnels under the storage from the crushing plant next door. Steps were taken on both sides to correct this dangerous situation.

4.4.2 Coinbra Paranaguá Explosion, Brazil, November 2001

This explosion occurred while loading corn onto a ship. There were no fatalities; however there were six injuries and extensive damage to the facility. The explosion started in the shipping bucket elevators that were in operation, and was most probably due to belt misalignment. This primary explosion expanded quickly throughout the whole facility. The first explosion caused dust within the facility to be placed into suspension in the air, thereby contributing to a series of subsequent explosions. The secondary explosion was so strong that all resistant structures collapsed, even rail cars were overturned, large pieces of concrete that weighed over five tons were blown 300 meters away, and the steel shipping tower was turned down to earth. The explosion was followed by fire, which ignited the grain and continued burning for nearly three weeks.

4.4.3 Aca San Lorenzo Explosion, Argentina,April 2002

This facility had a comprehensive housekeeping program that ensured dust accumulations were promptly and regularly cleaned. It included a brand new and highly efficient dust collection system, fitted with modern filters (bag-houses) with low pressure automatic cleaning. A thorough maintenance program was in place as well as a training program for employees and contractors on the hazards of handling and collecting dust.

The explosion occurred while loading soybeans onto a ship on a dry and sunny autumn day. The explosion left three dead, 19 injured, and caused massive destruction of the terminal. The facility had been loading a ship that was receiving simultaneously from three spots: from the horizontal silo, directly from trucks through the receiving pits, and directly from railcars. All three spots were connected through a tunnel that collected from the horizontal silo, passed through the truck reception pits, through the railcar pits, and continued to the shipping head house tower that supported the shipping bucket elevators. This head house was built on a concrete structure, but with no walls, that helped to stop and dissipate the destructive explosion wave.

An unknown ignition source ignited dust within the facility and resulted in a series of explosions that severely destroyed the heart of the port facility. The actual ignition source may never be known due to the damage that occurred in the tunnel beneath the horizontal silo, where the first or primary explosion started, and because the employee working in this area at the time of the explosion was killed. The below-surface tunnel was in the flat storage, connected to another collector tunnel, an underground avenue that led to the shipping tower. The collector tunnel collected grain from the truck reception area, from other tunnels under other horizontal silos, and from the railcar reception area arriving at the bucket elevator pits in the shipping tower.

The underground infrastructure was a long network of confined spaces that distributed and accelerated the propagation and intensity of the explosion. A second worker was killed while operating the railcar gates and a third was found dead three days later in the shipping bucket elevator pit, 15 meters below ground level, where the shipping tower stands.

When the ACA facility was rebuilt the original design concept of relying on extended lay-outs with no bucket elevators was changed. Current think is that the installation of bucket elevators will prevent the transmission of any primary explosion to the rest of the facility. The design is supported by several concepts of minimizing explosion risks, by:

- Eliminating tunnels where possible and instead using open galleries and catwalks that operate above ground, loaded by new bucket elevators at the end of each horizontal silo.
- Isolating the risks to certain sectors by installing the bucket elevators at the end of every tunnel, eliminating connections between tunnels, avoiding the propagation of the explosion.
- All the mechanical handling is now fitted with hazard monitors, controlling speed, belt misalignment, belt slip, plugging and maximum belt extension, with emergency stop.
- The elevator towers are open and made of steel. Elevator pits are also open.
- Reduction of environmental pollution (fugitive dusts) by replacing cyclones with low-pressure filters (bag houses). The current design incorporates thirty three large aspiration systems with filters (bag houses), that collect the dust emitted during operations, in different sections of the facility.
- The design includes a white mineral oil application system for dust-emission control.
- The design includes dust-suppression systems in two of the four ship loading tubes, with telescopic hoses, thereby minimizing the dust emission during loading of a vessel.
- The rebuilt system is designed for greater operational efficiency, accomplished by reducing belt speeds and increasing capacities, with wider belts to reduce dust generation. For example: previously, the shipping belt ran at a capacity of 1000 tph at 4.00 m/s. By comparison, in the new facility the same conveyors run at a capacity of 1,200 tph at 3.00 m/s.

4.4.4 Grain Elevator Dust Explosion in Minnesota, August 17, 2012

A combustible dust explosion damaged part of a grain elevator at Greenway Co-Op in Kasson, Minnesota. Greenway Co-Op has been providing agricultural products and services since 1930 including grain storage and custom grain drying. Employees at Greenway were moving grain at the time of the explosion, when there was some sort of spark that caused the dust to explode and blew the top part of the elevator off. Investigators speculated that there was a mechanical problem above the elevator. The explosion occurred at the top of an 80-foot leg that lifts corn to be loaded onto trucks and that the source of the heat might have been bearings in a moving part. The Greenway Kasson operation was shut down for 24 hours to ensure that another dust explosion did not occur. Initial damage estimates were as much as $100,000.

4.4.5 De Bruce Grain Elevator in Wichita, KS 1998

This incident resulted in killing seven workers and injuring several others. The ground shook was so hard that people believed that McConnell Air Force Base, close to the facility, was under missile attack. Windows were shattered for blocks in every direction. Flames shot hundreds of feet into the air and the smoke plume shocked city habitants

4.4.6 Grain Elevator Explosion in Kansas City, October 29, 2011

An explosion at a grain elevator in Kansas has claimed six lives on October 29th 2011. Three victims were found initially but unstable concrete and other damage forced crews to temporarily call off their search at the Bartlett Grain Co. facility in Atchison, about 50 miles NW of Kansas City. The first three casualties found were Bartlett Grain Co. workers aged 20, 21 and 24. The blast, which shook the ground in neighboring Missouri, highlights the dangers workers face inside elevators brimming with highly combustible grain dust at the end of the harvest season. The explosion sent an orange fireball into the sky, created a large hole in the side of the one of its concrete silos and blew a huge section of the grain distribution building roof off. Workers were loading a train with corn when the explosion occurred, but the cause was not immediately known.

4.4.7 Port Colbourne Elevator in Ontario, Canada, 1952

In the Port Colbourne elevator in Ontario, Canada, the explosion occurred in steel bins and the roof was blown off entirely. The ensuing fire also damaged the head house seriously. A grain dust explosion also seriously damaged the grain elevator no. 4A of the Saskatchewan grain pools, Sept. 24, 1952, where six men were killed and 14 injured. The primary explosion in a shipping bin was followed by a secondary explosion involving large quantities of dust, which had been allowed to accumulate in the building. The roof gallery above the bins was also destroyed.

4.4.8 Explosions at Various U.S. Facilities

One of the earliest and most serious accidents was the grain dust explosion of the Peavey terminal elevator at Duluth, Minn., in 1916. After the explosion, the cribbed grain bins caught fire, completely destroying the elevator. This was one of the worst roaring infernos in dust explosions.

An explosion in a starch / corn plant at Cedar Rapids, Iowa, in 1919 killed 43 people and one at a similar plant in Peking, Illinois, in 1924 resulted in 42 deaths.

A corn dust explosion occurred in the feed mill Wayne Feeds at Waynesboro, U.S.A., May 25, 1955. 3 men were killed, 13 were injured, and the violent explosion caused extensive property damage.

A grain dust explosion occurred in the workhouse of the South Chicago elevator in 1956. Because of the light steel frame construction of the roof gallery on top of bins, the explosion pressure was relieved so that the concrete bins below were not seriously damaged.

A severe explosion occurred in Kansas City, Mo., in 1958, when the Murray elevator was badly damaged. The head house of steel construction was completely shattered and its installations destroyed by fire.

4.4.9 Other Examples

A dust explosion, which was caused by the welding of a spout, excessively worn by the flow of grain, occurred in the Kampffmeyer grain silo at Albern near Vienna, July 4, 1960. The welding was performed in the elevator pit, when a spark ignited the dust in the running bucket elevator. The pressure wave of the explosion went up through the elevator shaft, ripping the casing of the elevator leg, and continuing up to the roof, causing severe damage to the building and machinery.

A flour dust explosion caused serious damage to the Sun flourmills in London on Aug. 7, 1965, 4 men died and 37 were injured when a giant blast shattered and set ablaze the mill building and a wheat storage silo of cribbed construction. The explosion is believed to have been initiated by welding a bin containing flour.

In Germany, a most violent dust explosion on Dec. 14, 1970, seriously damaged the grain silos at Kiel-Nordhafen on the Kaiser Wilhelm shipping canal, connecting the North and Baltic Seas. It was the worst accident of this kind in Germany. 6 men died and 17 were injured. The damage to plant, building and machinery is estimated at 10 million dollars.

A dust explosion occurred in the United States at Destrahan near New Orleans, where a Bunge Corp. terminal elevator with an 8,000,000- bushel capacity was badly damaged. The roof gallery above the storage tanks was entirely blown off. A checked belt conveyor reportedly caused it. The heat from the explosion and the resultant fire badly damaged the concrete storage bins and the adjacent workhouse.

4.5 Best Industry Practices

A good starting place for good industry practices to prevent dust explosions is the OSHA standard[14]. The Occupational Safety and Health Administration (OSHA) issued its Grain Handling Standard - *Title 29 Code of Federal Regulations* (CFR) Part 1910.272 in 1987 to protect workers exposed to fires and explosions. Excessive amounts of grain dust was one of the major causes of these devastating catastrophes that killed or maimed hundreds of workers. The standard is intended to protect workers from hazards faced while walking on or underneath accumulations of grain within a grain storage facility. These hazards include engulfment and entrapment in grain and grain handling equipment, which can result in asphyxiations, crushing injuries, and amputations.

In 1996, OSHA further amended the standard to protect employees whenever they enter a "flat storage structure" regardless of their point of entry. OSHA believes that this technical amendment will prevent from 2 to 4 additional fatalities annually and a similar number of traumatic injuries caused by mechanical devices such as augers.

The requirements of the standard apply to more than 250,000 workers at 24,000 grain elevators and mills that have been and continue to be exposed to fires, explosions, engulfment, and entrapment hazards. Engulfment and entrapment hazards have killed or maimed hundreds of workers. Note that a "flat storage structure" means a grain storage building or structure that will not empty completely by gravity, has an unrestricted ground level

[14] Grain Handling U.S. Department of Labor, Occupational Safety and Health Administration, OSHA 3103, 1996 (Revised)

opening for entry, and must be entered to reclaim the residual grain by using powered equipment or manual means.

According to the Bureau of Labor Statistcs[15], in 1993 and 1994 OSHA estimates that there have been more than 45 workers engulfed in gain and asphyxiated or crushed to death by grain augers. There are several provisions employers must follow to comply with the grain handling standard, including a requirements for hot work; entering bins, silos, tanks, and other storage structures; inside bucket elevator legs; preventive maintenance; housekeeping; handling emergencies; and training.

A permit system is required for workers performing hot work. Hot work includes electric or gas welding, cutting , brazing, or similar flame-producing operations. The permit is to ensure that the employer is aware of the hot work being performed—particularly, when performed by contractors—and that appropriate safety precautions have been taken prior to beginning the work. The standard does not require a work permit if the hot work is performed in the presence of the employer or the employer's authorized representative in an employer-authorized welding shop, or when work is conducted out of doors and away from the grain facility.

Employees must also be issued a work permit before they enter bins, silos, or tanks unless the employer or the employer's representative is present. Such permits will help employers maintain control over employee entry into these areas.

In addition to the permit system, employees should be thoroughly informed of the hazards associated with entry into bins, silos, tanks, and other structures. For example employees should never enter these areas from the bottom when grain or other agricultural products are hung up or stuck to the sides. Employees should be made aware that the atmosphere in bins, silos and tanks can be oxygen deficient or toxic. Consequently, employees must be trained in the proper method of testing, the atmosphere procedures to take if the atmosphere is to take if the atmosphere is found to be hazardous. The air inside the enclosure must be tested for oxygen content both before and during employee entrance, unless there is continuous natural air movement or forced-air ventilation in the space.

Provisions for ventilation are needed for supplement by the use of appropriate respirators if necessary. If oxygen levels are less than 19.5 percent or if concentrations of toxic agents are present in the air either exceed ceiling limits in OSHA's health standards or will have health affects that restrict

---

15 U.S. Department of Labor, Bureau of Labor Statistics, *Census of Fatal Occupational Injuries*. Washington. DC, 1993–1994.

an employee's abilities to effect self-rescue or obtain assistance or, if there is a combustible gas or vapor concentrations in excess of 10 percent of the lower flammable limit proper ventilation must be provided.

An employee must wear a body harness with a lifeline or use a boatswain's chair whenever entering a grain storage structure at or above the level of stored grain and the depth of stored grain poses an engulfment hazard. If the employer can demonstrate that the lifeline or boatswain's chair is not feasible or creates a greater hazard the employer must provide an alternative means of protection. Where employees work in a bin, silo, or tank, a trained and equipped observer must be present on the outside maintain communication with employees and provide help if needed.

The standard prohibits "walking down grain" to make it flow within or outside of the storage structure, or standing on moving grain. Also, all mechanical, electrical, and pneumatic equipment that presents a danger to employees inside grain storage structures must be de-energized and disconnected, locked-out and tagged, blocked-off, or otherwise stopped by other equally effective means or methods.

In addition, no employee is permitted to be in any location where an accumulation of grain on the sides of the storage structure or elsewhere could fall and engulf him or her.

4.5.1 Bucket Elevator Legs

The insides of bucket elevators are well recognized as potential ignition sources for primary explosions. To lessen these hazards, the standard requires that belts purchased after March 30, 1988, have a surface electrical resistance not exceeding 300 megohms. Bucket elevators must have an opening to the head pulley section and boot section to allow for inspection, maintenance, and cleaning; bearings must be mounted externally to the leg casing or the employer must provide vibration, temperature, or other monitoring of the conditions of the bearings if bearings are mounted inside or partially inside the leg casing. Also, elevator legs must be equipped with a motion-detection device that will shut down the leg when the belt speed is reduced by 20 percent or more of the normal operating speed. A belt-alignment monitoring device with an alarm to alert employees when the belt is not tracking properly is also required: alternatively, employers must provide a means to keep the belt tracking properly.

Bearing monitors, motion detection devices, and belt-alignment devices need not be installed if the employer equips bucket elevators with a fire and explosion suppression system capable of protecting the head and boot sections of the leg, or with a pneumatic dust control system; that will keep the dust concentrations inside the leg casing 25 percent below the lower explosive limit during operation.

Preventive Maintenance - Preventive maintenance is a very important aspect of any grain industry safety and health program. It is a must for controlling fuel and ignition sources and for keeping equipment functioning properly and safely. The OSHA standard does not require the employer to have a written preventive maintenance program but states that all mechanical and electrical equipment must be kept in proper operating condition. To do this the employer must annually inspect the mechanical and safety control equipment associated with dryers, grain stream processing equipment, dust collection equipment, including filter collectors, and bucket elevators. This equipment must be lubricated and maintained according to the manufacturers' recommendations or as determined necessary by prior operating records. Equipment that malfunctions or operates below designed efficiency must be promptly repaired or removed from service. Inspected or repaired equipment must how the date of inspection. The standard also requires procedures for locking out and tagging equipment to prevent the inadvertent application of energy or motion to equipment being repaired. serviced, or adjusted. All employees who repair, service, and operate the equipment must be familiar with the employer's locking out and tagging procedures.

Good Housekeeping - Housekeeping is an important part of any safety and health program especially in facilities where combustible material might accumulate. The standard request the employer to develop and implement a written housekeeping program to help eliminate these dangers. The program must include instructions for reducing dust accumulations on ledges, floors, equipment, and other exposed surfaces, and must identify "priority" areas in grain elevators that are known to be potential sources of ignition. These include floor areas within 35 feet (10.9728 meters) of inside bucket elevator legs, enclosed areas containing grinding equipment and enclosed areas containing grain dryers located inside the facility. The housekeeping program also must address the methods for removing grain spills from work areas. The use of compressed air to remove dust is permitted only when all machinery that presents a source of ignition in the area is shutdown and all other known potential ignition sources are removed or controlled.

Because grain dust is the main source of fuel for explosions in grain handling facilities, the standard allows a maximum accumulation of more than an 1/8-inch (0.3175 centimeters) in priority housekeeping areas of grain elevators, if dust accumulations exceed the 1/8-inch (0.3175 centimeters) action level in priority housekeeping areas designated means or methods must be initiated immediately to remove such accumulations.

The standard also provides for the employer to use alternative means to the 1/8-inch (0.3175 centimeters) action level where the alternative can be demonstrated to provide equivalent protection from explosions. This may involve additional treatment of the dust and/or the area of dust accumulation, such as spraying with oil or water. In addition, the use of oil additives such as white mineral oil in the grain flow, and changes in materials handling processes can also help reduce the accumulation of dust and make the dust less explosive.

Emergency Action Plan - Employers must develop and implement a written emergency action plan. The plan does not have to be written if there are fewer than 10 employees. This plan must include a distinguishable and distinct alarm system (especially for those employees who work indoors) and evacuation procedures, and must include employee training in emergency procedures. Employees must know where the nearest escape routes are and must be familiar with workplace maps that clearly show these emergency escape routes. In addition, at least two means of emergency escape from galleries (bin decks) are required in grain elevators. The employer must also designate a safe area outside the facility where employees can congregate after evacuation and must implement procedures to account for all employees after emergency evacuation has been completed. It is recommended that employers seek the assistance of local fire departments to preplan for emergencies and designate a means of contacting fire and rescue agencies under emergency conditions.

Training and Education - Training employees to recognize hazards associated with their jobs is an effective method for increasing overall safe operations. Employers are required to train employees in their work tasks annually or whenever changes in job assignments expose them to new hazards. New employees are to be trained prior to starting work. Employees assigned special or infrequent tasks, such as bin entry and the handling of flammable or toxic substances, must also be trained to perform these tasks safely. Training must include the following:

- General safety precautions associated with the grain facility as well as the recognition and prevention of hazards related to engulfment, mechanical devices, dust accumulations, and common ignition sources such as smoking
- Specific procedures and safety practices applicable to the job tasks including, but not limited to, clearing choked legs, and performing housekeeping, hot work, preventive maintenance, and lockout/tagout
- Training in emergency procedures

OSHA states that effective management of worker safety and health protection is a decisive factor in reducing the extent and severity of work-related injuries and illnesses and their related costs. To assist employers and employees in developing effective safety and health programs, OSHA published recommended *Safety and Health Program Management Guidelines (Federal Register 34 (I8):3908–3916, January 36,1989)*. These voluntary guidelines apply to all places of employment covered by OSHA. The guidelines identify four general elements that are critical to the development of a successful safety and health management program:

- Management commitment and employee involvement
- Worksite analyses
- Hazard prevention and control
- Safety and health training

The guidelines recommend specific actions under each of these general elements to achieve an effective safety and health program.

Continuous housekeeping and sanitation (regular cleaning of the elevator), and regularly scheduled bearing service should be top priorities at all grain elevators and at flour and feed mills. Many insurance companies insist on strict housekeeping, sanitation, and preventive maintenance at insured elevators. Grain, broken kernels and grain dust accumulate in the leg boots and should be cleaned out periodically. Some elevators install easily removable doors on leg boot side panels for quick, easy cleanout. The following are grain dust control and prevention procedures. All elevators and mills should be doing item number one, housekeeping and sanitation, for elevator safety and worker health, as well as for integrated pest management (IPM) purposes. It is the most important safety practice in any elevator or mill.

1. Maintain a rigorous housekeeping and sanitation program inside the grain elevator structure. Keep grain dust cleaned up in all working areas of the elevator.
2. Implement a weekly or bi-weekly (or as specified by the manufacturer) bearing lubrication program, based on the bearing manufacturer's specifications.
3. Use a food-grade mineral oil spray system on grain during transfer and loadout.
4. Install bearing temperature monitors on leg boot, head, and knee pulley shafts, on horizontal drag head and boot bearings, and on belt conveyor drive and idler bearings.

5. Install belt rub sensors inside bucket elevator leg casings to detect belt misalignment to prevent friction heating.
6. Maintain a periodic (weekly or bi-weekly) bearing temperature monitoring program. Document periodic bearing temperature readings and compare with previous readings. A substantial bearing temperature increase (10°F to 20°F or more in a week or two) may indicate bearing failure and the need to replace the bearing.
7. Replace steel cups with plastic cups in elevator legs.
8. Use anti-static belting material in legs and horizontal belt conveyors.
9. Install quick-opening cleanout doors on leg boot side panels for grain and dust cleanout.
10. Install dust aspiration systems at grain transfer points or ventilation systems in tunnels and galleries with open conveyors, and truck dump pits where dust accumulation is a problem.
11. Install dust aspiration or suction ventilation systems on inside enclosed legs and conveyors to keep suspended dust below MEC levels.
12. Clean out dust collectors and change filter bags at intervals recommended by the manufacturer.
13. Clean out dust cyclone collector holding bins at scheduled intervals.
14. Install dump pit baffles on truck dump pits to provide a major reduction in airborne dust during dumping operation.
15. explosion relief panels and devices in elevator design.
16. Install explosion proof electrical outlets and equipment in sensitive areas.
17. Train employees on the dangers and prevention of dust explosions.

A primary reason why grain elevator dust explosions continue to prevail is the fact that most preventive measures and best practices are voluntary standards. Private regulation in general tend to be weak or diluted, representing a kind of lowest common denominator of prevailing business practices. Many private standards tend to be largely defensive, motivated by the desire to forestall government regulation. Public standards, by contrast, are thought to be more stringent than their private counterparts, to the point of being seemingly unreasonable. Public regulation is also considered inordinately cumbersome, laden with procedural requirements and judicial appeals. These conceptions of public and private regulation are borne out substantially in the case of grain elevator safety. The private standard (NFPA 61B) is generally lax, reflecting industry's desire to avoid meaningful regulation of dust-control practices. The public sector, on the other hand, developed a more demanding standard but spent almost ten years in the process. The Office of Management and Budget has the opinion which it shares with the National Grain and Feed Association (NGFA) that the OSHA regulation is burdensome and unreasonable. As with so many regulatory issues, however, the relevant data are weak and inconclusive. Proponents and critics advance plausible, but divergent, arguments, each as good as its unsure assumptions.

Two popular explanations of public regulation—'capture' theory and 'life-cycle' theory—have been discredited for their incompleteness. Life cycle is a vivid metaphor, but it lacks causal content about the force and nature of the regulatory aging process. Similarly, capture theory has been challenged by the rise in social regulation, where government actions are opposed, not favored, by the regulated.

Several provisions in NFPA 61B are surprisingly strict. When judged by the criteria used to evaluate public regulation, a few NFPA provisions appear to be unreasonable. This suggests a previously unrecognized aspect of private regulatory behavior. Private standards-setting is not all lowest common denominator politics. There is a mixture of managerial behavior (marked by considerations of politics and economics) and technical behavior (permeated by the ethos of engineers). Managerial considerations explain the biggest weakness in NFPA 61B: favoring one regulatory approach to dust hazards (ignition control) to the complete exclusion of another technique (dust control). But some realms of private decision making are dominated by technical, more than managerial, influences. The technical provisions in NFPA 61B are the most stringent. Technical influences help explain why NFPA 61B was developed more than fifty years before OSHA considered writing a standard. This indicates how professional engineers alter the dynamics of private regulatory decision making.

It is important to recognize the grain elevators come in all sizes and shapes. Some are connected to processing facilities, others serve only as bulk storage. Bulk storage facilities are often grouped into three functional categories that roughly correlate with their size: country elevators (the smallest), inland terminals, and export terminals (the largest). Differences in function, products, and capacity have all been urged as reasons for avoiding safety standards of general application. Feed mills, for example, have a much better safety record than bulk grain storage facilities. Some so-called 'country elevators' handle grain only a few days a year, obviously minimizing the opportunities for accidents. The significance of these distinctions for safety regulation is unclear because throughput—the amount of grain moved through a facility—has a more direct bearing on safety than function or capacity.

Of all the hazards associated with grain handling, the most serious one is that the dust, generated whenever grain is handled, is easier to ignite and results in a more severe explosion than equal quantities of TNT. Dust explosions account for the vast majority of personal injuries and property losses in grain-handling facilities. There are thousands of small fires in these facilities every year; with estimates as high as 11,000. But the twenty to thirty explosions a year account for almost 80 percent of the property damage and 95 percent of the fatalities. Consider that the major explosions in 1977–78 coincided with tremendous increases in wheat exports (primarily to the Soviet Union). The comparatively low number of explosions in recent years, on the other hand, is attributable in part to decreases in grain sales. Fluctuations in trade patterns notwithstanding, the explosion problem remains the most serious hazard in grain elevators. This leaves us with a conundrum that we may expect a certain number of these incidents each year and largely influenced by market volume.

But as noted earlier, explosions in grain elevators are not caused by spontaneous combustion. To cause an explosion, the dust must be mixed with air to form a dense cloud. Layered dust only smolders upon ignition. The airborne dust concentrations must also be above the lower explosive limit (LEL), a condition so opaque that visibility is minimal and breathing is difficult. Concentrations above the LEL occur regularly inside most bucket elevators but seldom elsewhere in a facility. Grain elevator explosions are not one big blast, but rather involve a primary explosion and a series of secondary ones. The primary explosion generates shock waves throughout the elevator, often raising into suspension layered dust on walls, rafters, equipment, and the floor. Accumulations of as little as one-hundredth of an inch will propagate the flame from an initial explosion. In other words, layered dust provides the fuel to turn a primary explosion—often itself quite minor—into a major one.

Experts argue two basic schools of thought about how to approach the explosion hazard: one, eliminate ignition sources; two, control airborne and layered dust. Although it seems obvious that neither strategy should be pursued exclusively, the debate over safety standards is often between groups that have a strong preference for one approach over another.

Industry's inclination seems to focus on ignition sources. Ignition occurs most frequently in the bucket elevator—a continuous conveyor belt with equally spaced buckets (often metal) that elevates the grain and discharges it into a spout (refer back to figure 4.1). The top section of a bucket elevator, where the drive is located, is referred to as the "head." The bottom section, where grain enters the elevator, is known as the "boot." The "leg" connects the head and the boot.

Ignition sources are varied and often notoriously difficult to pinpoint. The USDA identified ten different probable sources, the largest group (four) being 'unknown'. Most studies agree that welding or cutting (also known as "hot work") is the largest known ignition source, accounting for perhaps 10 to 20 percent of all explosions. Other common ignition sources include electrical failure, overheated bearings, foreign metal objects sparking inside the leg, and friction in choked legs. Jogging the leg – i.e., trying to free a jammed bucket conveyor by repeatedly stopping and starting the driving motor—is a primary cause of friction-induced explosions.

Ignition-source control can take several forms. One is mechanical. Mechanisms including electromagnets and special grates can minimize the problem of metal objects entering the grain stream. Belt speed, alignment, and heat monitors can be used to detect hazardous conditions and shut down the equipment before suspended dust is ignited. The effectiveness of these devices varies, but quality has improved since their introduction to the grain-handling industry ten or fifteen years ago. Another approach to ignition control is behavioral. Employees are instructed not to jog the legs. Rules against smoking are strictly enforced. Permit procedures are instituted to ensure that hot work is done safely, and preventative maintenance schedules are instituted and implemented.

Ignition-source control has two limitations. First, there are countless potential ignition sources. The National Academy of Sciences (NAS) has reported the results of two surveys in which the ignition source remained unknown in over half the cases. Virtually every piece of equipment, as well as every grain transfer point, is a potential ignition source. Eliminating or controlling them all seems impossible. A concern with ignition control is that human error can play a big role. A common example is jogging jammed conveyors instead of inspecting and digging out the elevator boot. This problem has been recognized for many years, yet it persists largely unabated. Some operators of elevators encourage this time-saving but risky practice. While everyone agrees that good operating procedures are a good idea, they are also inherently difficult to enforce. Since operational breakdowns are inevitable, critics consider this loss control method inadequate by itself.

Housekeeping (dust control) is an issue that is largely within industry focus. Housekeeping practices, however, can be misleading to the extent that it conjures up only images of brooms and vacuum cleaners. Dust control is aimed at both airborne and layered dust. Airborne dust can be removed by various systems of aspiration, also called pneumatic (i.e., moved or worked by air pressure) dust control. Pneumatic dust-control systems have four major components: hoods or other enclosures; duct-work; a filter, or dust collector; and an exhaust fan. Layered dust can also be removed automatically, but in all but the largest facilities it is removed manually: with vacuum cleaners, brooms, compressed air, and, in some cases, water.

Both types of housekeeping (pneumatic dust control and layered dust removal) are controversial, but in the first case the disagreement is mostly technical, while in the second it is largely a matter of economics. The technical potential of pneumatic dust control is highly disputed. One grain elevator insurance company touts a system it claims can reduce dust concentrations in the bucket elevator below the lower explosive limit. The technology was not proven to the satisfaction of a 1982 NAS panel, but panel members placed a high priority on continued research in this area. Much of the industry considers such technology unavailable, while some segments claim it to be effective. This implies that the design of pneumatic dust-control systems is more of an art than an exact science. Engineers cannot simply take specifications for the desired concentration of airborne dust (or rate of accumulation for layered dust) and design a system that will perform accordingly. The technology and criteria for removing airborne dust is too uncertain with a large number of variables. Much depends on how the system is installed and maintained.

Layered dust is generally believed to pose much less of a technical problem. How to remove it effectively is not very controversial. For large facilities, dust removal is usually part of normal operations. Sweeping is done at least once per shift at all export facilities. But for many smaller facilities, removing layered dust means hiring additional labor and slowing down operations when they are most profitable. In general however, housekeeping and maintenance are often given low priority and are usually the first tasks postponed when there is a rush of business.

Insurance companies have been troubled with grain elevators since a series of big explosions in 1919 and 1921. Responding to calls for an industry-wide standard on grain elevator safety, NFPA appointed a Committee on Dust Control in Grain Elevators. Insurance interests have been major participants in NFPA's subsequent efforts. The committee, lacking sufficient information on certain aspects of the explosion problem, hired Underwriters Labs (another organization created by the insurance industry) to investigate methods of controlling floating dust in terminal grain elevators. The results of the UL study formed the basis for the dust-control provisions in the early versions of the standard, which also contains general operation and design provisions. It has been revised at least five times since then, although many of the provisions in the most recent version can be traced back to 1953. The most significant changes were in 1970, when NFPA added country elevators (formerly governed by a separate standard) to the scope of the standard, and in 1980, when NFPA responded to the threat of imminent government regulation by strengthening the ignition-control requirements for bucket elevators. The revision process began again with a meeting in July 1985 of the Technical Committee on Agricultural Dusts, and a revised version of NFPA 61B (hereinafter "61B") was adopted the following year.

The Agricultural Dusts committee, chaired for the past fifteen years by a representative of Continental Grain, has approximately two dozen members. Twelve to fifteen attend formal committee meetings, but all actions are subject to letter ballot by the full committee. The largest-segment of committee membership comes from the insurance industry—Industrial Risk Insurers, the Mill Mutuals, the Insurance Services Office of Nebraska, Kemper Insurance Co., and Factory Mutual Research Corp. are each represented. Two other organizations related to the insurance industry, UL and Johnson & Higgins (a brokerage firm), are represented on the committee along with Cargill, the country's largest grain company, and two major grain processors (Kellogg Co. and General Foods). Other representatives include those of a fumigant company, a fire equipment manufacturer, a manufacturer of grain-handling equipment, and several academics. Conspicuously absent from the committee roster is a representative from OSHA. OSHA's policy on participating in private standards-setting activities has varied by administration and by issue.

In its present form, 61B is a remarkably compact regulation, devoting no more than a page or two apiece to chapters on construction requirements, equipment, and dust control. Most provisions are as general as they are brief. "Extraneous material that would contribute to a fire hazard shall be removed from the commodity before it enters the [grain] dryer." "Boot sections [of the elevator leg] shall be provided with adequate doors for cleanout of the entire boot and for inspection of the boot pulley and leg belt."

NFPA 61B avoids almost all design details. The complex topic of explosion venting, for example, is addressed by reference to a separate NFPA venting guide that committee members agree is not particularly applicable to grain-handling facilities. Similarly, dust-control systems are mentioned, but there are no performance requirements or construction specifications.

In contrast, the National Academy of Sciences has published a 116-page guide to designing pneumatic dust-control systems.

The most significant provision in 61B is the basic housekeeping requirement. Unlike the standard later adopted by OSHA, which contains specific action levels and alternatives for dust control, 61B dispenses with the topic by stating simply that "dust shall be removed concurrently with operations." Whatever this means, the provision is not enforceable. OSHA learned that when it tried to use 61B in support of citations issued after the 1977 explosions. The commission that reviews OSHA citations was unwilling to rely on such an ambiguous provisions. In actuality, the housekeeping provision is not a requirement at all; rather it is more of a gentlemen's agreement to recognize the problem but leave its solution entirely to the individual operator.

The lack of specificity of the housekeeping requirement in 61B is not adequately justified by the reasons most commonly offered in its defense. Research sponsored by the NGFA indicates that a dust layer as thin as one-hundredth of an inch can propagate an explosion. The NGFA has been accused by its detractors of conducting this research precisely to bolster the scientific argument against any standard. But there rarely is a strong basis for resolving complex problems involving the trade-off between cost and safety. If this call for greater scientific certainty applied equally to all standards-setting, it would largely paralyze the effort. There are too many variables interacting and changing over time to expect anything resembling scientific certainty for each one.

Separate NFPA standards for country elevators were combined with those for other grain-handling facilities in 1973, when the committee decided that a distinction between types of grain elevators on the basis of capacity or shipping or receiving media is no longer practical. Similarly, the committee argued in July 1985 that motion switches should not be mandated on all bucket elevator legs because some country elevators do not realistically need them. The committee rejected the argument on the grounds that motion switches were generally a good idea and it would be impossible to identify in a standard those situations in which they are not necessary. The same could be said about dust-control requirements.

It is reasonable to ask why is the housekeeping provision in 61B so vague. Two factors other than the limitations of science and the diversity of facilities are at play. One is specific to this issue, the other indicative of a larger force affecting the development of private safety standards. There is general belief that good housekeeping simply is not the answer. To some, it is a matter of practicality. It would require excessive effort to keep an elevator clean enough to prevent explosions. Another concern is that voluntary standards can be used against a company in legal disputes. Most explosions lead to litigation. The two largest explosions in 1977 resulted in settlements of approximately $25 million each, and a jury recently applied the bane of the tort law—punitive damages—for the first time in a grain elevator case. The vaguer 61B is on housekeeping, the less powerful a weapon it would be after an explosion. Some of the vagueness, then, is simply an attempt to make the standard liability-proof. The ill-fated OSHA citations that relied on 61B are testimony to the effectiveness of this strategy.

Standard 61B can, has, and will be applied to situations other than brand-new elevators. When issuing citations, OSHA often refers to 61B without regard for the date the facility was constructed. NFPA 61B attempts to be unusually lenient in this regard. Unlike most building codes, it does not even require compliance in the case of major expansion or renovation—the vast majority of "new" construction. Moreover, the 1980 version also exempted from retroactive coverage a host of operational activities—such as hot-work procedures and housekeeping—that can be carried out without regard to the age or design of the facility.

4.6 OSHA Grain Handling Standard Audit Questionnaire

Despite the criticisms of the OSHA standard, their roadmap to preventing grain dust elevator explosions represents that best available practices to date. The Occupational Safety and Health Administration's Grain Handling Standard, 29CFR 1910.272, forms the heart of the safety standards grain elevator and feed mill operators should follow in providing a safe workplace for employees. The following is a self-audit checklist based on the standard[16].

4.6.1 Section (d) Emergency Action Plan

1. Has an emergency action plan that complies with the requirements of 1910.38(a) been developed and implemented?

   Yes __________ No __________ NA __________

[16] Prepared by Adam Schupp, safety consultant; N. Fort Myers, FL; 941-731-5947 (http://www.grainnet.com/pdf/59986.pdf)

4.6.2 Section (e) Training

1. Is appropriate job training provided to employees at least annually?

   Yes _______ No _______ NA _______

2. Is training provided when changes in job assignment will expose employees to new hazards?

   Yes _______ No _______ NA _______

3. Are current employees, prior to starting work, trained in at least the following:

   Yes _______ No _______ NA _______

   a. General safety precautions associated with the facility? Yes _______ No _______ NA _______

   b. Preventive measures for the hazards related to dust accumulations and common ignition sources? Yes _______ No _______ NA _______

   c. Procedures and safety practices applicable to their job tasks, such as choked legs, hot work, preventive maintenance, lockout/tagout, etc.? Yes _______ No _______ NA _______

   d. Special assignments, such as bin entry and handling flammable or toxic substances? Yes _______No _______ NA _______

4.6.3 Section (f) Hot Work Permit

1. Is a hot work permit required in all situations other than the three exceptions stated below? Yes _______ No _______ NA _______

   a. When the person who would normally issue the permit is present while the hot work is being performed.

   b. In welding shops authorized by the employer.

   c. In hot work areas authorized by the employer that are located outside of the grain handling structure.

2. Does the hot work permit certify that all of the requirements of 1910.252(a) have been implemented before hot work operations begin? Yes ______ No ______ NA ______

4.6.4 Section (g) Entry into Grain Handling Structures

Note: Section (g) applies to entry into bins, silos, tanks, and other grain storage structures. Entry through unrestricted ground level openings into flat storage structures in which there are no toxicity, flammability, oxygen-deficiency, or other atmospheric hazards is covered by Section (h) Entry into Flat Storage Structures.

1. Is a written permit issued for entry? Yes ______ No ______ NA ______

   a. Exception: A written permit is not required if the employer or his designated representative is present during entry operations.

2. Does the permit certify that the precautions contained in this section (Section (g)) have been implemented prior to entry? Yes ______ No ______ NA ______ Is all equipment (mechanical, electrical, etc.) that presents a danger to an entrant deenergized or isolated in a manner (such as lockout/tagout, disconnection, or blocking) that effectively protects an entrant? Yes ______ No ______ NA ______

3. Is the internal atmosphere in the space tested for the presence of combustible gases, vapors, and toxic agents when the employer has reason to believe they may be present? Yes ______ No ______ NA ______

4. Is the internal atmosphere in the space tested for oxygen content unless there is continuous natural air movement or continuous forced-air ventilation before and during entry? Yes ______ No ______ NA ______

5. If the oxygen level is less than 19.5%, or if combustible gas or vapor is detected in excess of 10% of the lower flammable

limit, or if toxic agents are present in excess of the ceiling values listed in Subpart Z, of 29CFR Part 1910, or if toxic agents are present in concentrations that will cause health effects that prevent entrants from effecting self-rescue or communication to obtain assistance, do the following procedures apply:

a. Ventilation is provided until the unsafe condition(s) are eliminated, and the ventilation is continued as long as there is a possibility of a recurrence of the unsafe condition(s) while the space is occupied? Yes ______ No ______ NA ______

b. The entrant wears an appropriate respirator if the toxicity or oxygen deficiency cannot be eliminated? Yes ___ No ___ NA ___

c. Respirator use is in accordance with the requirements of 29CFR 1910.134? Yes ______ No ______ NA ______

6. Is the practice of "walking down grain" strictly prohibited? Yes ______ No ______ NA ______

7. Is a lifeline with a body harness or a boatswain's chair used under the following two conditions:

   a. Whenever entry is made at or above the level of stored grain? Yes ______ No ______ NA ______

   b. Whenever an entrant walks or stands on or in grain that is deep enough to pose an engulfment hazard? Yes ______ No ______ NA ______

8. Is the lifeline so positioned and of sufficient length to prevent an entrant from sinking further than waist-deep in the grain? Yes ______ No ______ NA ______

   a. Exception: Where the employer can demonstrate that the protection required by (8) (a),(8) (b), and (9) above is not feasible or creates a greater hazard, the employer is required to provide an alternative means of protection that will prevent an entrant from sinking further than waist-deep in the grain. Note: When an entrant is standing or walking on a surface that the employer demonstrates is free from engulfment hazards, the lifeline or alternative means may be disconnected or removed.

b. Is an observer, equipped to provide assistance, stationed outside the space while entry operations are underway? Yes _______ No _______ NA _______

9. Are communications (visual, voice, or signal line) always maintained between the entrant and the observer? Yes _______ No _______ NA _______

10. Is rescue equipment that is specifically suited for the space being entered provided and in good condition? Yes _______ No _______ NA _______

11. Is the entry observer trained in rescue procedures and notification methods for obtaining additional assistance? Yes _______ No _______ NA _______

12. Are entrants prohibited from entering a space underneath a bridging condition or where a buildup of grain on the sides could fall and bury them? Yes _______ No _______ NA _______

4.6.5 Section (h) Entry into Flat Storage Structures

1. Are entrants who walk or stand on or in grain that is deep enough to pose an engulfment hazard equipped with a lifeline or some alternative means that will prevent the entrant from sinking further than waist-deep in the grain? Yes _______ No _______ NA _______ Note: When the employee is walking or standing on a surface that the employer demonstrates is free from engulfment hazards, the lifeline or alternative means may be disconnected or removed.

2. Whenever an entrant walks or stands on or is in grain that is deep enough to pose an engulfment hazard, is all equipment that presents a danger to entrants (such as as auger or other grain transport equipment) de-energized or otherwise isolated by such means as lockout/tagout, disconnection, blocking, etc.? Yes _______ No _______ NA _______

3. Are entrants prohibited from being underneath a bridging condition, or in any location where an accumulation of grain on the sides could fall and engulf them? Yes _______ No _______ NA _______

4.6.6 Section (i) Contractors

1. Are outside contractors informed of the following:

    a. The known potential fire and explosion hazards related to the contractor's work and work area? Yes _______ No _______ NA _______

    a. The applicable safety rules of the facility? Yes _______ No _______ NA _______

2. Are the applicable provisions of the facility emergency action plan explained to the outside contractor? Yes _______ No _______ NA _______

4.6.7 Section (j) Housekeeping

1. Has a written housekeeping plan been developed and implemented? Yes _______ No _______ NA _______

2. Have priority housekeeping areas been identified in the written housekeeping plan? Yes _______ No _______NA _______
Note: Priority housekeeping areas must include at least the following:

    a. Floor areas within 35 feet of inside bucket elevators.

    b. Floors of enclosed areas containing grinding equipment.

    c. Floors of enclosed areas containing grain dryers located inside the facility.

3. Are grain dust accumulations that exceed 1/8 inch in priority housekeeping areas removed immediately or, in lieu of removal, is equivalent protection provided? Yes _______ No _______ NA _______

4. If compressed air is used for cleaning, are all known ignition sources in the area shut down, removed, or isolated in some manner? Yes _______ No _______ NA _______

5. Does the written housekeeping program include procedures for the removal of grain and product spills? Yes ______ No ______ NA ______

4.6.8 Section (k) Grate Openings

1. Is the width of receiving pit grate openings (truck or rail) a maximum of 2–1/2 inches? Yes ______ No ______ NA ______

4.6.9 Section (l) Filter Collectors

1. Are filter collectors equipped with a monitoring device that will indicate a pressure drop across the surface of the filter? Yes ______No ______ NA ______
2. Are filter collectors installed after March 30, 1988, located as follows:
   a. Outside the facility. Yes ______ No ______ NA ______
   b. Located inside the facility and protected by an explosion suppression system. Yes ______ No ______ NA ______
   c. Located in an area inside the facility that is separated from other areas of the facility by construction having a one-hour fire-resistance rating and that is adjacent to an exterior wall and vented to the outside? Yes ______ No ______ NA ______

4.6.10 Section (m) Preventive Maintenance

1. Are regularly scheduled inspections of the mechanical and safety control equipment associated with dryers, grain stream processing equipment, dust collection equipment including filter collectors, and bucket elevators performed? Yes ______ No ______ NA ______
2. Is lubrication and other maintenance in accordance with the manufacturers' recommendations, or as determined by prior operating experience? Yes ______ No ______ NA ______
3. Is malfunctioning equipment, such as dust collection systems, overheated bearings, or slipping or misaligned belts on

inside bucket elevators, promptly repaired or removed from service? Yes _______ No _______ NA _______

4. Is a certification record maintained of each preventive maintenance inspection? Yes _______ No _______ NA _______
5. Have lockout/tagout procedures been developed and implemented in accordance with the requirements of 29CFR 1910.147? Yes _______ No _______ NA _______

4.6.11 Section (n) Grain Stream Processing Equipment

1. Is grain stream processing equipment, such as hammermills, grinders, and pulverizers equipped with an effective means of removing ferrous material from incoming grain? Yes _______ No _______ NA _______

4.6.12 Section (o) Emergency Escape Note: Applies only to grain elevators.

1. Are there at least two means of emergency escape from galleries (bin decks)? Yes _______ No _______ NA _______
2. Are tunnels in grain elevators that were in existence before March 30, 1988, provided with at least one means of emergency escape? Yes _______ No _______ NA _______
3. Are tunnels in grain elevators that were constructed after March 30, 1988, provided with at least two means of emergency escape? Yes _______ No _______ NA _______

4.6.13 Section (p) Continuous-Flow Bulk Grain Dryers Note: Applies only to grain elevators.

1. Are grain dryers equipped with automatic controls that will accomplish the following:

   a. Shut off the fuel in case of flame failure or interruption of air movement through the exhaust fan? Yes _______ No _______ NA _______

b. Will stop the grain from being fed into the dryer if excessive temperature occurs in the exhaust of the drying section? Yes _______ No _______ NA _______

2. Do grain dryers installed after March 30, 1988, comply with either (a), (b), or (c) below:

   a. Grain dryers are located outside the facility? Yes _______ No _______ NA _______

   b. Grain dryers are located inside the facility and protected by a fire or explosion suppression system? Yes _______ No _______ NA _______

   c. Grain dryers are located inside the facility and in an area that is separated from other areas by construction having at least a one-hour fire-resistance rating? Yes _______ No _______ NA _______

3. Note: Applies only to grain elevators. Does the facility have procedures that stipulate that bucket elevators shall not be "jogged" to free a choked leg? Yes _______ No _______ NA _______

4. Are belts and lagging purchased after March 30, 1988, conductive? Yes _______ No _______ NA _______

5. Are bucket elevators equipped with a means of access to the head section to allow for inspection of the head pulley, lagging, belt, and discharge throat of the elevator head? Yes _______No _______ NA _______

6. Is the boot section of bucket elevators equipped with a means of access for clean-out of the boot and for inspection of the boot, pulley, and belt? Yes _______ No _______ NA _______

7. Are leg bearings mounted externally to the leg casing? Yes _______ No _______ NA _______

8. Are those bearings not mounted externally to the leg casing provided with vibration monitoring, temperature monitoring, or other means to monitor their condition? Yes _______ No _______ NA _______

9. Are bucket elevators equipped with a motion sensing device that will shut down the leg when the belt speed is reduced

by no more than 20% of the normal operating speed? Yes _______ No _______ NA _______

10. Are bucket elevators equipped with a belt alignment device that will initiate an alarm when the belt is not tracking properly, or equipped with a means that provides constant belt alignment? Yes _______ No _______ NA _______

Note: Audit items (7) and (8) above do not apply to grain elevators having a permanent storage capacity of less than one million bushels, provided that daily visual inspection is made of the bucket movement and tracking of the belt. Audit items (5), (6), (7), and (8) do not apply to bucket elevators that are equipped with any system that will keep the dust concentrations inside the leg at least 25% below the lower flammable limit at all times during operation.

5

Coal Dust Explosibility and Coal Mining Operations

5.1 Introduction

Many facilities throughout the world handle coal, such as preparation plants. And/or rely on coal as a fuel, such as cement, lime factories, and power plants. Although coal can be handled safely and can be an efficient fuel, there exist explosion hazards which are accentuated as the particle size is reduced. Particle sizes of coal which can fuel a propagating explosion occur within thermal dryers, cyclones, baghouses, pulverized-fuel systems, grinding mills, and other process or conveyance equipment. In addition, the deadly combination of coal dust and methane gas in coal mining operations has led to some of the world's most catastrophic industrial accident resulting in unacceptable loss of life over the years.

This chapter provides an overview of coal dust explosions, safe handling operations, and coal mine safety practices. By way of a quick review, the reader should recall that there are three necessary elements which must occur simultaneously to cause a fire: fuel, heat, and oxygen; i.e. the three legs of the fire triangle. Removing any one of these elements eliminates the possibility of fire. For example, if there were very little or no oxygen present, a fire could not occur, regardless of the quantities of fuel and heat that were present. Likewise, if insufficient heat were available, no concentrations of fuel and oxygen could result in a fire.

For an explosion to occur, there are five necessary elements which must occur simultaneously: fuel, heat, oxygen, suspension, and confinement. These form the five legs of the explosion pentagon. Like the fire triangle, removing any one of these requirements would prevent an explosion from propagating. For example, if fuel, heat, oxygen, and confinement occurred together in proper quantities, an explosion would still not be possible without the suspension of the fuel. However, in this case, a fire could occur. If the burning fuel were then placed in suspension by a sudden blast of air, all five sides of the explosion pentagon would be satisfied and an explosion would be imminent.

Bearing in mind the three sides of the fire triangle (fuel, heat, oxygen) and the five sides of the explosion pentagon (fuel, heat, oxygen, suspension, confinement) are important in preventing fires and explosions at any facility, by eliminating the possibility of either suspension or confinement, an explosion cannot occur but a fire may occur. By eliminating the fuel, the heat, or the oxygen requirements, neither a fire nor an explosion can occur.

5.2 Coal as a Fuel

Coal, as a primary fuel, must meet several requirements in order to be explosive. These requirements are volatile ratio, particle size, and quantity. The volatile ratio is a value that has been defined and established by the former United States Bureau of Mines to evaluate the explosibility of coals based on large-scale tests in the Lake Lynn Experimental Coal Mine.

To calculate the volatile ratio, a proximate analysis must be performed in the laboratory on a sample of the coal. Such an analysis determines the volatile matter and fixed carbon quantities of the coal along with moisture and ash content. The volatile ratio is defined as the volatile matter divided by the summation of volatile matter and fixed carbon of the coal.

The calculation of the volatile ratio produces a value independent of the natural or added incombustible in the coal. It has been determined and widely reported that coals with a volatile ratio exceeding a value of 0.12 pose a dust explosion hazard. All bituminous coals fall into this category.

Anthracite coals, by definition, have a volatile ratio of 0.12 or less, and as such they do not present an explosion hazard. Note however that both bituminous and anthracite coals can be involved in fires, but only bituminous coals can be involved in explosions.

Another important property of the fuel is related to particle size. Experimental studies have shown that bituminous coal particles passing through a U.S. standard 20-mesh sieve can participate in a coal dust explosion. A 20-mesh sieve allows particles up to 841 microns or about 0.03 inch to pass – these are the largest particles that contribute to a coal dust explosion. As the particle size is reduced further, a more severe explosion hazard materializes. Typically, in pulverized-fuel systems, the coal is reduced to a particle size where more than 85% will pass a U.S. standard 200-mesh sieve with openings of 74 microns or about 0.003 inch. These coal dust particles require less energy or temperature to ignite and, since heat transfers more quickly between smaller particles, the pressure and rate of pressure rise during an explosion are accentuated.

A third requirement for explosibility is related to the quantity of coal dust available, known as the minimum explosive concentration (MEC). The reader may recall from previous chapters that the MEC is defined as the minimum quantity of dust in suspension that will propagate a dust explosion and generate sufficient pressure to cause damage. The MEC for bituminous coal is approximately 0.10 ounce per cubic foot or 100 grams per cubic meter. When pulverized coal dust at the MEC was dispersed in an entry, a cap lamp 10-feet within the cloud was not visible to observers standing in front of the dispersed dust. Also, a person cannot breathe in an atmosphere containing dust at the MEC. This amount of dust in the air is 25,000 times greater than the average concentration of respirable dust to which a coal miner may be exposed during an 8-hour shift.

A layer of pulverized coal dust at the MEC deposited on the floor of the Experimental Coal Mine in Bruceton, Pennsylvania averaged 0.005-inch thick. This thickness is almost unobservable. In other words, if footprints are visible in coal dust on the floor or the coal dust is seen on the walls of a plant, then there is a sufficient amount of coal dust at that location to propagate an explosion.

As with other dusts, the upper explosive limit of coal dust is not well-defined. Experiments reported by various investigators support that a coal dust loading of 3.8 ounces per cubic foot would propagate a low-velocity explosion and that a 5.0 ounces per cubic foot loading would quench itself within 10 feet of ignition.

As discussed further on, the presence of other flammable dusts or gases can lower the MEC of the coal, which increases the risk of fire and explosion. Conversely, the hazard can be lessened with the addition of ash, rock dust, inert gas, and any other inert material.

5.3 Heat and Energy

The heat requirements to complete the fire triangle or the explosion pentagon can be in the form of temperature or energy. The ignition temperature of a coal dust cloud decreases as the volatile content increases.

At high volatile contents, the ignition temperature of a coal dust cloud approaches a limiting temperature as low as 440°C (824°F). The ignition temperature of a Pittsburgh Seam coal dust cloud is fairly constant as the particle size increases to about 180 microns. Further increases in aerodynamic particle size result in rapid rise in the ignition temperature requirements. As the particle size decreases, the coal dust becomes easier to ignite.

The ignition temperature of a coal dust layer also decreases as the volatile content increases. At high volatile contents, the ignition temperature of a coal dust layer approaches a limiting temperature as low as 160°C (320°F). With dust layers on hot surfaces, the minimum ignition temperature decreases sharply as the thickness of the deposit is increased. This is attributed to the fact that thicker dust layers capture and hold heat more readily. The minimum ignition temperatures to ignite clouds of various ranks of coal are shown for illustrative purposes in figure 5.1. These should be considered very approximate.

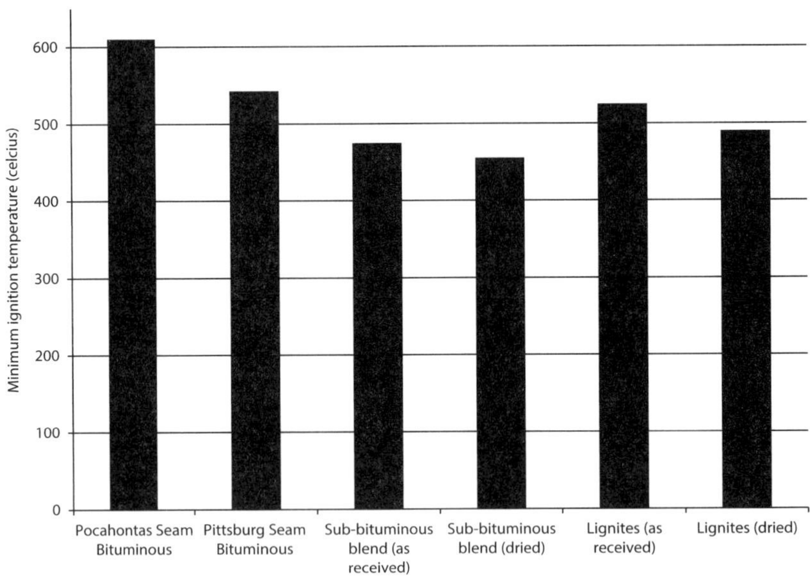

Figure 5.1 Range of minimum ignition temperatures reported from literature for coals of different types and rank.

Further, considerable variation in the minimum ignition temperature of coal dust layers have been reported with much lower values. In addition, the minimum ignition energy of coal dust varies with the oxygen content of the atmosphere, volatile content, and the amount of fine dust that will pass a U.S. standard No. 200-mesh sieve (74 microns). Coals are easier to ignite with higher oxygen contents, and/or higher volatile content, and/or in the amount of fine particulate coal. However, there is a limiting value of minimum ignition energy which varies for each coal.

All coal dusts should be regarded as being prone to ignition when exposed to the frictional sparks of badly maintained machinery or when they become contaminated with tramp metal. For mixtures of coal dust and flammable gas, the critical minimum ignition energy is that which affects the gas. When ignited, the gas releases sufficient energy to suspend and ignite a coal dust cloud.

As the volatile content of a coal increases, less oxygen is required to complete the fire triangle or the explosion pentagon. Less oxygen is also required as the rank of the coal decreases. Semi-anthracite has a very low volatile content and lignite is at least as volatile as high-volatile bituminous coals. However, at ambient temperatures, the oxygen content must be reduced to below 13% to prevent ignition of bituminous coal dusts with a strong ignition source.

5.4 Coal Dust Suspension, Confinement, Resuspension and Explosions

Suspension of fine coal dust particles in not a requirement for fires to occur; however, it is needed in order to complete the explosion pentagon. One should always consider the risk of explosion whenever coal dust is placed in suspension because, in most circumstances, there only needs to be an appropriate energy source available that can initiate an explosion. Furthermore, one may expect the possibility of multiple explosions. Consider an operation which has poor housekeeping practices and the external surfaces of machinery and floor areas have layers of dust. Multiple explosion scenarios are possible because of a domino effect, as illustrated in figure 5.2. Once the primary explosion has occurred, the blast waves from the primary explosion can disturb other layers of dust resulting in the formation of another ignitable dust cloud. Should this second cloud be ignited by an energy source, it in turn generates explosion blast waves which can cause another dust layer to become airborne, form a dust cloud and then be ignited.

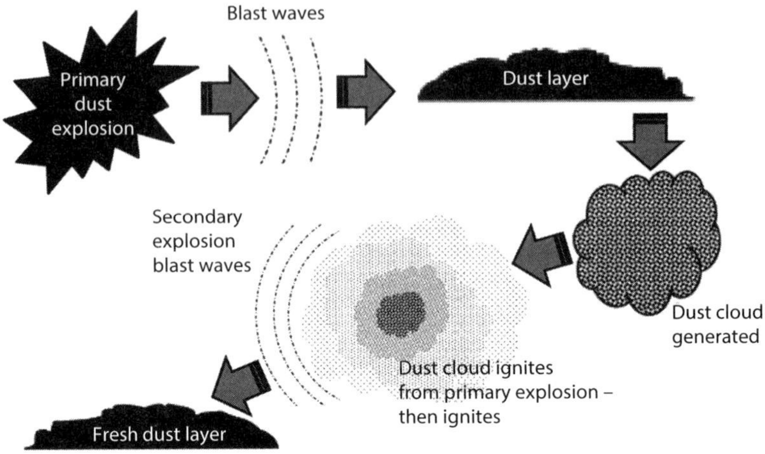

Figure 5.2 Illustration of a domino dust explosion effect.

If a coal dust layer on the floor is smoldering, an explosion is imminent if the layer is disturbed and the dust becomes a suspension in air. Under such conditions, the heat or energy needed to satisfy the fire triangle and the explosion pentagon are present.

The speed and duration of the moving air in an explosion is capable of dispersing additional coal dust from the floor, walls, overhead beams, and surrounding processing equipment. By way of comparison, a hurricane can cause substantial physical damage to structures at wind speeds of 150 to 200 miles per hour (230 to 290 feet per second). For most coal dust explosions, the air speed can exceed 200 miles per hour. In fact, a coal dust explosion will generally die out if the air speed is less than 100 miles per hour (150 feet per second).

As a point of reference, the maximum explosion pressure developed is about 90 psig for Pittsburgh Seam coal. The maximum rate of pressure rise for Pittsburgh Seam coal is 2000 psi per second. These reference values are important in predicting the violence or destructive powers capable of being generated when a particular dust is suspended and ignited. Since the maximum pressure is 90 psi for Pittsburgh Seam coal and the rate of pressure rise is 2000 psi per second, it is easily seen that only about 0.045 seconds elapse before the maximum pressure is realized. In a pulverized-fuel system using Pittsburgh Seam coal and designed to withstand 50 psi, vents must rupture within 25 milliseconds, otherwise pressures become excessive and equipment in the system is destroyed.

Once the explosion flame and pressures burst from the confinement into the plant area, a secondary explosion may be fueled by any additional dust suspended by the blast. This is why good housekeeping practices are so important to minimizing fugitive coal dust. If coal dust layers are eliminated then there would not be any fuel to allow a continuation of the explosion flame. The characteristic secondary explosion is responsible for the most damage to the plant itself. Also, the secondary explosion is usually responsible for the loss of lives or the serious injuries to personnel that occur.

Remember also that while confinement is not a leg of the fire triangle, it is an important leg on the explosion pentagon. The role of confinement is to keep the fine coal particles in close proximity after they are placed in suspension. Without the closeness, heat transfer could not occur rapidly enough to allow continued propagation. Without confinement, a propagating explosion is not possible, but rather, a large fireball with no appreciable forces associated with It is likely to materialize.

If an explosion is vented to the atmosphere outside of the plant or work area, confinement is eliminated and only part of the coal forced out of the vent will be burned, with the remaining unburned coal falling to the ground. As with the suspension leg of the explosion pentagon, if confinement is lost, the air speed will drop, additional coal dust will not be placed in suspension, and the explosion will extinguish.

5.5 Processing Equipment Explosion Hazards

Coal preparation serves several purposes. One important purpose is to increase the heating value of the coal by mechanical removal of impurities. This is often required in order to find a market for the product. Run-of-mine coal from a modern mine may incorporate as much as 60 percent reject materials. Various air pollution control devices are relied upon for the partial removal of pyrites with the ash to reduce the sulfur content of the coal. Ash content often must be controlled to conform to a prescribed quality stipulated in contractual agreements. Because of firing characteristics, it is often as important to retain the ash content at a given level as it is to reduce it. Freight savings are also a consideration since they are substantial when impurities are removed prior to loading. Also, the rejected impurities are more easily disposed of at the mine site remote from cities than at the burning site, which is usually in a populated area.

As coal ore leaves the mine, coal varies widely in size, ash content, moisture content, and sulfur content. These are all characteristics that can be controlled by preparation. Sizes range upward to that of foreign materials, such as a chunk of rock that has fallen from the mine roof or a metal tie;

large pieces of coal from a very hard seam are sometimes included. Ash content can range from three to sixty percent at different mines. Most of the ash is introduced for the roof or bottom of the mine or from partings (small seams of slate) in the coal seam. This ash is referred to as extraneous ash, and has a specific gravity greater than 1.80. The remaining ash is inherent in the coal. The density of coal increases with the amount of ash present.

The moisture content of the coal is also of two types. The surface moisture, that which was introduced after the coal was broken loose from the seam, is the easier to remove. This moisture is introduced by exposure to air, wet mining conditions, rainfall (in stockpiles), and water sprays. The remaining moisture, called "bed", "cellular", or "inherent" moisture, can be removed only by coking or combustion. This moisture was included during formation of the coal.

Foreign materials are introduced into the coal during the mining process, the most common being roof bolts, ties, car wheels, timber, shot wires, and cutting bits. Sulfur in coal occurs as sulfates, organic sulfur, and pyrites (sulfides of iron). The sulfates usually are present in small quantities and are not considered a problem. Organic sulfur is bound molecularly into the coal and is not removable by typical coal preparation processes. Pyrites generally are present in the form of modules or may be more intimately mixed with the coal.

Coal preparation plants remove only a portion of the pyritic sulfur; therefore the degree of sulfur reduction depends on the percentage of pyrites in the coal, the degree to which this is intimately mixed with the coal, and the extent of coal preparation. All of the materials described above are combined with the coal to form the ROM feed (the run-of-mine ROM coal). Coal, as referred to above, denotes the portion of the feed that is desired for utilization. Figure 5.3 shows the key features of a coal sizing circuit where various coal processing operations are performed. Each processing area where coal is handled and each piece of equipment in a manufacturing process poses individual hazards. Some common cases are now considered.

The raw coal for a pulverized fuel system is usually received from a variety of sources with particle sizes generally limited to approximately 2 inches or smaller. This raw coal is typically stored on an outside stockpile where it is moved around by heavy machinery such as frontend loaders. The fire and explosion hazards associated with a coal stockpile are usually limited to spontaneous combustion. Hot material must never be loaded into the pulverized-fuel system as there is a strong possibility of an explosion occurring within the pulverizer. In such a situation, all of the sides of the explosion pentagon could occur simultaneously. Hot spots should be

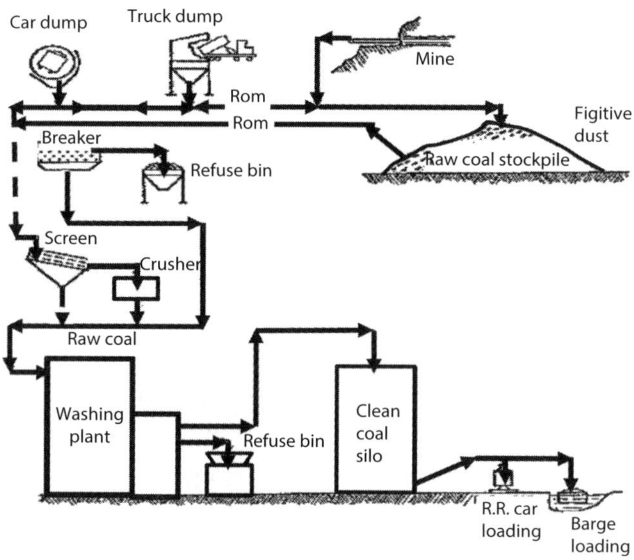

Figure 5.3 Illustrates key operations in coal sizing.

removed from the coal stockpile and spread about in thin layers allowing them to rapidly cool down.

If there are no hot spots in the coal, the front-end loader will load the coal onto a conveyor belt, which feeds a coal storage bin. These bins are usually equipped with mechanical sensors to detect high-level or low-level coal storage. There is also an emergency chute for unloading the bin in the event of a problem inside the bin. Coal in the bin may be susceptible to spontaneous combustion; however, some airflow is required to provide the oxygen necessary for heating. Thermocouples are sometimes located inside the bin to give warning of a fire, but carbon monoxide sensors would be more reliable for detecting an incipient fire. The raw coal empties from this bin onto a weigh scale. The weigh scale is a short conveyor belt that monitors the weight and the feed rate of the raw coal to the pulverizer. When any problems are detected in the system, the coal feed to the pulverizer is stopped completely.

A pulverizer is a size reduction machine; e.g., a hammermill. See figure 5.4 for the key features of this machine. Under normal operating conditions, coal is dropped from the weigh scale into a rotary airlock before it enters the pulverizer. A rotary airlock allows the coal and its inherent moisture to enter the pulverizer, but prevents any outside air from entering the system. Generally, the outside air has a higher oxygen content than the air circulating in the system and this additional oxygen could lead to completion of

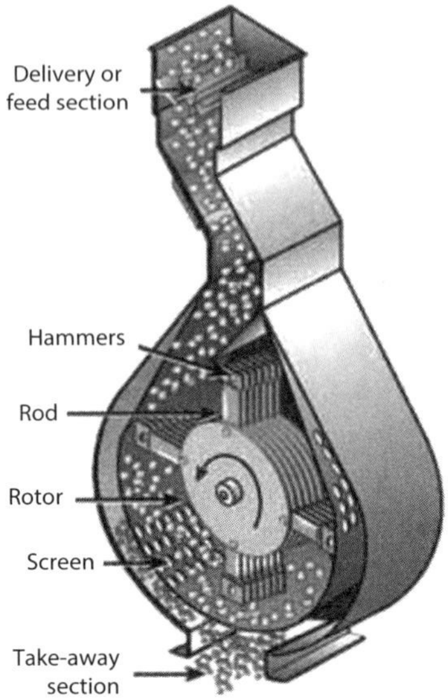

Figure 5.4 Shows key features of a hammermill.

the explosion pentagon and potential disaster. Coal that passes through the rotary airlock falls on the grinding table inside the pulverizer.

The coal feed rate and the size of the grinding table are variable. As the coal is being ground, hot air enters the bottom of the pulverizer and passes up through the pulverizer. The air is used for its drying and conveyance abilities. This hot air can come either from the clinker cooler or can come from the kiln hood. However, hot air from the clinker cooler is generally around 400°F as opposed to hot air from the kiln hood which can be much higher (typically 900 °F to 1200 °F). These elevated temperatures can lead to heating in any coal that has deposited along internal surfaces.

The main explosion hazard associated with a pulverizer is related to startup and shutdown procedures. When a system goes down under load, all the coal falls out of suspension. The internal surfaces are at elevated temperatures and the process of spontaneous combustion begins immediately. If the system is then restarted without full knowledge of internal conditions, an explosion could occur when the hot particles are suspended.

Another area of concern is the primary fan. The drive motor at the base of the mill can provide power for both the mill and the primary air fan, if the fan does not have a separate motor. The one advantage to having a single-drive motor is economical. The major disadvantage is that, in the event of a pulverizer shutdown, there is no way to pneumatically clear the coal out of the mill because the primary fan would also be down. This could lead to spontaneous combustion problems inside the mill that would make restarting the system hazardous. Before restarting, it must be verified that no hazardous conditions exist within the system. The primary fan has the role of transferring the coal particles into the kiln. If this fan is downstream from the pulverizer, it can exert a negative pressure, or suction, on the pulverizer. With higher pressures outside the pulverizer, typically nothing leaks out while a little air may leak in. However, a disadvantage to this configuration is the fact that pulverized coal will pass through the rotating blades of the primary air fan. This is generally not a problem unless an ignition source occurs within the fan. If on the other hand, the primary air fan is upstream of the pulverizer, then it exerts a positive pressure on the pulverizer. In this case, fine coal particles may find their way out of the mill through any small crack or fracture in the mill or through parts of the mill that are not well sealed. This leads to accumulations of coal dust in the plant and, if an explosion ruptures the pulverized-fuel system, it could use this coal dust as additional fuel for a serious secondary explosion.

Now let's consider the cyclone collector (See figure 5.5 for equipment profile). After the coal has been pulverized to a fine enough size, the

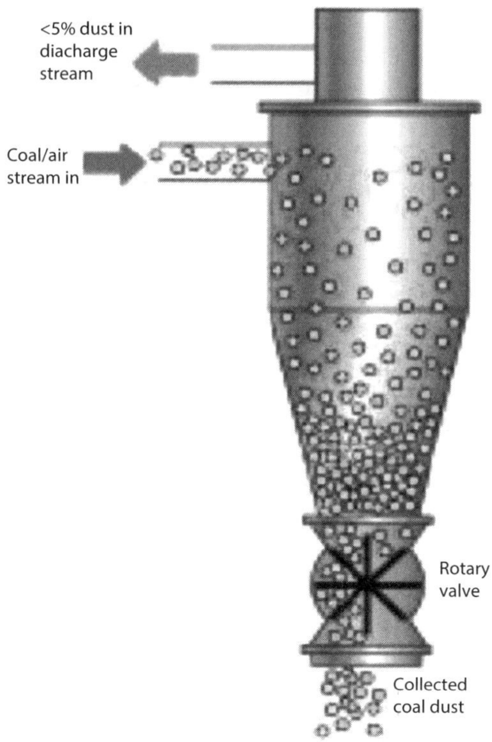

Figure 5.5 Shows details of cyclone separator/classifier.

circulating air lifts it out of the top of the pulverizer and through a classifier. When these fine coal particles pass out of the classifier, they may be transported through a duct lea ding to a dust cyclone. The coal-dust-laden airstream enters the cyclone where separation of the pulverized coal and circulating air is accomplished. The cyclone is designed such that the pulverized coal falls into the bottom of the cyclone, while the clean circulating air is allowed to pass out of the top of the cyclone.

Cyclones are capable of removing approximately 95% of the coal fines from the circulating air. The other 5% of coal fines will pass out of the top of the cyclone and continue through the system fan. A limited amount of coal is stored in the base of the cyclone for a short period of time. The rotary valve then feeds the coal into the airstream of the primary air fan. Immediately, the coal dust is blown into the kiln. Without any bulk storage of pulverized coal in the system, a shutdown of the pulverizer will stop the continuing coal feed to the kiln.

The clean air that flows out of the top of the cyclone passes into the system fan. Primarily, the system provides an airstream to transport fine coal from the pulverizer to the cyclone. Basically, the circulating air travels in a loop comprising the system fan, the pulverizer, the dust cyclone, and the connecting ducts. In this respect, the air provided by the system fan enters the pulverizer, but only that air passing through the primary fan enters the kiln (refer to figure 5.6 for an example of a kiln).

The pulverized coal does not pass through the blades of any operating fan. The fact that the system fan is a clean fan is an advantage. However, in the case of the dust cyclone, 5% or more of the coal fines are not removed from the air stream. This means that that accumulation of coal fines can occur in the closed loop, including the system fan. With sufficient fuel and

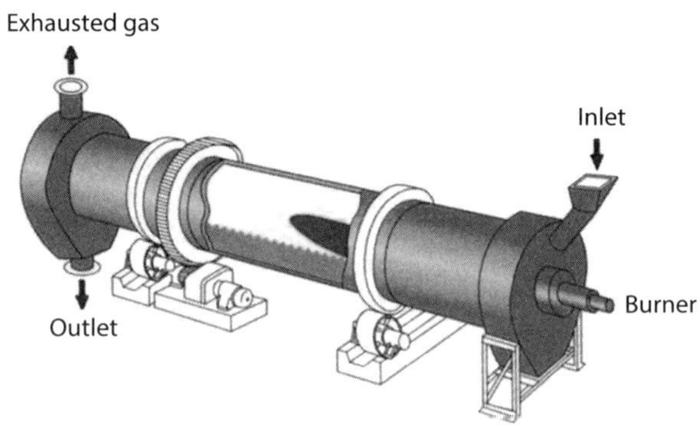

Figure 5.6 Example of rotary kiln.

oxygen in an area, only an ignition source need be present for a fire or explosion to occur. Along these lines, startup and shutdown are critical times where explosion hazards could exist. During shutdown, an accumulation of coal may begin to smolder in a state of spontaneous combustion. If this condition is not detected before startup, the smoldering coal could be placed into suspension by the system fan and the five sides of the explosion pentagon would be satisfied.

Along with cyclones, baghouses (see figure 5.7) are also a concern. The air that passes out of the top of the cyclone, with the 5% or more coal dust, is transported to a baghouse. A baghouse is designed with many filter-type bags hanging vertically, which are capable of removing the remaining coal dust from the circulating air. Baghouses or fabric filters are applicable for capture of fine particles of dust when the gases are at moderate temperatures, contain no sticky materials, and are nonexplosive. The dust-laden stream is passed through a finely woven or felted fabric on which a layer of dust serves as the filtering medium. As this dust layer thickens, the bag is "shaken" mechanically or by abrupt pressure changes to remove a portion of the filter cake. The filter is usually in the shape of a circular closed-end cylinder five to 12 inches in diameter and up to 30 feet long. Smaller filters are used to control emissions from bin and silo openings. The size of a filter installation depends on the amount of air and dust to be filtered. The ratio of air to filter cloth area depends on several variables; different services require different ratios or filtering velocities. The guidelines used for filtering of coal dusts are ratios of 6/1 for high dust loadings and elevated

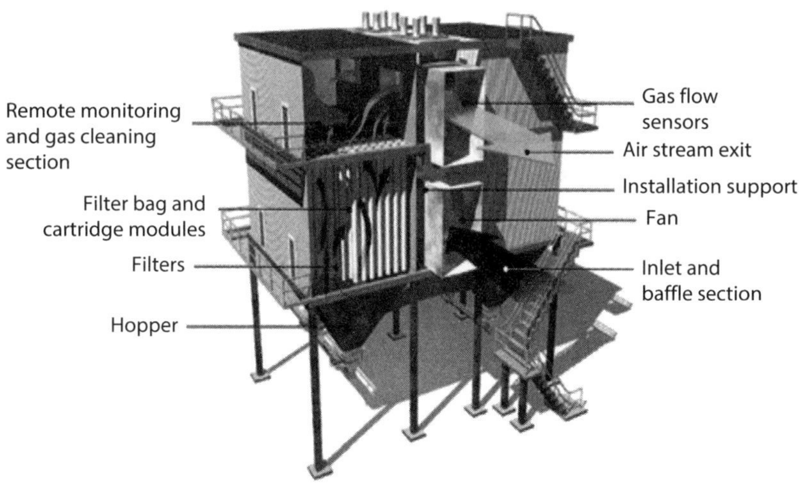

Figure 5.7 Features of a modern baghouse.

temperatures, progressing up to 8/1 for general dust loadings. Pressure differentials range from two to ten inches w.g. (water gauge).

In the baghouse, all the coal dust is captured in filter-type bags and the air circulating from the system fan is vented to the atmosphere. The primary advantage is twofold: first, all coal dust can be used as fuel; secondly, the air vented to the atmosphere carries a high-moisture content and is not recirculated through the system. This allows hot, dry air from the kiln hood or clinker cooler to be mixed with air from the system fan before entering the pulverizer. The moisture content of this air is lower than that which is vented to the atmosphere from the baghouse because it is not involved in coal drying until it enters the pulverizer.

After the coal-laden primary air passes through the primary air, the stream enters the kiln. Through a fan, the coal is blown through the burner pipe and directly into the kiln. The burner pipe is a long cantilever, which can extend 35 feet of more into the kiln, but is usually limited to 10–15 feet. The rotation and slope of the kiln cause the raw material to fall towards the lower end of the kiln where the burning zone is located. It is in this burning zone that the process of transforming the raw material to clinker takes place at around 2800°F. This temperature and the length of the flame are directly related to the volatile matter and the moisture in the coal.

Dryers – Dryers are a special class of processing equipment that if not carefully specified and designed can lead to explosions. Designing a thermal coal drying system is quite different than developing drying systems for most other materials. Considering the wide range of coal types and processing applications, system requirements and drying characteristics vary greatly. Design requirements are not only affected by the type of coal to be dried, but also the source of the coal and the specifics of each unique application.

Engineering systems to thermally dry disparate types of coal efficiently, safely and economically can be a challenge. Drying systems for a wide range of coal types and applications include:

- Run-of-mine coal
- Filter cake
- Coal fines from a prep plant
- Reclaimed waste coal
- Steam coal
- Coking or metallurgical grade coal
- Washed and sized bituminous coal
- Subbituminous or lignite coal prepped for gasification or liquefaction processes

Each of these coal types and application processes require special considerations and/or drying system features. For example, filter cake coals are very fine and typically hold a high volume of water. Therefore, a filter cake coal dryer must be able to handle high evaporative loads and high entertainment levels in the exhaust gases. Some coals, such as lignite, are subject to spontaneous combustion and explosions. A dryer for lignite should include explosion venting, a water deluge system and possibly an inert gas or a recirculated self-inerted process gas. Conversely, anthracite coal exhibits a high ignition temperature, rendering explosion vents or a deluge system unnecessary.

Figure 5.8 illustrates various types of fryer systems. There are many innovative designs for dryers; only a few examples are illustrated in the diagram.

Overheating of a thermal dryer may be caused by failure of a regulating valve (gas or oil fired) or other control and usually will produce an "upset" plume. The "internals" of dryers involve mechanical (sometimes moving) parts, which are subject to wear and damage. The refractory lining also is subject to scheduled replacement.

The second section of a thermal dryer is a large diameter cyclone, used to extract the particulate matter larger than 300 micrograms/m^3 from the gas

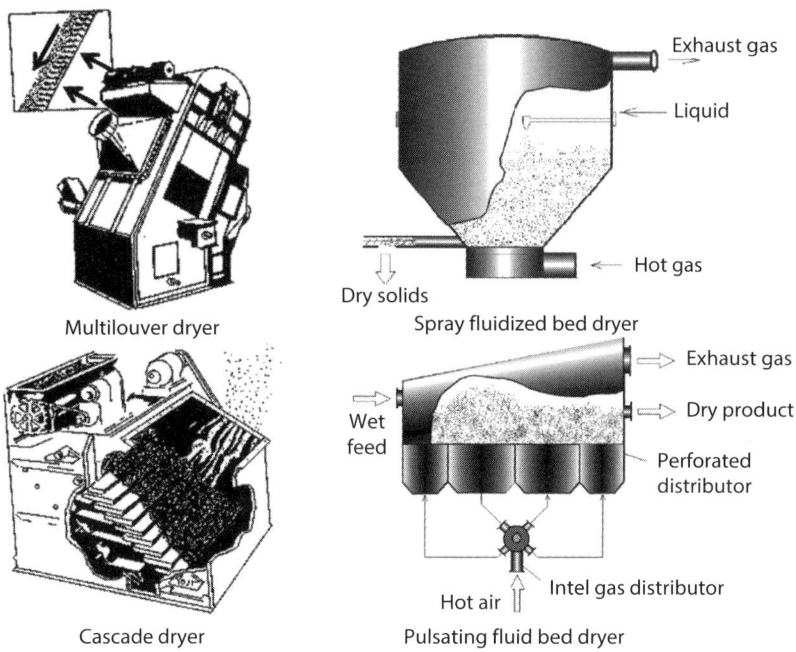

Figure 5.8 Various types of dryer configurations.

stream. This cyclone has no internal working parts, but the refractory lining is subject to wear. Volume of air, temperature, and pressure drop are the chief criteria for satisfactory or normal cyclone operation. Other causes of emissions are mechanical failure of the discharge feeder and wearing of seals.

Conventional fluid bed dryers are typically used for coals with a smaller size fraction, such as washed crushed coals or fines, and can also handle fairly coarse coal from 1/2" up to 1" with very high fluidization velocities. Lower rank coals (low rank subbituminous, PRB, or lignite coals) handle well in conventional fluid beds with larger size fractions because they tend to thermally breakdown while being dried. Conventional fluid beds offer the advantage of being able to be built much larger than vibrating fluid beds.

Conventional fluid bed thermal dryers with in-bed heat exchangers are employed for increased thermal input, for higher evaporative loads, or where lower process gas temperatures are being used, particularly to take advantage of lower temperature wastewater such as at a power plant. In-bed heat exchangers can also be used with steam as a heat medium. Alternatively, they can also be used with superheated high-pressure steam with a closed loop recirculated indirect heated, superheated atmospheric pressure steam as the process gas. This arrangement results in a very high efficiency inert thermal coal drying system for high volatile coals.

Convention flash dryers, with their high capacity and ability to handle very fine particles, are excellent for thermal drying of fine coals such as filter cakes or crushed coals (particularly those with high moisture contents and high evaporative loads). Moisture can typically be reduced to the inherent moisture level of the coal. Recirculating the process gas greatly enhances the efficiency, and can be beneficial for drying lower rank coals by helping to render the process inert and preventing potential explosions.

Tornesh flash dryers offer the advantage of longer retention time and higher drying efficiency for applications that require drying to lower moisture levels. Materials to be dried are fed into the feed chamber of the unit and begin to swirl in a cyclonic pattern like a tornado. As moisture is removed from the product, it becomes lighter and eventually is entrained in the swirling upward hot gas stream. This swirling motion creates a more torturous path for the process gas to travel, increasing the retention time in the drying chamber and heat transfer while lowering the moisture content. The system arrangement used for a specific application will depend on the type of coal being dried, the application requirements and the process requirements of the application. The specific arrangement/schematic will vary, not only based on the type of thermal dryer used, but also the choice of system auxiliary components and how they are interconnected. Such thermal coal drying systems are custom designed for the application. Typical arrangements may include:

- Systems that dry only
- Systems that dry and cool
- Systems that recirculate the cooling zone exhaust gas to the drying zone supply for improved efficiency
- Systems with closed loop recirculated process gas streams that are heated with a direct-fired process gas heater
- Systems with closed loop recirculated process gas streams that are heated with indirect heat
- Systems with recirculated self-inerted process gas streams heated by direct-fired process gas heaters
- Systems with recirculated indirect heated inert process gas streams, such as nitrogen, with in-bed heat exchangers

Direct-fired systems can be fired by a number of fuels, including but not limited to natural gas, propane, fuel oils, coal fines, stoker coal, blast furnace gas, coke oven gas, and other "low BTU" by-product gases. A leading designer and provider of custom designed dryer systems for coal handling operations is Carrier Vibrating Equipment, Inc. and Vibranetics Division (Louisville, Kentucky; Web site: www.carriervibrating.com).

Usually the last section of a thermal dryer is a scrubber, used to remove the fine portion of coal from the air stream. The scrubber may incorporate a bypass for emergency use, which may be partially open and permit particulate emission. The venturi scrubber is the type most commonly used. Its efficiency is a function of the volume, pressure drop, and temperature of the contaminated air and the volume and pressure of the scrubbing medium (water). Deviation from established norms can be checked by reference to records of previous plant inspections. Inspectors should examine the seals at manholes and duct collars for tight fit and should check the device for removal of solids and spent scrubbing liquids to ensure that it is not blocked partially open.

5.6 Coal Mining Operations and Safety

5.6.1 Overview

In the previous sections, the nature and problems associated with handling coal in relationship to dust explosibility were discussed. Coal mining operations pose additional risks of explosion because of the presence of methane, a highly flammable gas. In this section we shall examine the following topics:

- The explosibility of methane gas mixtures
- Effect of pressure and temperature on explosibility
- The amount of methane stored in coal
- Forecasting the methane emission rate
- Layering of methane at the mine roof
- Conditions when the recirculation of mine air becomes hazardous
- The importance of higher air velocity in preventing methane explosions
- Mine explosions, barometric pressure, and the seasonal trend in explosions
- Case studies
- Prevention, detection and suppression

5.6.2 Origins of Coal Bed Methane and Explosions

Coal seams form over millions of years by the biochemical decay and metamorphic transformation of plant materials. This coalification process produces large quantities of byproduct gases, such as methane and carbon dioxide. The amount of these byproducts increases with the rank of coal - it is the highest for anthracite, where for every ton of coal nearly 1,900 lb of water, 2,420 lb (20,000 cu.ft.) of carbon dioxide, and 1,186 lb (27,000 cu.ft.) of methane are produced. Most of these gases escape to the atmosphere during the coalification process, but a small fraction is retained in the coal. The amount of gas retained in the coal depends on a number of factors, such as the rank of coal, the depth of burial, the type of rock in the immediate roof and floor, local geologic anomalies, and the tectonic pressures and temperatures prevalent at that time.

These gases are contained under pressure and mainly adsorbed on the surface of the coal matrix, but a small fraction of gases is also present in the fracture network of the coal. Methane is the major component of gases in coal, comprising 80%–90% or more of the total gas volume. The balance is made up of ethane, propane, butane, carbon dioxide, hydrogen, oxygen, and argon.

Methane is a colorless, odorless, combustible gas that forms an explosive mixture with mine air in the concentration range of 5 to about 15% by volume. The maximum concentration of methane in mine air is restricted by law to 1 to 1.25% in all major coal-producing countries.

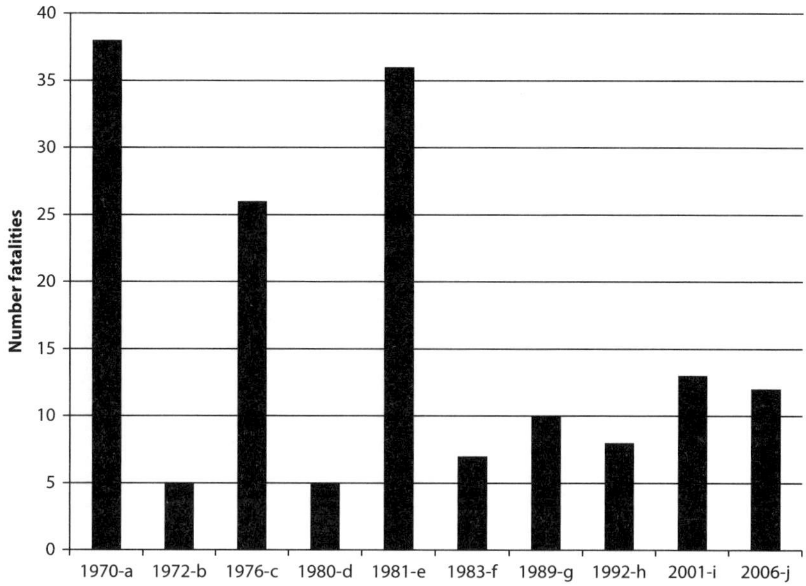

Figure 5.9 Examples of major U.S. coal mine disasters and numbers of deaths, 1970-a: No. 15 & 16 Mines, Hyden, KY; 1972-b: Itman No. 3 Mine, Itman, WV; 1976 –c: Scotia Mine, Oven Fork, KY; 1980 –d: Ferrell No. 17 Mine, Uneeda, WV; 1981 – e: Dutch Creek No. 1 Mine, Redstone, CO (15 deaths); No. 11 Mine, Kite, KY (8 deaths); No. 21 Mine, Whitewall, TN (7 deaths); 1983 – f: McClure No. 1 Mine, McClure, VA; 1989 –g: William Station No. 9, Wheatcroft, KY; 1992 – h: No. 3 Mine, Norton, VA; 2001 – i:Blue Creek No. 5 Mine, Brookwood, AL; 2006 – j: Sago Mine, Tallmansville, WV.

Coal mine explosions have resulted in some of the most horrific and tragic industrial accidents over the years, causing significant numbers of fatalities. Figure 5.9 provides but a few examples of explosions that have resulted in deaths in U.S. coal mining operations.

In the early 1900s, it was a common belief that most mine explosions were the result of unsafe practices that occurred throughout the mines. Early research, therefore, emphasized education of the mining industry. It was assumed that a more educated mining community would act in a way that would result in safer mines. In 1910, the Bureau of Mines opened an experimental mine in Bruceton, PA. From 1914 to 1925, the bureau focused almost exclusively on the causes of explosions and ways to prevent them.

Figure 5.10 illustrates conceptually how methane seeps through cracks in the coal bed, intermingling with air resulting in areas that are within the explosivity range of the gaseous mixture.

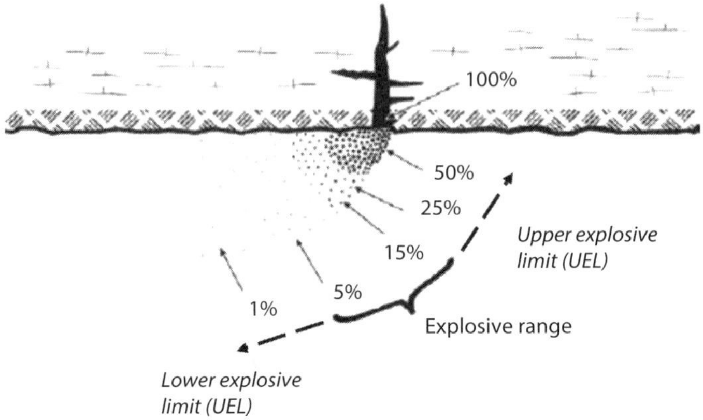

Figure 5.10 Illustrates methane gas seeping through a crack and being diluted by a moving air stream within a mine.

Methane gas has a tendency to pass through an explosive range as it seeps through a crack and travels into the mine environment. An effective mine ventilation system must be designed to ensure that the passage through the explosive range is as rapid as possible and that the volume of gas mixture in or above the explosive range is minimized. Even though methane-air mixtures under 5% are not considered to be explosive, worldwide experience with methane in mines has indicated that a considerable margin of safety must be provided.

Most explosions in coal mines occur when an explosive methane-air mixture is present. Since its inception in 1910, the U.S. Bureau of Mines and later the National Institute for Occupational Safety and Health have conducted research to help prevent methane explosions. Some of this methane research has been in direct response to underground explosions. Other research has been the result of applying new technologies to reduce the likelihood of underground explosions (for example, degasification with directional drilling).

An inert gas such as nitrogen or carbon dioxide cannot chemically react with methane. Inert gases can be added to an explosive methane-air mixture to make it nonexplosive. This is illustrated by the methane explosivity chart shown in figure 5.11.

Zabetakis[1] devised a simplified graphic method for determining the explosibility of mine atmospheres under mine-fire conditions from

[1] Zabetakis, M. G., Determining the Explosibility of Mine Atmospheres: Summary and Introduction, U.S. Bureau of Mines; Information Circular, 1959, IC7901

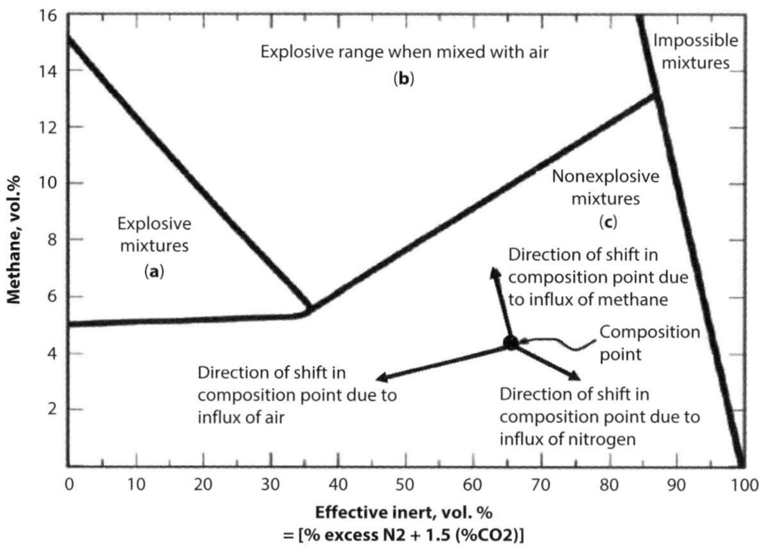

Figure 5.11 Methane Explosibility Diagram (after Zabetakis *et al.* [1959]).

mine- atmosphere composition data. Sometimes it is necessary to seal part of a mine during a fire to exclude air and thus extinguish the fire. Unfortunately, liberation of methane and formation of hydrogen and carbon monoxide in the sealed area may create an explosive or flammable atmosphere if enough oxygen is present to support combustion. On the other hand, the oxygen concentration may be reduced through combustion processes and the concentration of inert gases (nitrogen and carbon dioxide) increased enough to render the atmosphere within the sealed area non-explosive. Before an area is unsealed, it is extremely important to know if the sealed atmosphere contains flammable gases in proportions that might form explosive mixtures when air is introduced. If an explosive mixture forms and the fire is rekindled, a gas explosion will result. Various experimental methods may be used to determine the explosibility of a particular mine atmosphere; however, these methods usually are impracticable in the field. Regardless, the Zabetakis chart is useful from a general knowledge standpoint.

We must recognize that there are two main types of coalmine explosions: methane and coal dust. Methane explosions occur when a buildup of methane gas contacts a heat source and there is not enough air to dilute the gas level below its explosion point. Likewise, fine particles of coal dust in the right concentration that contact a source of heat can also be explosive.

Hybrid explosions consisting of a combination of methane and coal dust can also occur.

As noted, methane is formed as a byproduct of the coal formation. The methane that is adsorbed in the coal is released as the coal is mined or it migrates from surrounding sources above or below the coal seam through fractures created by the coal extraction process. Large fans are intended to circulate air in mines to provide ventilation to the working areas. These fans operate to dilute the methane to well below explosive levels (5–15%). Monitors installed on mining machines deactivate the machine when the methane concentration reaches 1%. Mines with excessive methane can remove the gas prior to mining by drilling drainage holes.

A common cause of methane face ignitions is the spark or hot streak created when cutting roof or floor rock with the mining machine. Applying directed water sprays behind the cutting bit to cool and limit the size of the hot streak can prevent these frictional ignitions.

Coal mines generally have active mining zones and areas that were previously mined. These mined-out areas can be abandoned areas that are no longer ventilated and are separated from the active mining areas by explosion-resistant structures called seals. Seals are designed to contain an explosion within the abandoned area.

Longwall mining is an example in which an "active" mined-out area called a "gob" is created behind the advancing panel extraction. This gob area becomes a reservoir for methane gas to collect, and may be difficult to fully ventilate because the overburden has collapsed and filled this area full of broken rock fragments. In this case, ventilation of the working face is critical to restrict the migration of methane from the gob and dilute that which escapes into the working area.

Methane is more readily ignitable than coal dust. In most cases, dust explosions are first initiated by methane explosions. The pressure wave or wind forces created by a small methane explosion can scour the mine entry, blowing coal dust up into the air. Once this dust is ignited, a chain reaction effect can occur and flame propagation can continue to spread over long distances consuming available fuel and oxygen and generating large volumes of toxic gases. Under constricting conditions such as blockage or restriction of the airway, both methane and coal dust explosions have the potential to accelerate and transition into the most severe situation, called a detonation.

Underground coal mining produces finely divided coal dust that deposits throughout the mine and serves as a source of combustible material for coal dust explosions. Limestone powder, known as rock dust, is spread throughout the mine workings on a regular basis. This rock dust serves to inert the coal dust when applied in the proper proportion. When explosions do occur, the dispersed limestone powder absorbs the heat generated from the explosion and will either stop the chain reaction or reduce the intensity of the explosion. It is critical that this inertization practice of rock dusting be consistent with the mining process. Even a thin layer of additional coal dust deposited on a previously rock dusted area can restore the explosive condition.

5.6.3 Longwall Mining

Methane control research by the National Institute for Occupational Safety and Health (NIOSH) has extended in recent years to applying reservoir modeling to predict methane inflows in gate road entries. Their studies support that emission rates in the gate roads decrease with the use of shielding boreholes and increased degasification time. Also, mining perpendicular to the face cleats were found to liberate more gas into the mine workings, whereby emissions were almost a linear function of Langmuir pressure and volume, and emissions were inversely related to sorption time constant. Subsequent simulations predicted changes in methane drainage using in-seam boreholes. The results showed that longer degasification times resulted in lower face emission rates. Pre-mining degasification produced more methane than that produced during panel extraction, a fact attributed to the already decreased methane content of the coal bed. NIOSH's[2] work concluded that longer pre-mining degasification periods would be more advantageous to the operator.

NIOSH reports that the industry trend toward increasing longwall face width can produce increased methane emissions from the face due to a higher volume of cut coal on the face conveyor. Permeability changes in the gob behind the longwall shields were also estimated using a NIOSH-developed numerical model. The results showed that permeability was highest near the edges and corners of the gob.

The following are some of the practical guidelines for controlling longwall coalbed methane offered in the NIOSH report:

---

[2] S. Schatzel, C. Karacan, , R. Krog, G. Esterhuizen, and G. Goodman, Guidelines for the Prediction and Control of Methane Emissions on Longwalls, NIOSH, Pittsburg, PA, March 2008

- NIOSH has recommended using shielding degasification boreholes to decrease emission rates by at least 25% for development entries. It recommends drilling the boreholes as close as practically possible (~27 m (90 ft)) to the entries and operate them for at least 6 months to achieve a 25%–50% decrease in emission rates.
- Equations for predicting methane emissions rates into gate road developments are presented by NIOSH. These relationships assume a supercritical longwall panel developed in the Pittsburgh Coalbed.
- For a longwall panel width greater than 305 m (1,000 ft), a trilateral borehole configuration is recommended for effective draining of methane. Short across-panel holes at close spacing can also be effective in degassing a panel. The required number of holes will depend on seam anisotropic reservoir conditions, but at least 12 holes for a 3,400-m (11,000-ft) long panel are recommended to achieve the same degasification as the trilateral configuration.
- If less than 12 months are available for pre-mining gas drainage, NIOSH recommends that degasification be continued until mining approaches the borehole. This approach maximizes the quantity of removed methane and reduces methane emission rates.
- To avoid shearer coal production delays, NIOSH recommends that continuous gob gas venthole (GGV) production be assured while GGVs are within about 500 ft of the working face. In many mines, the quantity of coalbed methane removed by a GGV is potentially 75% of the volume of gas emissions on the longwall face. A similar finding was observed by prior NIOSH research. If an operating GGV ceases producing gas, the gas that was being removed will enter the ventilation system.
- Assuming a well-caved gob, increasing the longwall face length by X% will increase the rate of methane emissions by, at most, two-thirds of X%.
- The length of the slotted casing section of a GGV will strongly influence its level of gas production. To effectively design the slotted casing section of a GGV, NIOSH recommends to (a) review the local geology to identify the location of gas-bearing units; and (b) set the top of the

slotted section at the highest gas-producing stratigraphic horizon.

- Completing a GGV into the caved zone is counterproductive and increases the likelihood of intermittent production from increased-width, supercritical panels. Therefore, the completion depth of GGVs should be at least 14 m (45 ft) above the top of coal for longwall panels, particularly in the Northern Appalachian Basin.
- Emphasize continuous GGV production since it will potentially produce 40%–50% more coalbed gas than GGVs operating intermittently.
- Increasing the longwall panel width increases the quantity of methane present because of the increased fractured reservoir volume. However, this increase does not enhance the performance of GGVs. As panel width increases, the effectiveness of GGVs completed near the tailgate margin will not extend as close to the headgate side. Drilling GGVs on the headgate side or near the panel centerline can produce coalbed gas from this portion of the panel.
- In favorable cases, GGVs drilled on the headgate side can be effective. Completion depths must isolate the borehole from communication with the ventilation network. These findings are based on supercritical panel designs.
- Mining-induced fracturing was observed to occur 24–46 m (80–150 ft) ahead of the mine face. Boreholes and exhausters should be installed before this occurs.
- Data were reviewed for GGV configurations completed from 7 to 32 m (24 to 106 ft) above the mined coalbed for supercritical panels in the Northern Appalachian Basin. It is recommended that GGVs be completed toward the top of this interval and be designed to include the Sewickley Coalbed. Permeability increases following undermining were dramatic, with increases of about 100–500 times the pre-mining values and instantaneous increases of up to about 1,000 times these values. These measurements did not differ significantly despite differences in borehole configurations. Fracture permeability pathways remain high to the mined coalbed toward the top of the described interval, yet the likelihood of drawing ventilation air into the borehole is minimized.

5.6.4 Controlling Explosion Risks at Coal Mine Working Faces

Kissel, et.al.[3] provide guidelines for controlling explosion risk at coal mine working faces. They note that at coal mine working faces, simultaneous application of three basic elements reduces the methane explosion hazard:

1. Adequate ventilation
2. Regular monitoring of gas concentrations
3. The elimination of ignition sources

Adequate ventilation is provided by using the mine entries to convey air for the long distances between the mine portal and the working sections (main ventilation systems) and then using line brattice or ventilation duct (face ventilation systems) to convey air the last hundred meters to the working face where coal is broken and removed. The air quantity provided is enough to safely dilute methane and the air velocity is enough to prevent layering.

Gas concentrations must be regularly monitored in accordance with regulations using knowledge of the circumstances under which the highest concentrations are likely to be found. Ignition sources must be eliminated by ensuring that electrical equipment does not ignite methane, that sparking from cutter picks is minimized, and that smoking by workers is strictly forbidden.

Risk-reduction studies using fault-tree analysis have shown that large reductions in explosion risk only result from *multiple* preventive actions. Kissel gives the example - a ventilation upgrade or a methane monitor upgrade by itself offers risk reductions under 50%. A risk reduction of 90% or more would typically require much more precautions. They note that other studies have shown that the everyday vigilance of those working underground is as important as engineering design.

Simultaneous application of three basic elements reduces the methane hazard at coal mine working faces. Large quantities of ventilation air circulated from mine portals to working sections through intake and return mine entries (main ventilation systems)and then to working faces using line brattice or ventilation duct (face ventilation systems). Regular

---

[3] F. Kissell, J. Tien, and E. Thimons, Methods for Controlling Explosion Risk at Coal Mine Working Faces, NIOSH TIC2 Number: 20032685, In: Proceedings of the 32nd International Conference of Safety in Mines Research Institutes, 28–29 September 2007, Beijing, China. Beijing, China: National Center for International Exchange & Cooperation on Work Safety (SAWS), 2007; 161–168

monitoring of methane gas concentrations with gas detection instruments, along with subsequent action when a threshold level is attained.

The simultaneous application of several elements is necessary because in the event that one fails, the others continue to ensure levels of safety.

As noted, methane gas is explosive when mixed with air in the range between 5 and 15 volume % of gas. The 5 volume % value is the lower explosive limit (LEL). Methane concentrations in air that are below the LEL are not explosive. When methane is emitted from strata, it is usually at high concentration. As it progressively mixes with air, the concentration will pass through the explosive range and down below the LEL. A proper ventilation system should supply sufficient fresh air to reduce all of the gas to far below the LEL as soon as the gas is emitted from the strata.

Main ventilation systems circulate air from the portal to working sections of the mine. In the US, main ventilation systems in coal mines always employ the mine entries to move air, at least one entry for intake air and at least one entry for return air. This is illustrated in figure 5.12. Ductwork is not used in a main ventilation system. The reliance on intake and return mine entries rather than ductwork makes possible the movement of large air quantities over long distances necessary for methane dilution.

Face ventilation systems carry air the last hundred meters to the working face where coal is broken and removed. U.S. practice is to employ either ductwork or line brattice for face ventilation. In most instances, the primary source of gas is at the working face, so it is vital to provide adequate ventilation air all the way to the face, that is, to the *last foot(30 cm).*

Kissel and coworkers describe two categories of face ventilation: exhausting and blowing, which are illustrated in figure 5.13. The exhausting system is the least efficient in clearing out methane gas from the face.

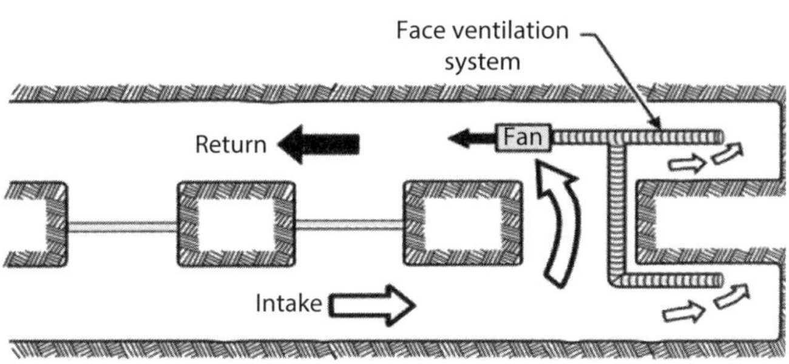

Figure 5.12 Illustrates how main and face ventilation systems interact to bring large quantities of air to working faces. Main ventilation systems use mine entries (intake and return) while face ventilation systems use ventilation duct (illustrated) or line brattice. Black arrows represent contaminated air (After Kissel).

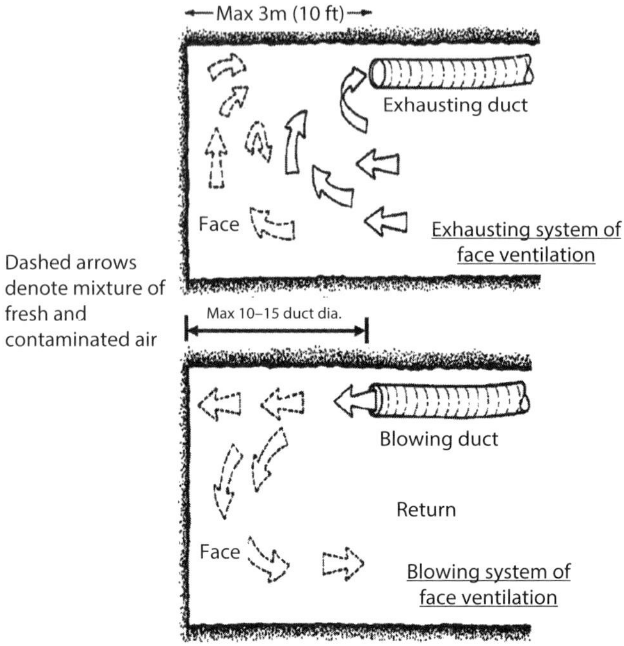

Figure 5.13 Illustrates exhausting and blowing systems of face ventilation (equipment not shown).

For example, the face ventilation effectiveness (FVE) of a 5-m^3/s airflow, 60-cm-diameter exhaust duct located 3m from the mine face is only about 0.10. This means that the concentration of methane measured one foot (30 cm) from the face is 10 times higher than the concentration in the air passing through the duct.

If the end of the exhaust duct is more than 3 m from the face, the FVE is even less. The end of an exhaust duct must always be 3m or less from the mine face unless other means are used to ventilate the face, such as venturi air movers powered by compressed air. Blowing face ventilation is reported to be better for clearing out gas than exhaust ventilation because the momentum of the air in a blowing jet carries it farther. However, blowing systems also lose effectiveness as the face-to-duct distance increases. Kissel, et.al. therefore recommend that the duct be kept as close to the face as possible, with the end of the duct not more than 10–15 duct diameters from the face. The area must also be kept free of obstructions that would prevent the emerging jet of air from reaching the face. Studies of blowing ventilation at coal mine faces show that the FVE for a 5-m3/s airflow, 60-cm-diam blowing duct at 6 m (10 duct diameters) is about 0.40, indicating that the concentration at the face is 2.5 times that in the return. Thus, although 5 m^3 of air emerges from the duct per second, only 2 m^3 of air actually reaches the face.

If the face emits 0.01 m^3 of gas per second, the average concentration in the immediate face area will be 0.01/2.0 or 0.5%, rather than 0.01/5.0 or 0.2%. Whether exhausting or blowing ventilation is employed, the end of the duct should be kept as close to the face as possible. If the face is drilled and blasted, keeping the ductwork in place is difficult. In some situations it is possible to move flexible ductwork forward and back on a trolley wire. Whatever method is used, when methane is present it is essential to keep the ventilation ductwork within the required face distance, regardless of cost or difficulty.

Another critical parameter to consider is air velocity. Low air velocities can lead to poor mixing between methane and air. Poor mixing in turn leads to fluctuations in the methane concentration that make an ignition scenario more likely. Bakke *et al.*[4] noted that a measurement of the methane concentration alone is an incomplete means of assessing the ignition hazard and that other measurable ventilation quantities might be important. Kissel reports a study of methane ignitions in U.K. coal mines that found that the probability of an ignition is determined by both the methane concentration and the densimetric Froude number (N_{Fr}). The Froude number is a dimensionless quantity related to the gas-mixing process in the presence of buoyancy forces. The expression for the Froude number is as follows:

$$\mathbf{N_{Fr} = u^2/[\,g(\Delta\rho/\rho)\,A^{1/2}]}$$

where:

u is the air velocity,

$\Delta\rho/\rho$ is the density difference between air and methane divided by the density of air, and

A is the cross-sectional area of the airway.

The data available to Bakke *et al.* from 123 ignitions on faces and gate roads at U.K. longwalls between 1958 and 1965 provided a basis for the correlation. Examination of the data indicated that the risk of an ignition was dependent on more than methane concentration alone and that it was possible to combine concentration and Froude number in one variable of

---

[4] Bakke P, Leach SJ, Slack A [1967]. On the general body concentration and other ventilation data as a measure of the average risk of ignition. Min Eng (U.K.) *Aug*:762–774

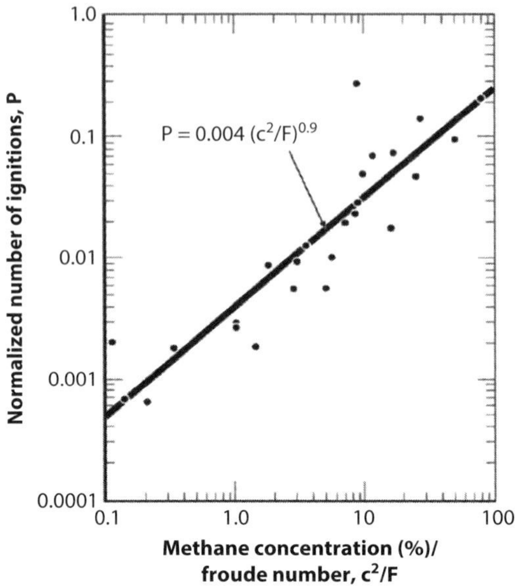

Figure 5.14 Correlation for the ignitions per year per gate road.

the form c^2/N_{Fr}. Figure 5.14 shows the normalized number of ignitions P (ignitions per year per gate road) versus c^2/N_{Fr} for the Bakke *et al.* data. The best fit to the data provides the following empirical correlation:

$$\mathbf{P = 0.004\ (c^2/\ N_{Fr})^{0.9}}$$

A high correlation was obtained, indicating that, absent other sources of mixing, the risk of ignition P does depend on the variable c^2/N_{Fr}. In most mines, A does not vary greatly compared to changes in c^2 and u^2. Also, the factor of 0.9 in figure 5.14 is close to unity. It follows that ignition risk varies with the quantity $(c/u)^2$. This departs from any notion that ignition risk depends on the concentration of c alone.

The findings described above have important implications for using higher air velocity to prevent methane explosions. In the absence of other means to promote turbulent mixing, raising air velocity is generally accepted as an effective way to reduce ignition risk. Higher air velocity promotes better mixing in addition to lowering the average methane concentration. High pressure (>10 bar) water sprays and auxiliary air movers (small fans or compressed-air venturis) that promote mixing can serve to reduce ignition risk.

It is also important to note that similar methane concentration levels, tunnels or mines with large cross-sectional entries and low air velocities have higher risk of ignition than those with small cross-sectional entries and higher air velocities. Both the lower velocity and higher area will subsequently contribute to reducing the Froude number.

5.6.5 Stratification

The term stratification, as applied here refers to a layering of methane at the mine roof. The density of methane is about half that of air. When methane is released at the mine roof it may form a buoyant layer that does not readily mix into the ventilation air stream. Such layers are believed to have been the source of many mine explosions. It is important to understand the circumstances that lead to the formation of methane roof layers and the methods used to dissipate them.

The formation of methane layers is may be attributed to inadequate ventilation. Raine[5] argued that a measurement of ventilation velocity is critical. He reported that under conditions of "normal firedamp emission," an air velocity of 0.5 m/s measured at the roof was enough to prevent layering. Kissel notes that most current-day estimates of the necessary velocity are close to this value. Aside from inadequate ventilation, there are other circumstances under which methane layers may occur. The location of airways adjacent to gobs are an example. UK experience during the 1960s with advancing longwalls, frequently traveled gate roads were directly adjacent to fresh longwall gob, where broken overburden provided a pathway for roof gas emissions.

It is essential that thorough gas monitoring be implemented in order to address methane layers. The literature reports that vigilance should be applied to ensuring the following guidelines:

- The air velocity measured at the roof level should be 0.5 m/s or less.
- The airway should be located next to a gob or should intersect a geologic anomaly, such as a fault, in order to serve as a conduit for gas.
- The mine roof (or tunnel crown) should not be within easy reach, so measurements at roof level are less apt to be carried out regularly.

[5] Raine EJ [1960]. Layering of firedamp in longwall working. Trans Inst Min Eng (U.K.) *119*(10):579–597

- The airway is equipped with cavities or roof-level obstructions to air movement.
- The airway should be inclined more than 5°. Airways inclined 20 to 30 degrees require 50% greater ventilation velocity if the air moves uphill.

Workers who test for methane layers should be instructed that the gas concentrations in these layers may fall outside of the accurate operating range of catalytic heat of combustion sensors. For accurate operation of these sensors, in most cases the concentration of methane must be below 8% and the concentration of oxygen must be above 10%.

Methane layers can be removed by increasing the ventilation velocity and reducing the gas flow by methane drainage. In instances where the methane source and layer size are limited, a less satisfactory but reportedly workable method is to use a (well-grounded) compressed air-powered venturi air mover or an auxiliary fan at each methane source to blow air at the source of the layer and disperse it. In either case, an aggressive sampling program is necessary to ensure safe conditions.

As noted, methane layering occurs as the gas is released at the mine roof. When methane is released at the mine floor or rib, the gas will readily mix into the ventilation air stream, losing its buoyancy. Hence, only the roof source produces a significant methane layer at the 2-cfm (1 l/s) rate.

5.6.6 Use of Portable Methane Detectors

The use of portable methane detectors is straightforward; however knowing where to make measures and proper interpretation of instrument readings can require additional expertise. NIOSH recommends that methane measurements be made as follows:

- Close to the methane source, where higher concentrations are more likely to be encountered.
- Close to the mine roof, where higher concentrations are more likely to be encountered.
- Where roof falls have created cavities in the top that may be ignored during routine checks.
- In areas where the dilution of methane is impaired, i.e., those that are poorly ventilated and those where air movement is blocked by equipment.
- While cutting is underway, because the methane release rate is higher as coal or rock is broken and the mining machinery advances.

- On the side of the entry that normally sees the highest concentrations. For exhaust ventilation systems, this is usually on the same side of the entry where the exhaust duct (or curtain) is located. For blowing ventilation, it is normally the opposite side of the entry from the blowing duct (or curtain).
- On mining machinery. In mines using continuous miners or longwall shearers, an additional methane detector is mounted on the equipment and wired to shut down the machinery if a threshold concentration is exceeded.

A quick glance back at figure 5.10 will remind the reader how methane is released and becomes diluted to safe levels. Methane generally enters a mine or tunnel as a localized source at high concentration. As illustrated in figure 5.10, the methane stream becomes diluted as it mixes with a moving air stream. As shown in the figure, methane enters through a crack in the coal. If no air enters the crack, the methane concentration in the crack can be close to 100% (well above the UEL); but as the methane emerges from the crack, it progressively mixes with and is diluted by the ventilation air. The progressive dilution of the methane gas results in the concentration changing from a very high level to perhaps as low as 1%. Any instrument reading depends highly on the location of the measurement; hence, of critical concern is how the instrument operator intends to use the reading to assess whether a hazard exists.

To address this issue, methane measurements should be taken at a distance of not less than 30 cm from the roof, face, ribs, and floor. If there is enough gas emanating from the crack (or other source) that causes the concentration to exceed safe limits at a 30-cm distance, then a hazardous condition must be assumed to exist.

If methane layering is likely, measurements must be taken at distances less than 30 cm because the thickness of such layers can be less than 30 cm. The degree of hazard resulting from a high-concentration layer of gas must be assessed from measurements of the size of the layer as well as the location and size of the source.

Care must be exercised by operators to avoid misinterpreting warning signs. It is not unusual to misinterpret a gas warning sign, especially in underground workings thought to have no gas. A common reason is that the gas flow can vary with the excavation rate. Consider that a mining machine begins to cut into an area of gassy ground, releasing methane into the ventilation air. The machine-mounted monitor on the mining machine senses this gas and shuts the machine down. After tracking down the source of the shutdown, the operator begins to search for gas with a handheld detector, but the worker hunting for gas cannot detect much gas because the emission dropped when the machine stopped. Under this scenario, one might conclude that the monitor on the machine is not working properly. Normal human reaction is given two instruments, one with bad news and the other with good news, the tendency is to believe the good news. It is crucial to recognize that when methane detection and monitoring instruments fail, they rarely give a false alarm or a false high reading – i.e., a gas meter rarely if ever indicates the presence of gas when there is none. The usual failure mode is to not register gas that is present. When any instrument registers gas, it is better to trust the reading and take appropriate precautions. In short – it is best to err on the side of safety.

Caution should also be exercised when successive methane readings vary more than they normally should. When the airflow is low or when measurements are taken close to the source of the gas, the methane will not be well mixed with the air. As such, a high reading may be detected in one area with a low reading just a few feet away. Such incomplete mixing may indicate that the ventilation air is deficient, which gives sufficient cause for suspecting that even higher concentrations of gas might be found nearby.

A good industry practice is to perform a *bump test* on every portable methane detector to ensure that it is working properly. Before every shift, the operator should briefly expose the portable detector to a known concentration of methane gas high enough to set off the methane alarm. The operator should record or note the reading to ensure that it is correct. A bump test is not a calibration, but a quick way to ensure that the most important functions of the instrument are operational.

In addition, the use of person-wearable methane monitors is a good practice. These are low-cost, light- weight instruments that operate continuously to warn miners of any gas accumulations. These units are typically equipped with visual, audible, and vibratory alarms that produce signals to warn the worker when a preset methane level is exceeded.

5.6.7 Summary of Monitoring Principles and Best Practices

Previously we covered the subjects of:

- Instrumentation
- Portable detectors
- Machine-mounted monitors
- Calibration of catalytic detectors
- Misinterpreting warning signs

It is worthwhile doing a brief review and adding some additional information. One of the critical points that has been highlighted in that the failure to properly sample for methane is a major contributing factor to methane explosion risk. Sampling errors are most likely to occur at mines or in tunnels where the presence of methane is not suspected or during non-routine tasks at mines or tunnels known to have gas.

There are actually quite a few models of gas detectors that are available to measure methane concentrations, as well as most of the other contaminant gases found in mines and tunnels. An example is the iTX Multi-Gas Monitor, a portable gas detector available from Industrial Scientific Corp., Oakdale, PA. This handheld instrument measures several gases simultaneously. The cost (2004) ranges from $1,300 to $2,200, depending on the number of gases measured. Similar instruments are available from other manufacturers. Most methane detectors used in mining use a catalytic heat of combustion sensor to detect methane and other combustible gases. These have been proven through many years of reliable operation. For detection of methane, proper operation of catalytic heat of combustion sensors is required. The major categories of methane detectors are:

- Portable Methane Detectors – these are hand-carried instruments that enable measurements to be made at any location. There are also "Indicating Detectors " which provide indications of gas at 0.25% methane and have an accuracy of at least 20% over most of the applicable range.
- Machine-Mounted Methane Monitors – these are mounted on certain types of mining machinery and operate continuously. They are designed to prevent the mining equipment from operating unless the methane monitoring system is functioning, a warning device that activates when the methane concentration is above 1.0%–1.5%, and a means to shut off power to the equipment when the methane concentration is 2.0% and above.

The proper use of portable methane detectors involves obtaining measurements:

- Close to the methane source where higher concentrations of methane are more likely to be encountered
- Close to the mine roof where higher concentrations are more likely to be encountered

- In regions where the dilution of methane is impaired - i.e., those that are poorly ventilated and those where air movement is blocked by equipment
- While cutting is underway because the methane release rate is higher as coal or rock is broken and the mining machinery advances

Consideration must be given to how methane is released and diluted to safe levels. Methane entering a mine or tunnel often enters as a localized source at a high concentration. Figure 5.10 depicts a cloud of methane being diluted into a moving air stream. As shown in the figure, methane enters through a crack in the rock. If no air enters the crack, the methane concentration in the crack can be close to 100%. However, as the methane emerges from the crack, it progressively mixes with and is diluted by the ventilation air. Suppose this progressive dilution reduces the concentration from 100% to 1%, as shown in the diagram. In this case, the instrument reading depends highly on the location of the measurement—a critical concern if one intends to use the reading to assess whether a hazard exists. This problem is handled by requiring methane measurements at a distance of not less than 12 in from the roof, face, ribs, and floor. If there is enough gas coming from the crack (or other source) to exceed statutory limits at a 12-in distance, then a hazardous condition exists.

In some cases, measurements must be taken at distances less than 12 in. For example, methane is lighter than air, so methane emerging from the roof can form a high-concentration layer along the roof of the mine (or crown of the tunnel). The thickness of such layers can be less than 12 in. Methane layers are more likely to form if the ventilation air velocity measured at the roof is 100 ft/min or less [see Raine 1960], if the roof has cavities [see Vinson *et al.* 1978], or if the roof has drill holes that serve as emission sources. Therefore, the degree of hazard resulting from a high concentration layer of gas must be assessed from measurements of the size of the layer, as well as the location and size of the source. This is why gas readings must be done by a qualified, competent person, as prescribed by MSHA and OSHA regulations.

Methane measurements in restricted spaces requires special considerations. A restricted space is one that cannot be readily entered to make a methane measurement. Examples of restricted spaces are a mineshaft that has been capped and a mine entry that has been closed to travel because of a hazardous roof. Lack of accessibility restricts both ventilation air and the opportunity to make a measurement. Simply making a measurement at the entrance of the restricted space is not adequate. Measurements must be made deep within and at the top of the restricted space. Measurements are needed at every location where an ignition source may be present or suspected.

Special consideration should be given to out-of-range gas concentrations in restricted spaces. Because some restricted spaces have little ventilation, the gas concentrations in these spaces may fall outside of the accurate operating range of catalytic heat of combustion sensors. For accurate operation of sensors, the concentration of methane must be below 8% and the concentration of oxygen must be above 10%. When measuring methane concentrations above 8%, instruments with catalytic heat of combustion sensors can act in a way that is misleading, responding with a rapid upscale reading followed by a declining or erratic reading. Such erratic instrument behavior should be a tipoff that very high, potentially explosive methane levels may be present.

Restricted spaces may also lack the 10% oxygen level necessary to ensure the proper operation of catalytic methane detectors. For example, the gas in exploration boreholes often contains little oxygen. If the instrument being used has a second sensor to measure the oxygen, an oxygen concentration < 10% will be indicated, thereby alerting the user that the methane reading may be incorrect. Even if the oxygen concentration < 10%, valid methane measurements are possible with other kinds of methane detectors. In this situation one should use a methane detector that operates by infrared absorption and/or use a catalytic methane detector that provides dilution sampling.

The term *dilution sampling* refers to adding a controlled quantity of ambient air to the sample in order to raise the oxygen content of the sample. If 1 L of sample gas is added to 1 L of ambient air, the oxygen level of the mixture will be adequate to operate a catalytic methane detector, and the true concentration of methane may be obtained by multiplying the detector methane reading by a factor of two. This is a useful example for an operator to remember.

We have already covered the importance of performing a bump test. Good industry practice is to:

- Perform a “bump test” on every portable detector to ensure that it is working properly.
- Perform the test before every shift by exposing the detector to a known concentration of methane high enough to set off the alarm. The operator should note the reading to ensure that it is correct.
- Remember that a bump test is not a calibration, but rather a quick way to ensure that the most important functions of the instrument are intact.

The use of machine-mounted methane monitors is critical. The disadvantage of portable handheld detectors is that a peak emission can be missed because readings at the appropriate locations are only taken at infrequent intervals. In contrast, machine mounted monitors operate continuously and can identify emission peaks and automatically shut off electrical equipment when the methane level is excessive. Machine-mounted methane monitors are usually mounted on mining and tunnel-boring machines. They are designed to have their readout display separated from the sensing head so that the readout is visible to the machine operator and the sensing head is placed in a location where methane is most likely to accumulate.

Figure 5.15 illustrates a methane profile map from a simulated continuous miner face. This diagram helps to illustrate the importance of placing the instrument's sensing head where methane accumulates. Proper placement of monitor sensing heads is crucial to the reliable detection of methane levels. The diagram shows a methane profile map measured from experiments at a full-scale simulated continuous miner face [see Wallhagen

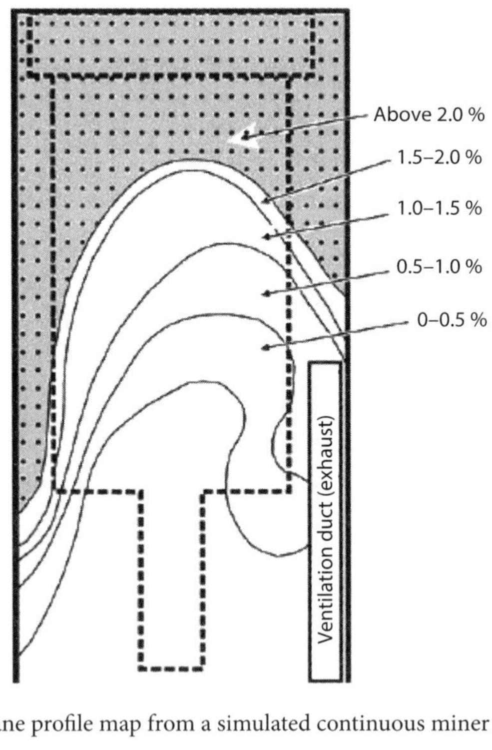

Figure 5.15 Methane profile map from a simulated continuous miner face.

1977]. A striking feature of such profile maps is the steep gradient in the methane concentration along the length of the machine. Thus, a distance of a foot or two forward or backward in the location of the sensing head will greatly change the indicated methane level. In the instance depicted, the sensing head should be as far forward as possible to measure higher methane levels. Inevitably, some tradeoffs are involved in picking the location, for a sensor head located too far forward will quickly become damaged or clogged with dust.

Another factor not to be overlooked is that there may be other flammable gases in addition to methane present in the mine or tunnel environment. Therefore, it is important to have an instrument that is calibrated to detect different gases. Methane detectors must be periodically calibrated with a known concentration of methane-air mixture that is injected into the instrument; but combustible gases other than methane are sometimes encountered. If the gas being sampled is different from the gas used to calibrate the instrument, the error in the instrument reading can be considerable.

Consider the following examples – (1) A tunnel being excavated under a leaking gasoline storage tank might contain gasoline vapor. Under these circumstances, a detector calibrated with a methane-air mixture will read low because higher molecular-weight gases (such as those in gasoline vapor) diffuse more slowly into the sensor element; (2) A methane-calibrated instrument is carried into a tunnel containing only gasoline vapor in the air, and the instrument reads 10% of the lower explosion limit (LEL). In this circumstance, the actual gasoline vapor concentration in air is about 20% of the LEL—twice the indicated reading.

Now consider a scenario where the opposite instrument error can occur. If the gas detector is calibrated with a higher molecular-weight gas such as pentane, and then carried into a tunnel containing only methane, and if it reads 10% of the lower explosive limit (LEL), the actual methane concentration is 5% of the LEL. Bear in mind that when operators are encountering methane, they should calibrate for methane; but if higher hydrocarbons are being encountered, operators should calibrate with a higher hydrocarbon, such as pentane or propane.

5.6.8 Estimating and Controlling Methane Concentration

Before any mining operations begin, thorough knowledge of the amount of methane stored in coal deposits must be known and well mapped out. Coal is the major source of methane gas in mines, with smaller amounts of methane found in oil shale, porous rock, and water. The methane that is stored in porous rock and water is of most concern during tunneling operations.

The amount of methane in coal is measured by using the "direct-method" test during exploration drilling from the surface, or it is estimated from the properties of the coal and the gas pressure or depth of the coal bed. In the direct method, a drill core of coal is brought to the surface. It is enclosed in an airtight container, and the methane emitted from the core is measured. The amount of gas that escaped the core as it was being brought to the surface can be calculated. Later on, the core is crushed and the residual gas given off during crushing is measured. Added together, these allow an estimate of the amount of gas in the coalbed.

When the direct-method results are not available, the amount of gas in coal may be roughly estimated from adsorption data. This estimate requires knowledge of the proximate analysis of the coal, assumes a standard moisture and ash content, and uses the hydrostatic head to estimate pressure.

For illustrative purposes, figure 5.16 summarizes literature reported upper range methane contents of coal versus depth and rank. When an active mine is nearby, the most effective way to forecast the methane emission rate for a mine under development is to use the emission rate from a nearby mine (or section) where similar mining methods are used under similar geological conditions. Corrections can be made for those factors

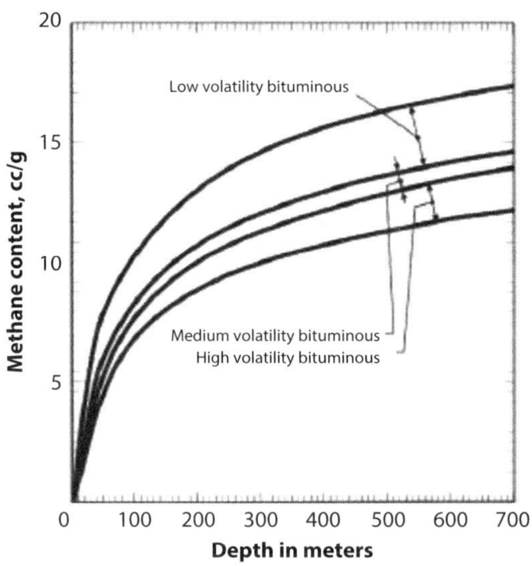

Figure 5.16 Estimated methane content of coal versus depth and rank. Values shown are an upper limit estimates.

that are likely to shift the emission rate. Such factors are differences in coalbed depth, differences in production rate, and geological anomalies such as faults. Some of these corrections are simple, if inexact.

Bear in mind that metal and nonmetal mines that encounter methane emissions require extra precautions against methane explosions. Most of incidents are probabilistic in nature. General literature reports that such mines with a return concentration exceeding 70 ppm of methane are inevitably classified as gassy. Although a measurement of concentration alone is not the complete methane story, a return concentration exceeding 70 ppm should serve as an alert to the presence of gas that has not yet shown itself in other ways.

Although discussed earlier, the subject of layering of methane at the mine roof is worth returning to. The density of methane is roughly half that of air, so methane released at the mine roof may form a buoyant layer that does not readily mix into the ventilation air stream. Such layers have been the source of many mine explosions, so it is important to understand the circumstances that led to the formation of methane roof layers and the methods used to dissipate them. Methane layers are largely a result of inadequate ventilation. A measurement of ventilation velocity is of most practical importance. Under conditions of "normal firedamp emission," an air velocity of 100 ft/min measured at the roof was enough to prevent layering. Most current estimates of the necessary velocity are close to this value. Estimating methane layers can be accomplished by using the dimensionless layering number, defined as follows:

$$\mathbf{L = U/[37(V/W)^{1/3}]}$$

Where:

L = layering number

U = air velocity in feet per minute

V = methane release rate in cubic feet per minute

W = entry width in feet.

Turbulent mixing begins at L > 2. Note that a value of L = 5 is necessary for adequate dilution.

Compared to the 100-ft/min criterion, the layering number concept is more difficult to apply because the methane release rate V is usually not known.

Thorough gas monitoring is a key to dealing safely with methane layers. Care in monitoring is particularly important if:

- The air velocity measured at the roof level is 100 ft/min or less
- The airway is next to a gob or intersects a geologic anomaly, such as a fault, that can serve as a conduit for gas

- The mine roof (or tunnel crown) is not within easy reach, so measurements at roof level are less apt to be carried out regularly
- The airway has cavities or roof-level obstructions to air movement

If the airway is inclined more than 5°, workers who test for methane layers should be aware that the gas concentrations in layers may fall outside of the accurate operating range of catalytic heat of combustion sensors. For accurate operation of these sensors, the concentration of methane must be below 8% *and the* concentration of oxygen must be above 10%. When measuring methane concentrations above 8%, instruments with catalytic heat of combustion sensors can act in a way that is misleading, responding with a rapid upscale reading followed by a declining or erratic reading. Such instrument behavior is a tipoff to the possible presence of high, possibly explosive methane concentrations.

Methane layers are removed by increasing the ventilation quantity and reducing the gas flow by methane drainage. In instances where the source and layer size are limited, a less satisfactory, but workable method is to use a (well-grounded) compressed air-powered venturi air mover or an auxiliary fan at each methane source to blow air at the source of the layer and disperse it. An aggressive sampling program is necessary to ensure safe conditions.

The reader may recall that earlier the concept of providing adequate mixing between methane gas and air for dilution purposes was discussed. It was noted that turbulent mixing conditions are favorable to providing adequate mixing and dilution. However; there is concern that methane gas can recirculate if inhomogeneous occurs. In this regard, recirculation leads to higher methane levels only when recirculated air *replaces fresh air.* A simple relationship to bear in mind is that the concentration of methane into a region = flow of methane into the region divided by the flow of fresh air into the region. Figure 5.17 provides an example of recirculation.

Figure 5.18 shows yet another example of recirculation, this one emanating because of a scrubber on the bit. The number of turnovers, 'n' depends on whether or not the scrubber is turned on, and if turned on, where the scrubber exhaust is directed. Turning on the scrubber raises the value of n, reducing the methane concentration in the zone. Directing the scrubber exhaust into the return (in this case, behind the exhaust line curtain) is the best exhaust configuration, yielding an nQ fresh air value over 4X higher and a zone methane concentration less than ¼ when compared to when with the scrubber off.

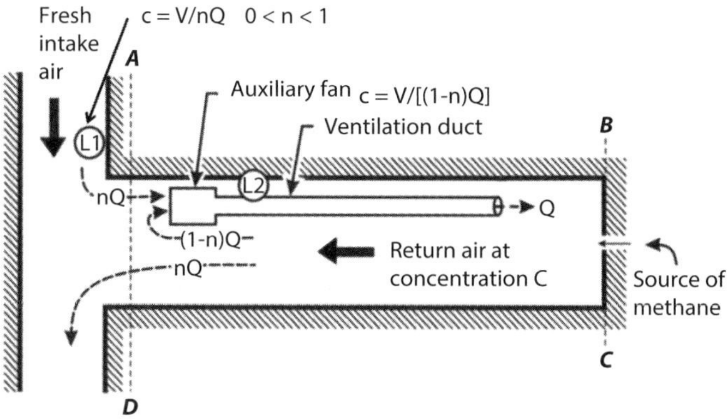

Figure 5.17 Illustrates the concern for the recirculation of gas streams.

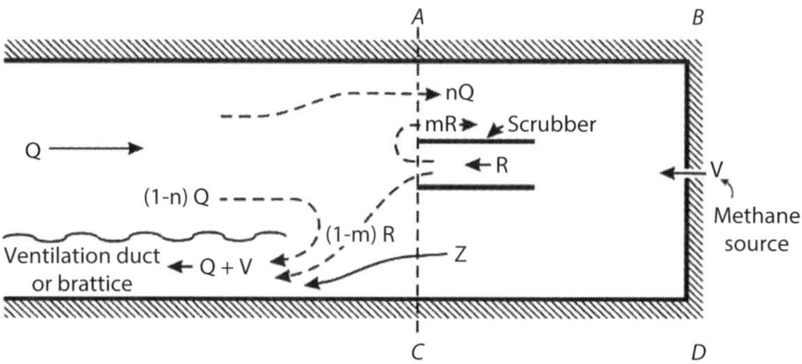

Figure 5.18 Illustrates recirculation caused by dust scrubber operation.

Another concern is for so-called *district recirculation*. An underground system moves air from a return airway back into an intake airway, thus raising the total air quantity in that portion of the mine in by the underground fan. Improved dust control can be a result, but there are some important caveats – the initial volume of fresh air to the district should be maintained. The recirculation fan should be placed far enough from the face for the dust to settle out, but close enough to the face to minimize stopping leakage. District recirculation systems will increase flow and pressure losses in the mine circuit, producing a small drop in main fan flow, and as such adequate monitoring and controls must be in place. Recirculation will raise the methane concentration only when recirculated air is substituted for fresh air. If the amount of fresh air entering a zone is unchanged, the methane concentration in the zone will be unchanged. At continuous miner faces, if the operation of a scrubber creates an airflow pattern that enhances the amount of fresh air entering the face zone, then the operation of the scrubber will lower the methane concentration (and vice versa).

The following are operating factors that impact methane dilution:

- *Scrubber air quantity* - Raising the scrubber air volume from 6,000 to 14,000 cfm produces a 23% decrease in methane concentration. The greatest decrease in methane concentration is at an intake airflow of 6,000 cfm, where raising the scrubber volume from 6,000 to 14,000 reduces methane levels by 38%.
- *Water sprays* - Effective scrubber operation depends on the air movement generated by the dust suppression water sprays. Turning off the spray system doubles the methane level. Also, switching from a conventional spray system to a directional spray system adds a 23% reduction in the methane level.
- *Exhaust direction* – Directing the exhaust toward the blowing curtain gives the highest methane levels. By comparison, directing the exhaust straight back lowers methane levels at the face by 30%. Directing the exhaust toward the return-side rib lowers methane an additional 25%, for a total decrease of 55%.
- *Clogging - Clogging of the flooded-bed filter panel or the scrubber ductwork will seriously* inhibit the methane dilution capacity of scrubbers. When a dust scrubber clogs, its air quantity declines.
- *Other operating factors* - Changing the line curtain airflow in the range between 6,000 and 14,000 cfm does not change the average face methane concentration. Also, methane levels do not increase when the line curtain airflow is less than the scrubber airflow (6,000-cfm line curtain versus 14,000-cfm scrubber), a situation that leads to a high amount of recirculation. Many years earlier, a study led to a conclusion that recirculation per se was not harmful, as long as a sufficient quantity of fresh air was provided by the line curtain.

Abnormally gassy faces may be ventilated with diffuser fans, high-pressure spray fans, or high volume scrubbers. Degasification (discussed further on) with horizontal or vertical boreholes is necessary if the section emits over 300 cfm of methane. Good industry practice is to use a diffuser fan for abnormally gassy faces. A diffuser fan is a small fan mounted on the continuous miner that directs an air jet at the working face. It can be used in conjunction with the exhaust line curtain. It was in fact the primary means of ventilating gassy faces before the development of the spray fan.

5.6.9 Managing Ignition Sources

Emphasis has been placed on ventilation methods and monitoring for methane gas. The potential for a methane ignition may be further reduced by dealing directly with the ignition source.

One obvious ignition source is faulty electrical equipment that has not been designed or maintained for ignition prevention. The other ignition source results from cutter bits striking rock. In this instance, abrasion from the rock grinds down the rubbing surface of the bit, producing a glowing hot metal streak on the rock surface behind the bit. The metal streak is often hot enough to ignite methane, causing a so-called frictional ignition. In the U.S., frictional ignitions are much more prevalent than those from faulty equipment.

It is generally recognized that the most important action one can take to reduce frictional ignitions is to replace bits regularly, thus avoiding the formation of wear flats on the bits. Frictional ignition with a mining bit always involves a worn bit having a wear flat on the tip of the bit [see Courtney 1990]. A small wear flat forms a small hot spot, which does not lead to an ignition, whereas a large wear flat forms a large hot spot that is more likely to cause an ignition.

Mining bits consist of a steel shank with a tungsten carbide tip. The steel is more incendiary than the tungsten carbide tip, so if the tip is worn off and the steel shank exposed, the chance of an ignition is much greater.

Good industry practice for reducing frictional ignitions included providing a regular change-out schedule to replace worn bits, providing bits with a larger carbide tip to reduce wear, and possibly changing the bit attack angle or the type of bit. Additionally, it is good practice to mount a water spray behind each bit, aiming the spray toward the location on the rock where the hot metal streak is expected. Anti-ignition back spray quenches the hot streak, reducing its temperature and the chance of a frictional ignition. Bringing water to the cutter head on continuous miners has been an engineering challenge. However, in recent years, practical (though expensive) water seals for continuous miner heads have been developed. As a result, a few "wet-head" continuous miners equipped with anti-ignition back sprays have been installed in U.S. coal mines with a history of frictional ignition problems. Figure 5.19 illustrates the concept of

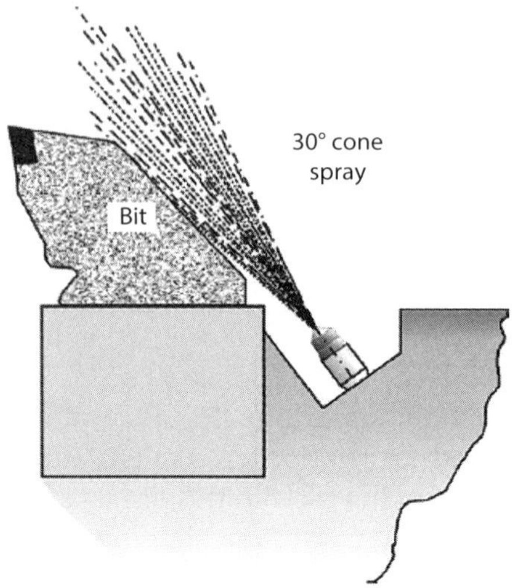

Figure 5.19 Anti-ignition back sprays for quenching the hot streak on the bit.

using anti-ignition back sprays for quenching the hot streak reduces temperature and the chance of a frictional ignition.

Proper spray nozzle selection, nozzle placement, and operating pressure of anti-ignition back sprays are important if the full hot-streak quenching potential is to be realized. If the spray density is too low or if too much water is wasted in wetting the back of the bit, then quenching effectiveness suffers.

5.6.10 Case Study – The Massey Mine Disaster

On April 10, 2010, twenty nine miners were killed as a result of an explosion that occurred at Massey Energy's Upper Big Branch mine in Montcoal, West Virginia. At the time of the explosion, Massey was mining two major sections of the Upper Big Branch Mine. At one section, they had a longwall machine operating and it was nearing the end of that area. At the same time, they were using a continuous miner to prepare another area for longwall mining. Initial ignition of methane gas took place at the tailgate section of the longwall.

Various news reports along with OSHA statement show that an initial methane ignition burned for 60–90 seconds before it set off a secondary powerful coal-dust-fueled explosion that tore through miles of underground tunnels, killing the 29 miners. Sparks given off from faulty bits on the longwall mining machine ignited a pocket of approximately 13 cubic feet of methane gas, which burned for 60 to 90 seconds.

The burning methane ignited a massive amount of highly explosive and untreated coal dust, which was pervasive throughout the mine. It carried the blast throughout miles of underground pathways that twisted and turned, killing miners over two miles away from the point of ignition. At least two miners were at the tailgate section of the longwall where the methane ignition started, but had time to telephone miners operating the longwall machine to tell them to power off the machine. They then ran about 400 to 500 feet before being killed by the coal dust explosion.

Longwall mining involves the digging of two parallel tunnels into the coal seam 1,000 feet apart, more than the length of three football fields (refer to figure 5.20). One of these tunnels is called the headgate and the other the tailgate. At the far end of the tunnels, a third tunnel is dug connecting the two; this is the mine face. The shearer is placed on a track running between the headgate and tailgate, running up and down the track cutting into the coal. Mining is done towards the entrance of the mine. Hydraulic jacks support steel covers that hold up the roof. As the coal is dug, the tracks, shearer and jacks are all moved forward. As the longwall miner digs through the coal seam and moves forward, the roof caves in behind the miner.

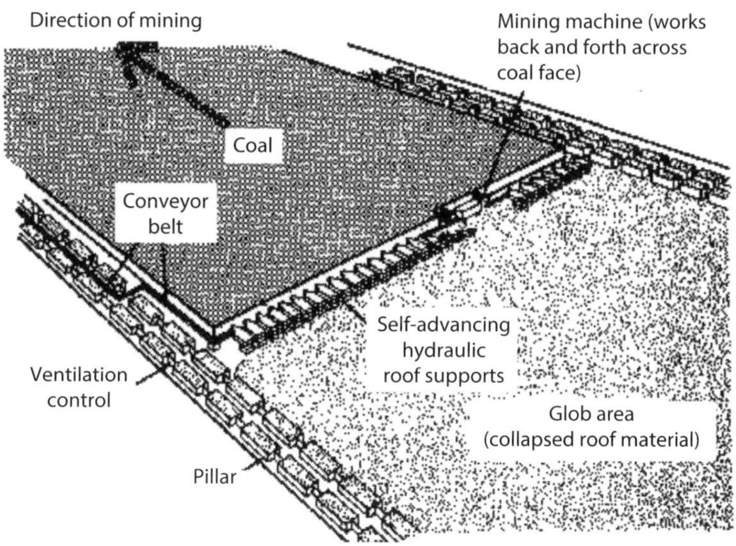

Figure 5.20 Illustrates longwall mining practice.

The incident has been attributed to Massey's failure to provide safe industry practices in four areas: ventilation, maintaining the bits and shearer, spraying water, and preventing the buildup of coal dust.

Ventilation - Ventilation is needed both to bring fresh air to the miners and to dissipate and prevent the buildup of methane gas in the mine. Methane gas is naturally produced in similar geological processes that produce coal and often seeps out into mines. Also, miners often run into pockets of gas embedded in a coal seam.

In the months and weeks leading up to the explosion MSHA investigators had cited Massey multiple times for failure to maintain proper ventilation. Among those citations were having only about half the needed fresh air, as well as redirecting the exhaust air from one section of the mine into the air intake of another.

About 30 minutes before the deadly blast, the fire boss called to the surface to say that the air was so poor and thick with coal dust that he couldn't see to get out of the mine. His friend and fellow fire boss, who was on the surface, testified that he was on his way into the mine to help out when the explosion occurred.

Bit Replacements - Drill bits need to have a carbide tungstun tip to prevent sparking as they strike rock. Investigators believe that a spark ignited methane gas, which in turn set off a secondary coal dust explosion that killed the miners. The bits are composed of a steel shank with a carbide tungsten tip. It is important that this tip be in place as a safety measure, because when a carbide tungsten tip hits rock it does not spark the way steel does. The shearer had many broken bits. The seam being mined at the mine was so narrow that the longwall machine was hitting sandstone continuously. Examination of the shearer showed that most of the carbide tungsten tips were broken off the bits, leaving exposed steel to strike the rock and coal, causing a shower of sparks.

Water Spray - Next to each bit on the shearer, a water sprayer is attached to spray water onto the bit and coal as it digs. The water helps to keep the bits cool, keep down coal and rock dust and extinguish any sparks that come off the bits. Tests on the sprayers show that many were not working, some having been removed or purposefully blocked. Investigators reported that many of the sprayers were missing or not working.

Rock Dusting - Another step in preventing underground explosions is to control coal dust in the air and ground. Coal dust must be removed from along conveyor belts and pulleys that generate heat. "Rock dusting" must also be completed throughout the mine. "Rock dusting" is the spreading of ground-up rock dusk, like chalk, among the coal dust, which prevents the coal dust reaching explosive concentrations. Tests of over 18,000 samples taken throughout the UBB mine after the disaster showed that 80% of the coal dust had not been rock dusted properly. Logs recorded by miners as they did their pre-shift inspections showed repeatedly that coal dust was building up throughout the mine, but that rock dusting had been completed only in a few instances. Of 62 pre-shift inspections done on the day of the explosion, and in the four days leading up to it in different parts of the mine, all of the records showed the need for rock dusting to be done. Rock dusting was performed in only 9 areas as a result of the inspection. All of this untreated coal dust turned a small methane ignition into a powerful and deadly disaster.

Lack of Inspections and Enforcement - MSHA investigators had issued more than 600 citations over the previous 18 months. MSHA was well aware of the problems with ventilation, the buildup of coal dust, and should have seen the state of the longwall and other mining equipment. Yet they took no action to shut the mine and protect the miners. A team of MSHA investigators had been in that area of the mine less than two weeks before the disaster. Joe Main defended the role of MSHA investigators and said there was an internal review only, but gave no indication of whether it had started and when or if its findings would be made public.

While commentaries are generally out of place in technical publications, this incident is an exception. Massey's failure to follow good industry practices and standards resulted in an ultra-hazardous working environment that was responsible for 29 deaths. Since the inception of OSHA in 1970, it has been the obligation of every industrial operator and owner to ensure safe working conditions for workers. This is known as OSHA's *General Duty Clause*. Massey's practices in this instance were beyond being irresponsible – they reflect gross negligence and a flagrant disregard for standards and the lives of its employees.

5.6.11 Other Case Studies

The Chemical Safety Board has assembled a number of case studies from its own and NFPA investigations over the years concerning dust explosions in processing equipment for coal handling operations. Table 5.1 provides a sampling of some of the case studies reported. It is worthwhile for the reader to read through these to gain perspective on the nature and extent of the types of explosion and fire incidents that have occurred with coal processing and then consider the vulnerability of the operations they have to deal with.

Table 5.1 Case studies involving coal handling (data assembled and reported by the Chemical Safety Board on its web site).

Date	City, State	Fatality	Injury	Description	Equipment Involved	General Facility Type	Ignition Source	Work Activity
1984	Brilliant, OH	1	4	One of several fuel pipes that carry fuel to the boiler furnace became very hot. When the system was shut down, an explosion occurred in the other fuel pipes. Coal dust was possibly ignited by the hot fuel pipe. The workers were preparing to cool the fuel pipe with a fire hose. Employees #1, 2, 4 and 5 were on the H R deck, unit 1. Employee #3 was on the H 8 deck, unit 1, 15 feet below the H R deck. The employees received burns and suffered from dust inhalation.	Boiler furnace	Electric Services	Hot surface	
1984	unknown, CA	0	0	An explosion occurred in a coal bin. Fire was initiated by spontaneous combustion. Carbon dioxide application from the top was insufficient to prevent the development of an explosive atmosphere in the confined space above the coal. When a smoldering fire burns for a considerable time without being detected, the buildup of volatile gases can produce an explosive mixture resulting in an explosion. There was extensive damage to the bin top and to the conveyor housing feeding the bin.	Coal bin	Stone, clay, glass, & concrete products	Spontaneous combustion	
1984	Hayden , CO	1	2	An explosion occurred in the ambient air intake ductwork of the hayden station, unit 2b mill (coal pulverizer). One worker was struck by a flying air shroud and killed. The cause of the explosion is unknown. A citation for lack of training was recommended.	Coal Pulverizer	Electric Services	Unknown	Crushing

Date	City, State	Fatality	Injury	Description	Equipment Involved	General Facility Type	Ignition Source	Work Activity
1984	Ravenswood, WV	2	0	Crushed coal tar pitch had jammed in a "brady" conveyor system. Four employees had gone to the basement of the building and removed a section of the pipe. They were going to increase air pressure in the line to purge the material by blowing it out at the point where the pipe had been disconnected. This was a common practice and was the second time it had been done on this same shift. When the line clears itself the entire building is engulfed in coal tar dust. The area had not been cleaned in over two weeks, allowing an unusual amount of pitch to build up in the basement. While one of the foremen was at the air control valve, pressurizing the system to blow out the plug, a cloud of dust and a ball of fire came upstairs and out of the basement. There was instantly a flash fire in the building. Then a major fire broke out as the pitch residue, which was everywhere, began to burn. The ignition source for the explosion and fire was most probably a broken incandescent light bulb.	"brady" conveyor belt	Primary metal industries	Hot surface	

(Continued)

Table 5.1 (*Cont.*)

Date	City, State	Fatality	Injury	Description	Equipment Involved	General Facility Type	Ignition Source	Work Activity
1985	Glenrock, WY	0	3	A primary air fan, a motor, and a coal pulverizer system running at maximum capacity overloaded, causing the system to shut down. The seal air pressure in the pulverizer unit overcame the decaying primary air fan pressure, causing the coal dust to back flow into the primary air fan shroud which had been running at 500°F. The coal dust coming in contact with the hot metal caused gasification and spontaneous combustion. The inboard side of the fan shroud separated and a large fire ball erupted, burning the three employees who were nearby.	coal pulverizer; air pressure fan; air fan shroud	Electric Services	Hot surface	Crushing
1985	unknown, NM	0	0	An explosion occurred in the coal grinding mill circuit during a scheduled shutdown of a cement kiln. Property damage and disruption of production resulted. Physical damage was limited to the 914.4-mm diameter hot air supply pipe which had separated at the welded joint, bent and distorted coal screw feeder covers, the hot air inlet damper (badly bent), and minor distortion of the cold air inlet control damper.	Grinding mill	Stone, clay, glass, & concrete products	Unknown	Unknown

Date	City, State	Fatality	Injury	Description	Equipment Involved	General Facility Type	Ignition Source	Work Activity
1986	Becker, MN	0	2	A lead worker and four other employees were assigned the task of changing dust filter bags on a coal dust collector at an electric power generation plant. Using their employer's energy control procedures, they isolated the system and set up the job. The employees were using two portable 500-watt quartz lights--one on the work platform and one at floor level. The light at floor level had a broken lens. While this light was on, one of the employees noticed coal dust falling on the light and realized that that was an explosion hazard. He reached for the light, and it exploded,.	2 portable quartz light devices; dust collector	Electric Services	Hot surface	
1994	Owensville, IN	0	22	A smoldering coal fire in a turbine generator hopper was being monitored. Fire hoses were pumping water down 19-inch-wide floor openings, which were normally covered, into the hopper, which was in a bunker room on the south side of a coal tripper. The fire escalated. Two yard supervisors were at the bunker room door handling fire hoses. Four yard workers were using the fire hoses to pump water into the hopper. At 9:00 AM, just as a township fire fighter walked through the east door of the ninth floor bunker room, an explosion occurred.	turbine generator hopper	Electric Services	Flame	Extinguishing fire

(Continued)

Table 5.1 *(Cont.)*

Date	City, State	Fatality	Injury	Description	Equipment Involved	General Facility Type	Ignition Source	Work Activity
1998	Hammond, IN	0	17	A coal dust explosion ignited at a Tripper, and propagated throughout the coal handling areas. An explosion and fire occurred on a coal fired electric generating station. The cause of the incident may have been due to coal dust and at the time of the incident an outside contractor was vacuuming in the area where possible ignition may have occurred.	Tripper, Vacuum	Electric Services	Unknown	Coal processing
1999	Dearborn, MI	6	30	A blast occurred while 3 maintenance men were working on the boiler. The blast could have occurred by pressure building up inside the boiler or a spark that ignited gas or coal-dust that fuels the massive machines.	boiler	Electric Services	Flame	Maintenance
2002	Marquette, MI	1	0	A coal-fired boiler was being brought on-line. The coal feeding system had not been purged properly. An excessive amount of coal dust left in the system ignited, causing an explosion.	coal-fired boiler, coal feeding system	Electric Services	Fire	
2003	Rothschild, WI	0	0	A coal dust explosion Monday forced plant to shut down one of its electrical generators				

Date	City, State	Fatality	Injury	Description	Equipment Involved	General Facility Type	Ignition Source	Work Activity
2003	Muscatine, IA	0	1	Coal dust fire at electrical power generating facility.				
2004	Lockport, IL	0	0	A fire erupted in a steel storage bunker holding coal. There have been three bunker fires at the plant in the past year.	steel storage bunker	Electric Services		
2004	Red Rock, OK	0	0	Explosion and flash fire resulted in the coal handling system of the plant. Coal dust was the source of the fire.	Coal handling system	Power genera-tion	Unknown	
2004	Wood River, IL	0	0	A mix of coal dust and welding heat caused a fire aboard a barge unloading facility at a power station. The fire started with an explosion at an excavator crane; several employees were doing preventative maintenance welding to the crane just moments before the fire started. The fire then progressed up into the engine compart-ment and into the cabin.	excavator crane, welding equipment	Electric Services	Unknown	

5.6.12 Application of Rock Dusting

Rock dust has been used for nearly a century as a precautionary measure to protect against dust explosions. It is generally agreed that the effectiveness of rock dust lies in its ability to be simultaneously dispersed with coal dust, and, by serving as a heat sink, thus preventing flame propagation. Most leading coal-producing nations have similar requirements, some more stringent and some less stringent than those enforced in the United States. A partial listing of these requirements is given in table 5.2. Passive barriers have been deployed in most leading coal-producing nations to provide supplemental protection against coal dust explosions. Conveyor belt entries have received emphasis. Barriers are designed to quench an explosion immediately on arrival at the location.

Cashdollar and co-workers[6] report on an extensive investigation resulting in recommendations for the application of rock dust to reduce the risks of explosions in mining operations. They note that the workings of a bituminous coal mine produce explosive coal dust for which adding rock dust can significantly reduce the potential for explosions. Based on their findings, guidelines have been established by the Mine Safety and Health Administration (MSHA) about the relative proportion of rock dust that must be present in a mine's intake and return airways.

Current MSHA regulations require that intake airways contain at least 65% incombustible content and return airways contain at least 80% incombustible content. The higher limit for return airways was set in large part because finer coal dust tends to collect in these airways. Based on extensive in-mine coal dust particle size surveys and large-scale explosion tests, NIOSH recommends a standard of 80% total incombustible content (TIC) be required in the intake airways of bituminous coal mines in the absence of methane.

MSHA inspectors routinely monitor rock dust inerting efforts by collecting dust samples and measuring the percentage of TIC, which includes measurements of the moisture in the samples, the ash in the coal, and the rock dust. These regulations were based on two important findings: a survey of coal dust particle size that was performed in the 1920s, and large-scale explosion tests conducted in the U.S. Bureau of Mines' Bruceton Experimental Mine (BEM) using dust particles of that survey's size range

---

[6] K. Cashdollar, M. Sapko, E. Weiss, M. Harris, C. Man, S. Harteis, and G. Green, Recommendations for a New Rock Dusting Standard to Prevent Coal Dust Explosions in Intake Airways, National Institute for Occupational Safety and Health Office of Mine Safety and Health Research Pittsburgh, PA, Report of Investigations 9679, May 2010

Table 5.2 International rock dusting standards.

Country	TIC %	Volatiles %	Methane %	Comments
Australia Queensland	85–80 (return) 85–70 (intake)	— —		85% TIC ≤ 200 m from the face 80% TIC > 200 m from the face 85% TIC ≤ 200 m from the face 70% TIC > 200 m from the face Supplemental protection—barriers
Australia NSW	85–70 (return) 80–70 (intake)	— —		85% TIC ≤ 200 m from the face 70% TIC > 200 m from the face 80% TIC ≤ 200 m from the face 70% TIC > 200 m from the face Supplemental protection—barriers
Canada (Nova Scotia)4	75 (intake) 80 (return)	— —	<1 >1	
Czech Republic	80 (intake/return) 85 (intake/return)	— —	<1 >1	Supplemental protection—barriers
Slovakia	80 (intake/return) 85 (intake/return)	— —	<1 >1	Supplemental protection—barriers
Germany	80 (intake/return)	—		Supplemental protection—barriers

(*Continued*)

Table 5.2 (*Cont.*)

Country	TIC %	Volatiles %	Methane %	Comments
Japan	78 (intake/return) 83 (intake/return)	35 35	<1 >1	Specific requirements depend on ash, moisture and volatile content, the gassiness of the seam, and the fineness of the rock dust used.
Poland	70 (intake/return)	>10 >10		70% in "non-gassy" roadways 80% in "gassy" roadways Supplemental protection—barriers
South Africa	80 (intake) 80 (return)	— —		80 % TIC ≤ 200 m from the face 65% TIC > 200 m from the face 80% TIC for 1000 m from the face Supplemental protection—barriers
United Kingdom	50 (intake/return) 65 (intake/return) 72 (intake/return) 75 (intake/return)	20 27 35 >35		Supplemental protection—barriers
United States	65 (intake) 80 (return)	— —	1.0 / 0.1 0.4 / 0.1	Add 1% TIC / 0.1% methane Add 0.4% TIC / 0.1% methane

to determine the amount of inerting material required to prevent explosion propagation. Mining technology and practices have changed considerably since the 1920s, when the original coal dust particle survey was performed. Also, it has been shown that as the size of coal dust particles decreases, the explosion hazard increases. Given these factors, NIOSH and MSHA conducted a joint survey to determine the range of coal particle sizes found in dust samples collected from intake and return airways of U.S. coal mines.

Results from the survey show that the coal dust found in today's mines is much finer than in the mines of the 1920s. This increase in fine dust is presumably due to the increase in mechanization. In light of this comprehensive dust survey, NIOSH conducted additional large-scale explosion tests at the Lake Lynn Experimental Mine (LLEM) to determine the degree of rock dusting necessary to abate explosions.

The tests used Pittsburgh seam coal dust blended as 38% minus 200 mesh and referred to as medium-sized dust. This medium-sized blend was used to represent the average of the finest coal particle size collected from the recent dust survey. Explosion tests indicate that medium-sized coal dust required 76.4% TIC to prevent explosion propagation. Even the coarse coal dust (20% minus 200 mesh or 75 μm), representative of samples obtained from mines in the 1920s, required approximately 70% TIC to be rendered inert in the larger LLEM, a level higher than the current regulation of 65% TIC.

Given the results of the extensive in-mine coal dust particle size surveys and large-scale explosion tests, NIOSH recommends a standard of 80% TIC be required in the intake airways of bituminous coal mines in the absence of methane. The survey results indicate that in some cases there are no substantial differences between the coal dust particle size distributions in return and intake air courses in today's coal mines. The survey results indicate that the requirement of 80% TIC in return airways is still appropriate in the absence of background methane.

Factors that can influence the amount of admixed rock dust required to make coal dust inert include coal and rock dust particle size distribution, coal dust volatile content, and the additional presence of methane. Much knowledge has been obtained from experimental mine and laboratory dust explosion research. Investigators have examined the effects of rock dust inerting requirements, the minimum explosible coal dust concentrations, the effect of volatile matter on the explosibility of coal dusts, the effect of the size of coal and rock dust particles, and the effect of background methane in full-scale experimental mines and in laboratory test vessels. Further research evaluated the effects of pulverized versus coarse coal particle size, coal volatility, extinguishment, and pyrolysis mechanisms. The consensus of these studies is that dust particle size emerges as the single most influential factor controlling coal dust explosion propagation. To determine compliance with regulations, inspectors from MSHA periodically collect samples of deposited dust from various areas in a mine. The MSHA laboratory determines TIC and compares it with the TIC requirement. This TIC requirement is based on a mean coal particle size of 20% minus 200 mesh and assumed to be constant throughout the intake entries. The size of the coal dust component is not measured by MSHA laboratories as part of the explosibility assessment.

5.6.13 Methane Degasification

Coal seams are classified as to the degree of gassiness. Table 5.3 reports the general classification scheme, with typical values reported for high-volatile bituminous coals.

All coal seams are gassy by definition, but they vary in their degree of gassiness, i.e., gas content per ton of coal. The depth of a coal seam and its rank are good indicators of its gassiness, but direct measurement of gas content is highly recommended.

Methane drainage must be performed when the ventilation air cannot dilute the methane emissions in the mine to a level that is below safe limits. The practice is known as degasification. It is economically feasible to handle specific methane emissions from a mine up to 1,000 ft^3/ton with a well-designed ventilation system. At higher specific emission rates, a stage is reached where ventilation cost becomes excessive or it becomes impossible to stay within statutory methane limits with mine ventilation alone. In other words, there is a break-even point between the costs of ventilation versus degasification (see figure 5.21).

There are a number of advantages of coal seam degasification. These include:

- Reduced methane concentrations in the mine air, leading to improved safety

Table 5.3 Gassiness of coal seams.

Category	Depth (for high volatile bituminous coals), ft	Gas Content (cu.ft per ton coal)
Mildly Gassy	<600	<100
Moderately Gassy	600–1,200	100–300
Highly Gassy	1,200–3,000	300–700

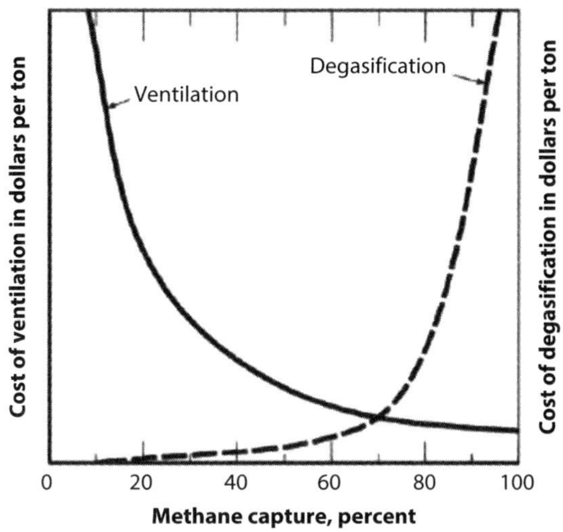

Figure 5.21 Optimum point for methane drainage.

- Reduced air requirements and corresponding savings in ventilation costs
- Faster advance of development headings and economy in the number of airways
- Improved coal productivity
- Additional revenue from the sale of coal mine methane
- Additional uses of degasification boreholes, such as water infusion to control respirable dust
- Advance exploration of coal seams to locate geological anomalies in the longwall panel

Underground mining is done in two phases: (1) development and (2) pillar extraction. Development work involves the drivage of a network of tunnels (entries) into the coal seam to create a large number of pillars or longwall panels to be mined later. This drivage is done with a continuous mining machine. This machine cuts and loads coal into a shuttle car, which in turn hauls and dumps the coal onto a moving belt. The coal travels out of mines on a series of belts and is finally brought to the surface via a slope or shaft. Figure shows a typical longwall panel layout in a U.S. coal mine. All methane produced during the development phase of mining is from the coal seam being mined. Methane is emitted at the working face as well as in the previously developed areas. All emitted methane is mixed with ventilation air, diluted to safe levels, and discharged on the surface. Methane drainage during or prior to development becomes necessary if the development headings will experience a high rate of methane emissions. This is called pre-mining methane drainage. Horizontal drilling of longwall panels prior to mining also falls into this category.

The ultimate goal of coal seam degasification is to reduce the gas content of the coal seam below 100 ft^3/ton prior to mining and the capture of at least 50% to preferably 75% of the post-mining emissions. Methane drainage techniques are comprised of:

- Horizontal in-seam boreholes
- In-mine vertical or inclined (cross-measure) boreholes in the roof and floor
- Vertical wells that have been hydraulically fractured (so-called frac wells)
- Short-radius horizontal boreholes drilled from surface

Figure 5.22 illustrates the two most commonly used variations of degasification with in-seam horizontal boreholes. Success of the technique is predicated on good coalbed permeability (≥5 mD). In highly permeable coal seams, e.g., the Pittsburgh Seam of the Appalachian Basin, nearly 50% of the in situ gas can be removed by this technique prior to mining. The major drawback of this technique is that only about 6 months to a year—the time between development and longwall extraction— is available for degasification.

In-mine inclined or vertical boreholes are comprised of short vertical or long inclined boreholes that have been drilled from an existing mine (or roadways expressively driven for this purpose) to intersect other coal seams in the gas emission space, allowing for the seams to be degassed prior to mining. Success depends on high permeability. A better way to degas these coal seams lying in close proximity to each other is to use vertical frac wells.

Vertical frac wells are best suited to highly gassy, deep, low permeability coal seams where it takes several years prior to mining to adequately degas the coal. The wells are drilled from the surface on a grid pattern over the

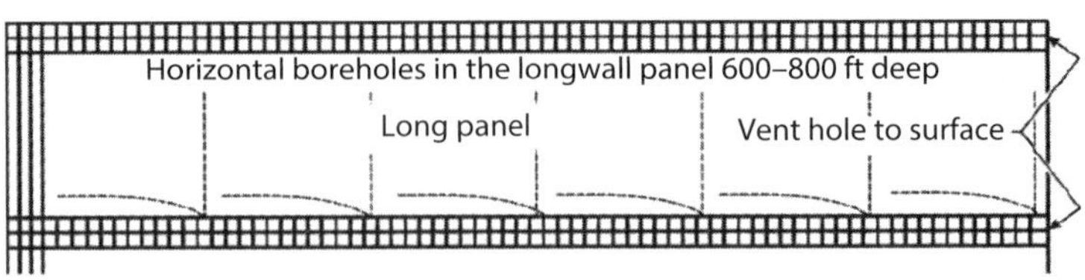

Figure 5.22 Illustrates horizontal in-seam boreholes. The drawing shows the longwall panel methane drainage method.

entire property or only on longwall panels to intersect the coal seam to be mined in the future.

Vertical wells drilled into the coal seam seldom produce measurable amounts of gas without hydraulic stimulation. High-pressure water (or other fluids) with sand are pumped into the coal seam to create fractures (see figure 5.23). The fluid (water) is then pumped out, but the sand remains, keeping the fractures open for gas to escape to the well bore. Under ideal conditions, if the vertical frac wells are drilled more than 5–10 years in advance of mining, 60%–70% of the methane in the coal seam can be removed prior to mining. Vertical frac wells have been very successful in the Appalachian and San Juan Coal Basins of the United States. They have also been attempted in the United Kingdom, Germany, Poland, China, and Australia, but met with only limited success.

Postmining Methane Drainage Techniques include: *Packed cavity method and variants, Cross-measure borehole method, Superjacent method; Vertical gob well method.*

The packed cavity method has historically been used in Russian coal mines. Early methods of methane control consisted of simply isolating the worked-out area in the mine using packed walls, partial or complete stowing, and plastic sheets or massive stoppings. A network of pipeline that passed through these isolation barriers was laid in the gob, and methane was drained using vacuum pumps. Figures 5.24 and 5.25 show typical layouts for caving and partially stowed longwall gobs.

The *Cross-measure borehole method* is a popular method of methane control on European longwall faces. A typical layout for a retreating longwall face shown in figure 5.26. Boreholes 2–4 inches in diameter and

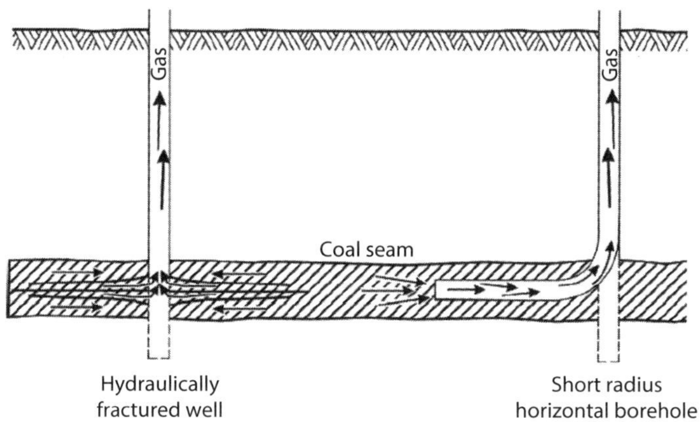

Figure 5.23 Illustrates premining methane drainage from surface.

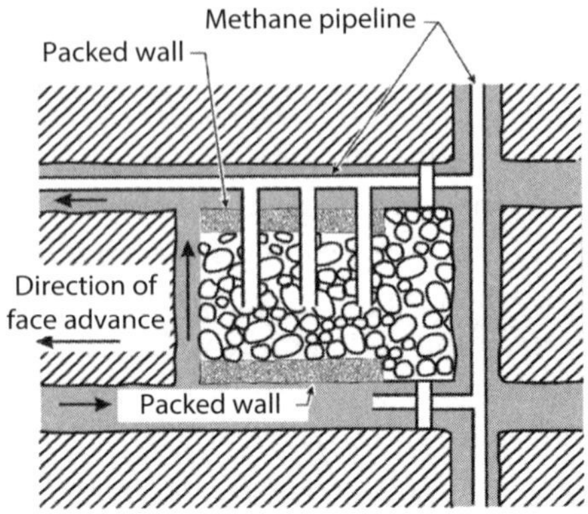

Figure 5.24 Illustrates methane drainage by the packed cavity method.

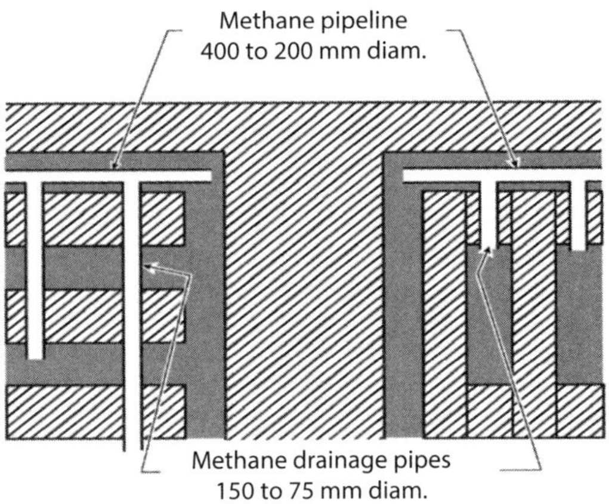

Figure 5.25 Illustrates partially filled longwall gob - *Packed cavity method.*

about 80 ft apart are drilled from the top gate to a depth of 60–500 ft. The cross-measure borehole method is generally more successful for advancing longwall panels than for retreat faces. The flow from individual boreholes is typically 20 ft^3/min, but can occasionally reach 100 ft3/min for deeper holes. Sealing of the casing at the collar of the borehole is very important and is usually done with quick-setting cement. Sometimes a liner (a pipe of smaller diameter than the borehole) is inserted in the borehole and sealed at the collar to preserve the production from the borehole even when it is sheared by rock movements.

The *Superjacent method* is used mainly for retreating longwall faces in highly gassy seams in French mines. See figure 5.27. A roadway is driven

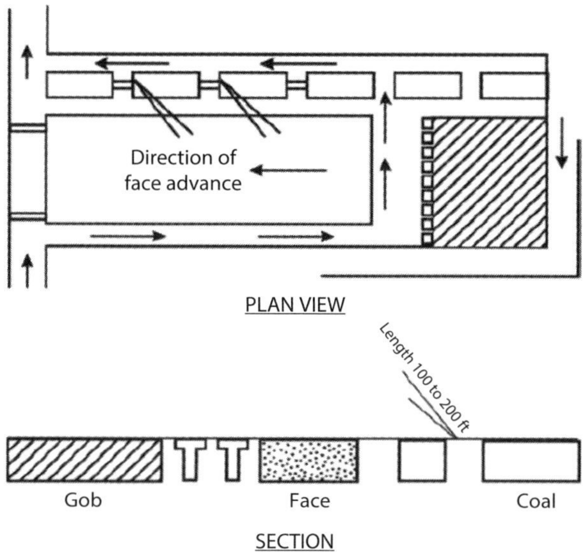

Figure 5.26 Illustrates methane drainage with cross-measure boreholes.

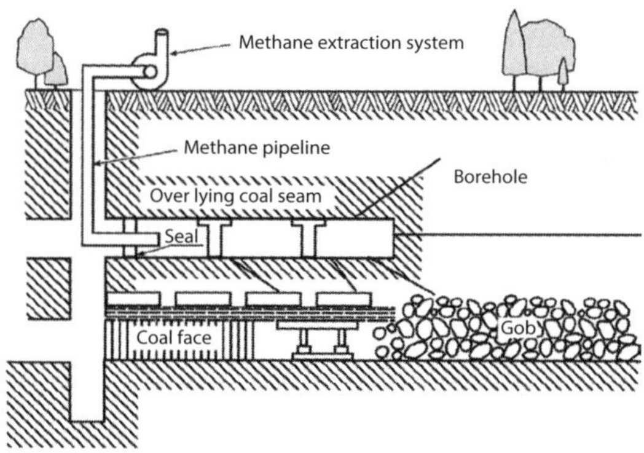

Figure 5.27 Illustrates the methane drainage by the superjacent method.

70–120 ft above the longwall face, preferably in an unworkable coal seam. The roadway is sealed, and vacuum pressures up to 120 in w.g. are applied. To improve the flow of gas, inclined boreholes in the roof and floor are drilled to intersect with other gassy coalbeds. If the mining scheme proceeds from the top to the bottom seams in a basin, the entries in a working mine can be used to drain coal seams at lower levels. Methane flow from these entries is high, averaging 700–1,000 ft3/min for highly gassy seams. About 50% of total emissions have been captured using this method.

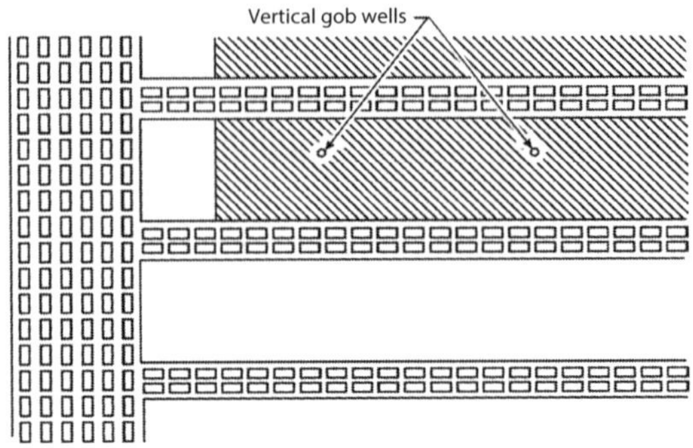

Figure 5.28 Illustrates methane drainage by vertical gob wells (Ventilation controls are not shown).

The *Vertical gob well method* has historically been the most commonly used in longwall mining in the U.S. (see figure 5.28). Methane emission rates from gobs vary depending on the geological conditions, but deep-seated longwall gobs (e.g., those in the Pocahontas No. 3 Seam in Virginia and the Mary Lee Seam in Alabama) produce methane in the range of 1,800–18,000 ft^3/min. Multiple entries are driven to develop longwall panels so that necessary air quantities can be delivered to the longwall faces via the mine ventilation system. In many cases, however, some type of additional methane control becomes necessary. The most popular method is to drill vertical boreholes above the longwall prior to mining. Depending on the length of the longwall panel (typically 10,000 ft) and the rate of mining, 3 to 30 vertical gob degas boreholes are needed. The first hole is usually within 150–500 ft of the start line of the longwall face. The borehole is drilled to within 30–90 ft from the top of the coal. The casing is cemented through the fresh water zones near the surface, and a slotted liner is provided over the lower open section to prevent closing of the hole by caving. These boreholes are completed prior to mining.

5.6.14 Prevention, Early Detection and Fire Suppression

An explosion or fire in a coal mine is considered a majorly hazardous event. Safe practices are essential to control risks to the health and safety of people, (and to control risks to plant, infrastructure and the environment) from the event of a fire in a coal mine. This may be done by providing guidance in undertaking a risk management approach to fire hazards and by recommending and enforcing standards which should be followed.

All fire risk scenarios in a coal mine must be identified through a fire risk assessment, with appropriate risk controls being implemented to prevent, detect and suppress any fire (that may occur in a coal mine) such as to minimize risks to health and safety of people.

Guidelines should be used by coal operators and mine owners when:

- Undertaking fire hazard risk assessments to fulfill OHS obligations
- In developing mine emergency systems and major hazard management plans for fire and explosion
- In reviewing the adequacy of current fire risk controls, including following a fire event
- When auditing the mines fire and explosion management plan and/or emergency management system

In devising and applying best practices, it is important to be clear on definitions. The following are some important terms that all workers should be versed in:

- Combustibility - The ease with which a substance or material is able or likely to catch fire or burn
- Combustible Gas - A gas that will burn when it is within its flammable range
- Combustible Liquid - A liquid that has a flash point above 610 C
- Combustible Metal - A metal that will burn under certain circumstances
- Competent Person - For any task means a person who has acquired through training, qualifications or experience, or a combination of them, the knowledge and skills to carry out that task
- Designed flow rate - Means the flow rate that the fire water reticulation system is designed to deliver at each fire hydrant
  - Note: This should be determined by the fire risk assessment, but is typically 10 l/s at the hydrant, which allows sufficient flow for two fire hoses to operate simultaneously
- Residual pressure - The remaining water pressure measured at a point within a system at the designed flow rate
- Safety of people - Applies to any person who may be harmed by the effects of a fire and includes:
  - Machine operators
  - People fighting a fire
  - People in the vicinity of a fire

  - People in/bye or down stream of a fire who may be affected by the fire effects, refer clause
  - Shall - Indicates a statement that is 'strongly recommended'
- Should - Indicates a statement is 'recommended'
- Static pressure - The pressure in the line at no flow
- Volatility - The ease with which a substance will evaporate
- Dynamic pressure - Means the water pressure under flow conditions
- Explosability - The ease with which a substance or device may suddenly produce a volume of rapidly expanding gas
- Fire hydrant - An assembly installed on a branch from a water pipeline, which provides a valved outlet to permit a supply of water to be taken from the pipeline for fire fighting
- Fire depot - Means a depository of fire equipment readily available and suitable for connecting to an adjacent inbye fire hydrant. Note: Typically this provides for a quick response to fight any fire in the fire risk area that should be covered by the adjacent fire hydrant.
- Fire station - Means a depository of fire equipment suitable for dispatch to underground parts of a coal mine from an area on the surface of the mine adjacent to the transport entry to the mine. Note: Typically this provides for additional fire equipment support for a fully developed fire.
- Fire substation - Means a depository of fire equipment strategically located in specific districts throughout the mine. Note: This provides for a quick response for additional fire equipment support to that provided at the fire depot.
- Flammability - The ease with which a substance is capable of catching fire
- Flash point - The lowest temperature at which a substance produces enough vapor to ignite and burn when an ignition source is applied but will stop when the source is removed
- Ignitability - The ease with which a substance or material is able or likely to catch fire or burn
- Flammable range - The range of concentrations, expressed as a percent by volume, over which a gas or vapor is flammable
- Must - Indicates mandatory legislative requirements, (i.e. a requirement of an Act or Regulation)
- Operator - Means the Coal Operator

Fire risk management is the responsibility of the mine owner/operator. Fire risks should be controlled by:

- Eliminating activities which have potential to generate explosive atmospheres or otherwise minimize through provision of adequate ventilation. Note: In a coal operation, such activities may include but be not limited to – use of cutting equipment, underground coal mining, working in confined spaces, etc.
- Eliminating potential ignition sources, and providing other measures necessary to control the risks.
- Regular monitoring to ensure effectiveness of control measures taken with respect to potential ignition sources.

An emergency management system *MUST* address:

- Fire officers and fire teams
- Supply and maintenance of fire fighting equipment
- Workforce training in the use of fire fighting equipment
- Provision and supply of water storage and reticulation
- Emergency fire procedures
- Reporting and replacement of damaged fire equipment

An emergency management system *MUST* also address:

- A fire and explosion (major hazard) management plan to provide for regular assessment of the fire and explosion risk
- Control measures that effectively manage identified risks
- Assessment of risk of fire and the early detection and control of fire on any conveyor belt in a reclaim tunnel or other enclosed space

Fire risks should be minimized to the lowest level reasonably practicable and controlled in the following order:

- Eliminate the fuel and/or ignition sources, if practical
- Substitute the fuel and/or ignition source to one of a lesser hazard
- Segregate the fuel and ignition source, (isolating the hazard)
- Use engineering means to:
    - Minimize the risk of initiating a fire
    - Provide systems for the early detection of fires
    - Provide systems to suppress and extinguish a fire

- Develop emergency procedures for early warning, fire suppression and the safe egress of people
- Use of personal protective equipment (PPE) (note, a combination of methods may be required to minimize the risk to the lowest level reasonably practicable)

5.6.14.1 Fire Risk Management Plan

The fire and explosion (major hazard) management plan must state how the health and safety of people who work at or are affected by the coal operation will be protected from a fire hazard. The fire and explosion management plan should be based on a risk management approach to safety, and should be an integral part of the mine's Emergency Management System (EMS). Essential elements of Fire Risk Management include:

- Fire hazard identification – fuel sources, ignition (heat) sources, oxygen sources
- Fire risk assessment and risk control
- Safe systems of work
- Consultation with all stakeholders
- Provision of adequate information, instruction, training and supervision on fire hazards and fire risk controls
- Monitoring, review and revision of the fire risk management process

5.6.14.2 Consider Fuels – Common Sources

In performing the risk assessment, take into consideration the following:

- Check for the presence of Class A fuels involving carbonaceous solids, which normally burn under the formation of a glow such as (but not limited to):
    - Environmental - coal dust, coal stockpiles and other coal deposits such as longwall goafs (spontaneous combustion)
    - Plant materials – non-metallic materials such as textiles, rubber, plastics, timber, urethane, electrical insulation, fabrics, tyres, carbon, etc.
    - Operational materials – rags, paper, timber, plastics, etc.
- Class B fuels involving flammable and combustible liquids such as (but not limited to):
    - Plant – diesel, petrol, hydraulic oil, brake fluid, greases

  - Operational materials - such as solvents, degreasers, alcohol, fuel, etc.
- Class C fuels involving combustible gases such as (but not limited to):
  - Environmental - methane, hydrogen, carbon monoxide, etc.
  - Plant - fuel vapor, LPG, etc.
  - Operational materials - such as acetylene, hydrogen, etc.
- Class D fuels involving combustible metals such as magnesium, aluminium, lithium, sodium, potassium and their alloys
- Class E fires involving electrically energized equipment e.g. fuels of any class

5.6.14.3 Consider Ignition (heat) Sources

Heat energy – such as diesel engines, exhaust systems, turbochargers, pumps, heat exchangers or radiators, brakes, hydraulic systems, hot work (cutting and welding), contraband in underground mines (naked flame, arcing), etc.
Electrical energy – such as electrical arcing, static electric discharge causing arcing/sparking, cables overheating, lightning, transformers, optical energy
Mechanical energy – such as frictional heating (from failed component surfaces rubbing or bearings), frictional sparking (from failed components, impact damage, grinding, continuous miner or shearer picks), strata failure (piezo-electric or frictional)
Chemical energy – such as spontaneous combustion of coal, contraband, chemical reactions (chemical used on site reacting in process or with another chemical), stockpile heating's, etc.
Pressure energy – such as shotfiring
Radiant energy – such as welding, overheating of motors, heated surfaces

5.6.14.4 Consider Fire Hazard Locations

Surface plant infrastructure:

- Surface parts – office buildings, workshops (hot work areas,), flammable material storage areas (pressurized flammable gas, fuels, oils, solvents, greases, etc.), bush fires, electrical switch rooms, overhead power lines

- Preparation plants – multi-storey buildings and structures fire hazards may pose different risks depending on the level above ground, (hot work, plant/machinery failure, coal dust spillage/accumulations, flammable materials)
- Rail and road delivery terminals – plant/machinery, haulage equipment, coal dust spillage/accumulations
- Coal handling parts – Reclaim tunnels (coal spillage/accumulations, coal dust, methane, conveyor belts, feeders, flammable materials, hot work); materials handling bins (coal dust, methane); and stockpiles (heating, mobile plant)

Open Cut or sections of open cuts:

- Open cut parts – travelling roads, the cut itself
- Mobile plant - trucks, excavators, drag lines, shovels, dozers, etc.
- Fixed plant - conveyor belts; coal handling machinery
- Storage areas - explosive magazines, fuels
- Electrical transformers and switch rooms

Underground parts:

- Face parts – longwall, continuous mines, bolting plant
- Outbye parts – travelling roads, drifts, methane drainage lines
- Return parts – returns, sealed areas, goaf areas, auxiliary/ventilation fans
- Mobile plant – LHD, man transport
- Fixed plant – Conveyor belt systems, feeders, crushers bins
- Electrical – transformers, gate end boxes, cables, battery charging stations

5.6.14.5 *Consider Hazardous Areas of plant*

- Mobile sources
- Conveyor belt systems – coal spillage/accumulations, coal dust, brakes, idler/pulley failure, fluid couplings, belt rubbing, belt/pulley slippage, static charge
- Underground face equipment – frictional ignition of methane, bearing failures, mechanical component failures,

mechanical friction, hose failures, fluid couplings, coupling failures, accumulations of coal, accumulations of oil/grease, cooling system failures
- Electrical – overheating, short circuit

General considerations in a fire risk assessment must take into consideration:

- All potential fuel sources
- All potential ignition sources
- All potential fire risk areas and fire risk areas of plant used in the mine
- All possible fire risk scenarios (what can happen?, when/where can it happen?, why/how can it happen?) For each fire risk area
- Fire effects
- Fuel properties and oxidation
- Maximum reasonable consequence and likelihood for each fire risk scenario
- Risk to health and safety of people.
- The risk to property, production and the environment
- Required measures to control the fire risk to the lowest level reasonable practicable
- Information requirements to employees and contractors on the mine site
- Instruction and training requirements
- Emergency management system requirements
- Whether the recommendations should be adopted or rejected.

Recommended References

Bakke P, Leach SJ [1962]. *Principles of formation and dispersion of methane roof layers and some remedial measures.* Min Eng (U.K.) *Jul*:645–669.

Bakke P, Leach SJ [1965]. *Turbulent diffusion of a buoyant layer at a wall.* Appl Sci Res *15*(sect A):97–135.

Bakke P, Leach SJ, Slack A [1967]. *On the general body concentration and other ventilation data as a measure of the average risk of ignition.* Min Eng (U.K.) *Aug*:762–774.

Bielicki RJ, Kissell FN [1974]. Statistical analysis of methane concentrations produced by incomplete mixing of methane and air at a model coal mine working face. Pittsburgh, PA: U.S. Department of the Interior, Bureau of Mines, RI 7987.

CEC [1985]. Investigation of methane emission and ignition risk at coalface machines. Final report on the European Coal and Steel Community (ECSC) research project 7225 – 11/08/068, October 1982 to September 1985, National Coal Board, Commission of the European Communities.

CFR. Code of Federal Regulations. Washington DC: U.S. Government Printing Office, Office of the Federal Register.

Chilton JE, Taylor CD, Timko RJ [2003]. Evaluation of Iyoni II methanometers. Proceedings 30th International Conference of Safety in Mines Research Institutes, Johannesburg, South Africa

Chilton JE, Taylor CD, Hall EE, Yantek DS [2005]. Evaluation of person-wearable methane monitors. Proceedings of the 8th International Mine Ventilation Congress, Brisbane, Queensland, Australia, July 2005, pp.189–195.

Courtney WG [1990]. Frictional ignition with coal mining bits. Pittsburgh, PA: U.S. Department of the Interior, Bureau of Mines, IC 9251. NTIS No. PB91110072.

Creedy DP, Phillips HR [1997]. Methane layering in bord-and-pillar workings. Final project report, project COL 409. Braamfontein, Republic of South Africa: Safety in Mines Research Advisory Committee (SIMRAC) (www.simrac.co.za).

CSA [1984]. Combustible gas detection instruments. Section 5.3.1, CSA standard C22.2 No. 152–M1984 (product ID 2001201). Mississauga, Ontario: Canadian Standards Association.

Davies AW, Isaac AK, Cook PM [2000]. *Investigation of a coal mine explosion and relevance of risk assessment.* Trans Inst Min Metall *109*(May-Aug)(Sect A):A61–A69.

Kissell FN, Banfield JL, Dalzell RW, Zabetakis MG [1974]. Peak methane concentrations during coal mining: an analysis. Pittsburgh, PA: U.S. Department of the Interior, Bureau of Mines, RI 7885. NTIS No. PB233424.

Kissell FN, Goodman GVR [1991]. Preventing tunnel methane explosions: what's most important. In: Proceedings of the Fifth U.S. Mine Ventilation Symposium (Morgantown, WV, June 3–5, 1991). Littleton, CO: Society for Mining, Metallurgy, and Exploration, Inc., pp. 605–610.

Kissell FN [2006]. Handbook for Methane Control in Mining. National Institute for Occupational Safety and Health, DHHS (NIOSH) Publication No. 2006–127, 184 pp.

McNider T, Grygiel E, Haynes J [1987]. Reducing frictional ignitions and improving bit life through novel pick and drum design. In: Proceedings of the Third Mine Ventilation Symposium (University Park, PA, October 12–14, 1987), pp. 119–125.

McPherson MJ [2002]. The Westray mine disaster. In: De Souza E, ed. Proceedings of the North American/Ninth U.S. Mine Ventilation Symposium (Kingston, Ontario, Canada, June 8–12, 2002). Lisse, Netherlands: Balkema, pp. 3–12.

Phillips HR [1996]. Identify methods to reduce the risks of frictional ignition hazards. Final project report, project COL 226. Braamfontein, Republic of South Africa: Safety in Mines Research Advisory Committee (SIMRAC) (www.simrac.co.za).

Raine EJ [1960]. *Layering of firedamp in longwall working.* Trans Inst Min Eng (U.K.) *119*(10):579–597.

Richmond JK, Price GC, Sapko MJ, Kawenski EM [1983]. Historical summary of coal mine explosions in the United States. Pittsburgh, PA: U.S. Department of the Interior, Bureau of Mines, IC 8909. NTIS No. PB83184226.

Schroeder WE, Kissell FN [1983]. *Face ventilation analysis by probability plotting of transformed methane concentration data.* J Mine Vent Soc S Afr *36*(11):109–116.

Titman H, Roberts AF, Brookes FR [1965]. The accumulation of methane in roof cavities. Safety in Mines Research Establishment (U.K.), research report No. 235.

Vinson RP, Thimons ED, Kissell FN [1978]. Methane accumulations in coal mine roof cavities. Pittsburgh, PA: U.S. Department of the Interior, Bureau of Mines, RI 8267. NTIS No. PB277919.

Wallhagen RE [1977]. Development of optimized diffuser and spray fan systems for coal mine face ventilation. Waltham, MA: Foster-Miller Associates. U.S. Bureau of Mines contract No. H0230023. NTIS No. PB277987.

6

Preventing Fires and Explosions Involving Metals

6.1 Introduction

Most metals are combustible to a varying degree, depending on the physical conditions. Many will undergo dangerous reactions with water, acids, and certain other chemicals; and some metals are subject to spontaneous heating and ignition. The hazard of an individual metal or alloy varies depending on the particle size and shape that is present.

This chapter provides information on combustible metals, their properties and some common sense guidelines for safe handling. Before discussing the subject in some depth, a few terms are worth reviewing. These definitions are useful to remember when dealing with metals in the laboratory or in a manufacturing environment.

The first term worth discussing is the class of metals known as *alkali metals*. Any of the six chemical elements in the leftmost group of the periodic table constitute these materials. These metals form alkalis when they combine with other elements. Because their atoms have only one electron in the outermost shell, they are very reactive chemically; more specifically, they react rapidly, even violently, with water, forming numerous compounds, and are never found free in nature. The most reactive class of alkali metals are sodium, potassium, lithium, rubidium and cesium. Upon reaction with water they generate hydrogen, and are easily ignited.

The term *combustible metal* refers to any metal, composed of distinct particles or pieces, regardless of size, shape, or chemical composition that will burn.

The term *combustible metal dust* refers to any finely divided metal 420 μm (microns) or smaller in diameter (that is, material passing a U.S. no. 40 standard sieve) that presents a fire or explosion hazard. Any time a combustible dust is processed or handled, the potential for explosion or fire exists. The degree of hazard will vary depending on the type of combustible dust, conditions, the amount of material present, and processing methods used.

The term *chips* refers to particles produced from a cutting or machining that are not oxidized and that are not diluted by noncombustible materials. Chips vary in ease of ignition and rapidity of burning, depending on their size and geometry.

The term *ribbon* refers to a piece of metal that is less than in. (3.2 mm) in two dimensions or less than an inch in single dimension. These sizes are generally thought of as a powder.

The term *pyrophoric material* refers to a substance capable of self-ignition on short exposure to air under ordinary atmospheric conditions.

Dispersions of alkali metals in organic solvents present special concerns. In addition to the water reactivity/pyrophoricity due to the reactive metal, the solvent presents the concerns of flammable or combustible liquids and vapors.

6.2 Combustibility Properties of Metals

The following section provides some general information about various metals that could be encountered in fires and explosions. If not properly identified and handled, combustible metal fires may present unique and dangerous hazards to personnel, facilities and firefighters. Generally combustible metals involved in fires generate extreme temperatures ranging from 5,000°F to 8,500°F. These temperatures result in the dissociation of water to its basic elements of hydrogen and oxygen. The application of CO_2 has similar results with dissociation of carbon and oxygen.

These ultra-high temperatures present extreme hazards to experienced firefighters entering a structure where a combustible metal fire may be burning, with rapid heat buildup and fire spread characteristics. The temperatures encountered with a combustible metals fire will far exceed those for which structural and proximity fire gear are tested and will quickly burn through material if burning metal deposits on it.

Burning combustible metal also generates high levels of smoke compared to other combustible materials. Usually the smoke is a white to grayish color if just the metal is involved. The smoke will also appear to be exiting the structure or building at a much higher velocity than one would expect in normal situations because of the high burning temperatures of the involved product.

Powders, dusts, chips, and swarfs (metallic particles and abrasive fragments removed by a cutting or grinding tool) of alkali earth or transitional metals present the greatest hazards, with powders and dusts posing greatest concern. In these forms, the metal has a greater surface area and a high explosion potential should they become airborne in a natural environment or secondary to attempts to extinguish the fire. Aluminum powders have the highest KST rate (a measurement of inherent explosive power) of all the combustible metal dusts.

As a rule, larger products of alkali earth and transitional metals such as bars, ingots, heavy castings, and thick plates or sheets are impossible to ignite; in most cases, they will self-extinguish when the heat source is removed.

Fires involving large quantities of combustible metals are impossible to extinguish and are extremely difficult to contain if not caught in the incipient stage and controlled with a Class D extinguishing agent appropriate for the metal involved. In most cases, the heat generated by a metal fire will make it impossible to apply an appropriate agent to control the fire once it is beyond the incipient stage.

Table 6.1 provides a summary of the categories of combustible metals. The metals identified in bold, italic are ones which are most likely to be encountered by first responders to fire incidents.

As noted by the table, combustible metals fall into four categories. Alkali metals are water-reactive under non-fire and fire conditions. Applying water to alkali metals will result in fire and extreme reactions. In a non-fire situation, alkali earth metals, transitional metals, and other metals do not present the same hazards as alkali metals; but when burning, these metals can become extremely dangerous if water is applied to extinguish fire. It is also important to note that most metals (such as iron) in a fine enough particle size will ignite and burn as well.

Aluminum (Al): Aluminum is used both as a commercially pure metal and as an alloy. In finely divided powder or dust form, aluminum and its

Table 6.1 Categories of Combustible Metals.

Alkali Metals	**Alkali Earth Metals**	**Transitional Metals**	**Other Metals**
Lithium	Magnesium	Hafnium	Aluminum Powder
Sodium	Beryllium	Niobium	Aluminum Flake
Potassium	Calcium	Titanium	
Rubidium	Strontium	Tantalum	
Cesium	Radium	Zirconium	
Radium			

alloys are combustible in air and present a serious combustion explosion hazard. Aluminum will react violently with many chemicals. Aluminum particles and smaller turnings will react with water to form hydrogen gas which is highly flammable and explosive in favorable concentrations. The rapid vaporization of water in contact with molten aluminum can result in rapid phase transformation explosions. Halogenated extinguishing agents should not be used.

Aluminum dust is considered to be highly explosive. Rocket fuel is made of aluminum. The NFPA used to have a standard dedicated to aluminum. NFPA 651: Standard for the Machining and Finishing of Aluminum and the Production and Handling Aluminum Products, covered the hazards of ignition and explosions in the manufacture of light metal flake powder or paste and atomized light metal granules, or dust of any light metal alloy that is explosive in an environmental atmosphere. However there are other combustible metals and in 2002 NFPA 651 was incorporated in NFPA 484.

NFPA 484: Standard for Combustible Metals updates the safety requirements for any metal that is considered a combustible metal, including aluminum, lithium, magnesium, niobium, tantalum, titanium, and zirconium. NFPA 484 also includes requirements for the production, processing, finishing, handling, storage and use of all metals and alloys that are in a form capable of combustion or explosion. For more information specifically on aluminum refer to chapter 6, in NFPA 484. There are specific statements on ductwork in section 6.1.9.8 confirming ducts "shall be constructed of non-ferrous spark resistant metal" and "plastics or other nonconductive ducts ... shall not be used." These statements eliminate galvanized steel as an option as galvanized steel is not spark resistant. Additionally, chapter 12, in NFPA

484, section 12.2.5.8 states "Duct seams shall be oriented in a direction away from personnel." This statement eliminates spiral duct as an option due to the nature of the spiral manufacturing process.

Antimony (Sb): Specific gravity, 6.62; melting point, 1166°F (630°C). This metal ignites and burns in air above 780°F (416°C).

Barium (Ba): Specific gravity, 3.5; melting point, 1300°F (704°C). When in contact with water, it liberates hydrogen but usually without ignition.

Beryllium (Be): Specific gravity, 1.85; melting point, 2400°F (1,316°C). A synonym for beryllium is glucinum. It decreases the combustibility of molten aluminum and magnesium. As a powder, ignition may occur if the metal is heated to about 1,115°F (600°C). Burning occurs with an intense flame, but can be extinguished by water. Beryllium powder should be kept away from air and moisture and stored in tight containers, preferably under argon gas. Beryllium and its compounds are very toxic and contact with skin or inhalation of dust or fumes should be avoided.

Bismuth (Bi): Specific gravity, 9.80; melting point, 520°F (271°C). This metal burns in air with a bluish flame when vaporized and oxidizes rapidly when molten. At red heat, bismuth decomposes steam. It reduces carbon dioxide, but does not react with nitrogen or hydrogen.

Calcium (Ca): Specific gravity, 1.55; melting point, 1562°F (850°C). In contact with water, calcium liberates hydrogen but usually without ignition. It is considered pyrophoric under some conditions and is normally shipped in lump form under argon.

Cadmium (Cd): Specific gravity, 8.65; melting point, 610°F (321°C). It oxidizes when heated, giving off dense brown fumes that decompose steam above 750°F (400°C). As a powder it decomposes hot water but without igniting.

Chromium (Cr): Specific gravity, 7.19; melting point, 3407°F (1875°C). It presents a moderate dust explosion hazard under favorable conditions of partial size dispersion and ignition source.

Copper (Cu): Specific gravity, 8.96; melting point, 1981°F (1083°C). Copper is not known to produce explosive properties when in a finely divided form.

Hafnium (Hf): Specific gravity, 13.36; melting point, 4,032°F (2,222°C). Fines (fine particulate matter) are pyrophoric. The metal burns with little visible flame but with a high rate of heat release. Chemical reacts with water to form hydrogen, which may ignite spontaneously. Its hazards are similar to those of zirconium. Crushing and sizing should be carried out under inert gas.

Iron (Fe): Specific gravity, 7.86; melting point, 2802°F (1540°C). Iron presents a dust hazard under favoring conditions of particle size and dispersion in air. It can be easily ignited in the form of dust or steel wool or as fine turnings or chips containing oil. Structural steel has a specific gravity of 7.83 and melting point of 2605°F (1430°C). When heated above 600°F (315°C), it begins to lose strength rapidly.

Lead (Pb): Specific gravity, 11.34; melting point, 621°F (327°C). At high temperatures lead volatizes and burns with white flame. At red heat, it is rapidly oxidized by air and at white heat by steam.

Lithium (Li): An alkali metal. Specific gravity, 0.53; melting point, 356°F (180°C). As the hydride, lithium offers the most concentrated method of storing and transporting hydrogen. Ignition and burning occur when lithium is heated to about 356°F (180°C). It reacts less vigorously than sodium with water or air, and usually does not ignite. It reacts strongly with chlorinated, fluorinated, and brominated organic compounds, halogens, and sulfuric acid. In the presence of moisture, lithium reacts exothermically with nitrogen at ordinary temperatures. Above the melting point, it rapidly forms the nitride. Near its melting point, it ignites in air and burns with a characteristic intense, brilliant white flame. Following treatment with nitric acid, it may explode on very light impact or friction. Lithium also reacts violently with hydrogen peroxide.

Magnesium (Mg): An alkali metal. Specific gravity, 1.74; melting point of 1202°F (650°C). Its ignition temperature is near the melting point, although ignition of some forms may occur at lower temperatures. As a dust cloud or in ribbon form, magnesium can be ignited almost instantly. Loose shavings ignite fairly readily. It is less easy to ignite the surface of a compact pile of chips. Magnesium fines wet with oils may ignite spontaneously. Fines wet with acids, water, water-soluble oils, or oils containing fatty acids will generate hydrogen. Powders form explosive mixtures with air that may be ignited by a spark. Fines will also react with chlorine, bromine, iodine and oxidizing agents.

Manganese (Mn): Specific gravity, 7.43; melting point, 2246°F (1230°C). As a dust cloud it can be ignited at 840°F (449°C) in air.

Molybdenum (Mo): Specific gravity, 10.2; melting point, 4750°F (2620°C). Molybdenum powder reacts vigorously with water vapor at 1300°F (704°C). It presents a slight dust hazard under favoring conditions of particle size, dispersion, and strong ignition source.

Plutonium (Pu): Specific gravity, 19.6; melting point, 1,170°F (632°C). It is pyrophoric, highly toxic (particularly in the oxide form when burning), and radioactive. Chips, turnings, and fine particles may ignite spontaneously. It is easily ignited by friction.

Potassium (K): An alkali metal. Specific gravity, 0.86; melting point, 145°F (63°C). Potassium is similar to sodium in its pyrophoricity, chemical activity, and hazards although somewhat more reactive. It is violently reactive with sulfuric acid and most halogens. It will detonate in contact with liquid bromine.

Silicon (Si): Specific gravity, 2.33; melting point, 2588°F (1420°C). A silicon dust cloud can be ignited in air at 1425°F (775°C). Pure silicon metal dust has been shown to be highly explosive under certain conditions.

Sodium (Na): An alkali metal. Specific gravity, 0.97; melting point, 208°F (98°C). Sodium metal is most often used to remove excess water from flammable liquids during the distillation process. It is also used in the production of other sodium-based compounds and in the reduction of organic esters. Sodium metal reacts violently with water to produce hydrogen gas and sodium hydroxide. The exothermic heat of this reaction can lead to auto-ignition of the hydrogen gas and/or the metal itself and create a severe fire or explosion hazard. Due to its low melting point, explosions of molten sodium may occur during reaction with water. Residual liquid present following the reaction contains a high percentage of sodium hydroxide, a caustic, corrosive material.

In finely divided form, sodium ignites spontaneously in air. Molten sodium reacts with most gases and liquids except the noble gases and nitrogen, and reacts vigorously with hydrogen to form the hydride. Solid sodium reacts strongly with water, alcohol, polyhalogenated hydrocarbons, halogens, acidic oxides, sulfuric acid, mercury, and certain alloys of lead, tin, zinc and bismuth.

Strontium (Sr): Specific gravity, 2.6; melting point, 1386°F (752°C). In contact with water, it liberates hydrogen readily, but usually without ignition.

Tantalum (Ta): Specific gravity, 16.62; melting point 5425°F (2996°C). It presents a moderate dust explosion hazard under favorable conditions of particle size, dispersion, and ignition source.

Tellurium (Te): Specific gravity, 6.24; melting point, 846°F (452°C). It presents a moderate dust explosion hazard under favoring conditions of particle size, dispersion and ignition source.

Thallium (Tl): Specific gravity, 11.85; melting point, 572°F (300°C). At red heat it decomposes water, producing hydrogen. It is very toxic.

Thorium (Th): Specific gravity, 11.6; density, melting point, 3,090°F (1,700°C). Thorium is pyrophoric and radioactive. As a dry powder it has a low ignition temperature. Thorium powder is shipped under helium or argon gases in special containers. When poured through air, it is subject to ignition by electrostatic spark. It should be handled cautiously with a non-sparking spoon or scoop. Containers and tools should be grounded. Ignition has occurred due to chemical reaction between finely divided thorium and water at ordinary temperatures.

Tin (Sn): Specific gravity, 7.29; melting point, 450°F (232°C). Its dust cloud can be ignited in air at 1165°F (630°C).

Titanium (Ti): Specific gravity, 4.51; melting point, 3040°F (1670°C). At red heat, 1300°F (704°C), the metal actively decomposes steam. Following strong nitric acid treatment, the metal may explode by the light impact or friction of handling with tongs. The powder can be ignited in pure carbon dioxide above 1260°F (682°C), in nitrogen above 1475°F (801°C), and in air at 626°F (330°C) to 1094°F (590°C). Powdered titanium immersed in water or wet with water at ordinary temperatures has been ignited by chemical reaction.

Tungsten (W): Specific gravity, 19.2; melting point, 6115°F (3380°C). As fine powder, tungsten may be pyrophoric. Hydrogen reduced powder may retain some absorbed hydrogen, presenting a dust explosion hazard. Tungsten reacts violently with molten nitrates, nitrites, and peroxides.

Uranium isotopes 233 and 235, plutonium, and thorium are radioactive, and present the same fire and explosion hazards as other combustible metals. They have been known to ignite spontaneously. As coarse particles, metals in this group offer little hazard; but in powder form, they present severe dust explosion possibilities. Some burn readily in pure nitrogen or carbon dioxide gas, and explode on contact with oxidizing agents. Some heat spontaneously in air and must be stored and handled under a nonreactive liquid or gas. Chips and other fine particles ignite easily, and special care is needed to prevent the ignition of chips by the friction of cutting. Fine particles of these metals should be shipped and stored in approved DOT containers.

When molten or when burning, these metals react vigorously with water, foam, carbon dioxide vaporizing liquid and dry chemical extinguishing agents. Applied directly to the burning material, these agents will intensify burning and may cause an explosion. Approved dry powder extinguishing agents, dry sand, graphite chips, limestone, or talc are used to smother fires.

Uranium (U): Specific gravity, 18.68; melting point, 2,071°F (1,132°C). The fire hazard of metallic uranium is similar to but more severe than magnesium. In finely divided form, it is pyrophoric. Spontaneous fires have occurred under argon, under vacuum, and underwater. The high temperature reaction of uranium with steam is very violent. Radioactivity and products of combustion present a health hazard. Uranium powder should be handled in an inert atmosphere. Chips and turnings may be stored if completely immersed in water with ventilation to remove evolved hydrogen. Partial submersion or slight moistening accelerates spontaneous heating.

Zirconium (Zr): Specific gravity, 6.51; melting point 3,326°F (1,830°C). Fine particles in dust layers or clouds are pyrophoric, and may be ignited by heat, static or friction. Particles can be ignited in nitrogen gas above 986°F (530°C) and in carbon dioxide above 1,040°F (560°C). Zirconium samples may explode during or following treatment with strong nitric acid or carbon tetrachloride. Samples have also exploded while being dissolved in a mixture of sulfuric acid and potassium acid sulfate. Particles form explosive mixtures with barium nitrate, potassium chlorate, and other oxidizing materials. In dry powder form, zirconium or zirconium-copper alloys in glass containers may explode by impact or friction if the container breaks. It ignites more easily than magnesium and is slightly less hazardous than uranium. Zirconium powder is commonly handled wet because it is then more difficult to ignite, although once ignited it will burn more violently. At least 25 percent moisture by weight should be present.

Zinc (Zn): Specific gravity, 7.13; melting point 786°F (419°C). In the form of dust, in contact with moisture, alkaline solutions, or acetic acid, zinc will heat spontaneously to ignition. Large pieces of zinc are difficult to ignite but once ignited will burn strongly. In oxygen, oxidation of the metal takes place rapidly at 300°F (149°C) and ignition occurs at 930°F (499°C) with a bluish flame. Steam is decomposed on contact with zinc at 660°F (349°C). Acids or caustic soda also liberate hydrogen on contact with zinc. In air, zinc burns to the oxide, forming white or bluish smoke. As a dust cloud, it can be ignited at 1,110°F (600°C).

6.3 Explosion Temperatures

The Pittsburgh Research Laboratory of the National Institute for Occupational Safety and Health[1] conducted a study of the explosibility of various metals and other elemental dusts dispersed in air, with a focus on the experimental explosion temperatures. The dusts studied included boron, carbon, magnesium, aluminum, silicon, sulfur, titanium, chromium, iron, nickel, copper, zinc, niobium, molybdenum, tin, hafhium, tantalum, tungsten, and lead. The dusts were chosen to cover a wide range of physical properties - from the more volatile materials such as magnesium,

[1] Kenneth L. Cashdollar' and Isaac A. Zlochower, Explosion Temperatures of Metals and Other Elemental Dust Clouds, Pittsburgh Research Laboratory, National Institute for Occupational Safety and Health, http://www.cdc.gov/niosh/mining/UserFiles/works/pdfs/etoma.pdf

aluminum, sulfur, and zinc to the highly "refractory" (very low volatility) elements such as carbon, niobium, molybdenum, tantalum, and tungsten.

The investigators conducted flammability studies in a 20-L chamber, using strong pyrotechnic ignitors. The experimental data obtained included the minimum explosible concentrations, maximum explosion pressures, and maximum explosion temperatures. A unique multi-wavelength infrared pyrometer was used to measure the temperatures. The measured maximum explosion temperatures ranged from – 1,550°K for tin and tungsten powders to – 2,800°K for aluminum, magnesium, and titanium powders. The measured temperatures were compared to the calculated, adiabatic flame temperatures. In general, the dusts whose experimental temperatures were closer to the adiabatic values were those with the finer particle sizes, those that were more easily vaporized, and/or those that were intrinsically more reactive.

The U.S. Bureau of Mines[2] studied the explosibility of metal dusts using a cylindrical 1.2-L Hartrnann chamber with an electric spark ignition source. Other investigators used a nearly-spherical 20-L chamber and much stronger pyrotechnic ignitors to study the explosibility of metal and other elemental dusts (see Hertzberg, et.al[3], 1991; Hertzberg, et.al.[4], Cashdollar[5]).

Various investigators measured explosion characteristics for the metal and other elemental dusts which included minimum explosible concentrations (lean flammable limits), maximum explosion pressures, and rates of pressure rise. The explosions observed were rapid deflagrations, not detonations. The infrared temperatures and other experimental explosion data provide evidence to evaluate the general mechanisms of homogeneous versus heterogeneous combustion of these elemental dusts.

6.4 Dry Powder (Class D Fires)

The National Fire Protection Association (NFPA) 484, *Standard for Combustible Metals*, states that it is extremely important to conduct a good

---

[2] Jacobson, M., Cooper, A. R., and Nagy, J. (1964). Explosibility ofMetal Powders, U.S. Bureau of Mines RI 6516,25-pp

[3] Hertzberg, M., Zlochower, I. A., and Cashdollar, K. L. (1991), Explosibility of Metal Dusts, Combustion Science and Technology, 75, 16 1-1 65.

[4] Hertzberg, M., Zlochower, I. A., and Cashdollar, K.L. (1992), Metal Dust Combustion: Explosion Limits, Pressures, and Temperatures. In: Twenty-Fourth Symposium (International) on Combustion, Pittsburgh: The Combustion Institute, pp. 1827 – 1835.

[5] Cashdollar, K. L. (1994). Flammability of Metals and Other Elemental Dusts. Process Safety Progress, 13, 139–145.

size-up by identifying the combustible metals involved, the physical state of the metals (e.g., shavings, chips, fine dust, etc.), the location relative to other combustible materials, and the quantity of the product involved. NFPA 484, A.13.3.3.10.3, states that the application of a wet extinguishing agent (particularly water hose streams) accelerates a combustible metal fire and could result in an explosion.[6] This is due to the water reacting with the combustible metal and giving off highly flammable hydrogen gas and oxygen. This conversion of water into hydrogen has a heat value (British Thermal Units per pound (Btu/lb)) of about 2.8 times that of gasoline, assuming 100 percent conversion of the hydrogen in the water. This equates to flowing 42.8 gallons per minute (gpm) of gasoline on the fire for every 100 gpm of water.

NFPA 484, A.13.3.3.5, states that the following agents shall not be used as extinguishing agents on a combustible metal fire because of adverse reactions or ineffectiveness: water, foams, halon, carbon dioxide, nitrogen (except on iron, steel, and alkali metals, excluding lithium), and halon replacement agents. Thus, in lieu of using a wet extinguishing agent, primarily water, it is recommended that a bulk dry extinguishing agent compatible with the product involved, such as dry sand, dry soda ash, or dry sodium chloride, be used. In most cases for large fires beyond the incipient stage, the application of a dry agent is not feasible. In these cases the best approach is to isolate the material as much as possible, protect exposures, and allow the fire to burn out naturally. Thorough training is a must to properly identify and handle these unique fires. Businesses that manufacture, use or store combustible metals, and fire departments with combustible metals in their jurisdiction, should review Chapter 13 of the National Fire Protection Association (NFPA) 484: Standard on Combustible Metals.

Class D fires only involving combustible metals such as magnesium, sodium (spills and in depth), potassium, sodium-potassium alloys, uranium and powdered aluminum require the use of special extinguishers. Class D fires are caused by burning metals that combust easily on contact with air, such as magnesium and lithium. Such specialized fires require a specialized metal powder fire extinguisher to deal with them, and will be required in scientific labs or where manufacturing processes involve the risk of metal fires. For example, aluminum dust or swarf can catch fire, so any process involving cutting, drilling or milling aluminum holds potential

---

[6] NFPA [2009]. NFPA 484: Standard on Combustible Metals. Quincy, MA: National Fire Protection Association.

risk. Generally the extinguishing agent is referred to as DRY POWDER. These extinguishers should be identified by a star containing the letter "D", if color-coded, the star is yellow.

The following is a list of combustible metals that require a type D fire extinguisher:

aluminum (powder)	neodymium
aluminum phosphide	nickel catalyst (Raney)
beryllium (powder)	phosphorus
calcium	phosphorus pentasulfide
calcium carbide	potassium
cerium	potassium hydride
cesium	potassium-sodium alloys
gallium arsenide	Red-Al
gallium phosphide	rubidium
lithium	sodium
lithium aluminum deuteride	sodium aluminum hydride
lithium aluminum hydride	sodium bis(2-methoxyethoxy) aluminum hydride in toluene
lithium aluminum hydride bis (tetrahydrofuran)	sodium borohydride
lithium amide	sodium borohydride cobalt-doped
lithium borohydride	sodium borohydride on alumina
lithium hydride	sodium hydride
lithium tetraphenylborate tris(1,2-dimethoxyethane)	strontium
lithium tri-tert-butoxyaluminohydride	titanium (powder)
lithium-6 deuteride	zinc (powder)
lutetium	zinc phosphide
magnesium and magnesium alloys	zirconium (powder)
magnesium hydride	

Agents that are typically used as extinguishing agents on combustible metal fires include dry sand, dry soda ash, dray calcium chloride, and a Class D extinguisher.

The following agents are not recommended by experts to be used as extinguishing agents on combustible-metal fires because of adverse reactions or ineffectiveness:

- Water
- Foams
- Halon
- Carbon dioxide
- Nitrogen (except on iron, steel, and alkali metals, excluding lithium)
- Halon replacement agents
- ABC or BC dry chemical extinguishers

There is an exception to the rule in that A: B: C dry-chemical and B: C dry-chemical extinguishers, although not permitted be used as an extinguishing agent on a combustible-metal fire, can be used on other classes of fires in the area where combustible metals are present. It is important to note further that the fire-extinguishing agent expellant gases must be compatible with the combustible metal.

An incipient fire occurring while the metal powder is in slurry form shall be permitted to be fought using listed Class B extinguishing agents, except that halogenated extinguishing agents shall not be used. An incipient fire occurring in semi-wet material or filter cake shall be fought using a listed Class B extinguishing agent. Where Class B extinguishing agents are used to extinguish fires involving solvent-wetted aluminum, the residual material shall be immediately covered with dry sand, with dry inert granular material, or with other listed Class D extinguishing agent, and the entire mass shall be allowed to cool until it reaches ambient temperature. When the material has cooled and it has been determined that there are no hot spots, the covered material shall be carefully removed for disposal. The material shall be handled in covered containers.

An incipient fire should be ringed with a dam of dry sand, with dry material that will not react with the metal being extinguished, or with a listed Class D extinguishing powder in accordance with the manufacturer's instructions. The application of the dry extinguishing agent should be conducted in such a manner as to avoid any disturbance of the combustible-metal dust, which could cause a dust cloud.

The use of pressurized extinguishing agents should not be permitted on a combustible-metal powder fire or chip fire, unless applied carefully so as not to disturb or spread the combustible-metal powder or chip fire. Only listed or approved Class D extinguishing agents or those tested and shown to be effective for extinguishing combustible-metal fires should be permitted.

Additionally, fire-extinguishing agents compatible for the hazards present should be made available in metal-scrap storage areas. Fire-extinguishing agents compatible for the hazards present should be available in metal-powder storage areas.

The two best known dry powder extinguishing agents for controlling combustible metal fires are graphite and sodium chloride (salt). More recently the U.S. Navy has developed a new agent specifically for lithium fires. It should be noted that the term "dry powder" should not be confused with "dry chemical" which is normally associated with extinguishing agents suitable for use on flammable liquid fires. Some "dry chemicals" may create an explosive reaction when applied to combustible metal fires.

One commercial extinguishing agent is Amerex G-Plus Dry Powder[7] which is a black, non-toxic material consisting of finely graded, granular graphite (carbon). It is normally applied to a metal fire using a spark proof scoop or shovel. The manufacturer recommends that the powder be applied with a minimum disturbance, adequately covering the fire area to a depth of approximately two inches. If the burning material is on a combustible surface, the fire should be covered with dry powder as previously noted. A one or two inch layer of powder should then be spread on a nearby area (preferably a non-combustible surface) and the burning material shoveled onto this layer with more powder added as needed.

Graphite is generally considered to be an excellent conductor of heat (heat sink). When applied to a combustible metal fire, the temperature is reduced below the metal's ignition temperature. The finely ground particles of the powder also act as a smothering agent by packing and effectively sealing the fire hazard from the air.

Practice drills should be conducted to train first responders of any of the three types of dry powder extinguishing agents. Fires in some combustible metals and combustible metal alloys generate by-products which constitute a health hazard. It is important that the MSDS provided by the material supplier be consulted for recommended breathing and clothing protection. In the U.S., the following OSHA requirements should be

[7] http://amerex-fire.com/learning-center/tech-tips-4/

followed: CFR 29 Part 1910.157 (6) "The employer shall distribute portable fire extinguishers or other containers of Class D extinguishing agent for use by employees so that the travel distance from the combustible metal working area to any extinguishing agent is 75 feet (22.9 m) or less. Portable fire extinguishers for Class D hazards are required in those combustible metal working areas where combustible metal powders, flakes, shavings, or similarly sized products are generated at least once every two weeks."

CFR29 Part 1910.158 (g) Training and education states: (1) "Where the employer has provided portable fire extinguishers for employee use in the workplace, the employer shall also provide an educational program to familiarize employees with the general principles of fire extinguisher use and the hazards involved with incipient stage fire-fighting. (2) The employer shall provide the education required in paragraph (g)(1) of this section upon initial employment and at least annually thereafter."

The following pages contain excerpts taken from NFPA 49 – "Hazardous Chemicals Data" and provide additional information on several hazardous metals. NFPA 49 and the "Hazardous Material Handbook" are valuable reference guides for combustible metals. An MSDS also contains fire-fighting information from the prime manufacturer of the metal.

- Aluminum (Dust or Powder) Description: Silvery colored powdered metal. Fire and Explosion Hazards: Forms explosive mixture in air. Bulk dust when damp may heat spontaneously. Hazard greater as fineness increases. Reacts with some acids and caustic solutions to produce hydrogen. Life Hazard: Respiratory and eye irritant only. Fire Fighting Phases: Do not use water. Smother with suitable dry powder. Usual Shipping Containers: Cans, barrels and boxes. Storage: Protect containers against physical damage. Keep dry and isolate acids, caustics and chlorinated hydrocarbons. Separate from oxidizing materials. Avoid storage near combustible materials. Remarks: Electrical installations in Class II hazardous locations, as defined in Article 500 of the National Electrical Code, should be in accordance with Article 502 of the Code; Class II, Group E electrical equipment should be used in atmospheres containing aluminum dust. This hazardous powder may be produced by some metal spraying operations. See Code for the Processing and Finishing of Aluminum (NFPA No. 65), Code for the Prevention of Dust Explosions in the Manufacture of

Aluminum Powder (NFPA No. 651) and National Electrical Code (NFPA No. 70).

- Beryllium – (Dust or powder) Description: Silvery material, resembling aluminum powder. Fire and Explosion Hazards: Forms explosive mixture in air. Hazard greater as fineness increases. Reacts readily with some strong acids, producing hydrogen. Life Hazard: Extremely toxic respiratory poison and eye irritant. If introduced under skin through cuts or punctures, may develop slow-healing ulcers. Personal Protection: Wear special protective clothing. Fire Fighting Phases: Do not use water. Smother with suitable dry powder. After exposure to beryllium fire, personnel should bathe carefully; all equipment and clothing should be washed down and clothing should be laundered separately from other non-contaminated material and clothing. Usual Shipping Containers: Steel and fiber drums. Storage: Protect containers against physical damage. Keep dry and isolate from acids, caustics and chlorinated hydrocarbons. Separate from oxidizing materials. Remarks: Electrical installations in Class II hazardous locations, as defined in Article 500 of the National Electrical Code, should be in accordance with Article 502 of the Code; and electrical equipment should be suitable for use in atmospheres containing beryllium dust.
- Calcium - Description: Silvery, soft metal. Tarnishes to grayish white on exposure to air. Fire and Explosion: When heated to its ignition temperature, calcium burns quiescently in air without displaying any tendency to melt or flow. Finely divided calcium exposed to moist air may ignite spontaneously at room temperature. Calcium reacts with water to form hydroxide and hydrogen, but the reaction is much less violent than with sodium. Mixed with air, the liberated hydrogen may present an explosion hazard. As discussed below, explosion or violent reaction may take place if care is not exercised in selecting extinguishants applied to a calcium fire. Life Hazard: Solid material will cause skin and eye burns since it reacts with moisture to form caustic. Similarly the fumes from burning calcium are highly irritating to skin, eyes and mucous membranes. Personal Protection: Wear self-contained breathing apparatus; wear goggles if eye protection is not provided. Fire Fighting Phases: Do not use water or halogenated

hydrocarbons such as carbon tetrachloride. Carbon dioxide and dry chemical are ineffective. Use dry graphite, soda ash, powder sodium chloride or appropriate metal fire extinguishing powder. Usual Shipping Containers: Hermetically sealed cans and drums. Storage: Protect containers against physical damage. Keep away from water or locations where water may be needed for fire in other storage or in the building itself. Avoid high temperatures. Store under kerosene or other neutral oil. Never store under halogenated hydrocarbons. A detached fire-resistant building is recommended for quantity storage.

- Calcium Carbide (CaC_2) Description: Grayish-black irregular lumps.Fire and Explosion Hazards: Not flammable in dry state but produces acetylene gas on contact with water or moisture. Will generate sufficient heat on contact with small amount of water to ignite acetylene gas formed. Life Hazard: Dust is an eye and respiratory irritant and can cause skin burns. Fire Fighting Phases: Do not use water, vaporizing liquids or foam. Carbon Dioxide is ineffective. Smother with suitable dry powder. Usual Shipping Containers: Steel drums and cans. Storage: Protect against physical damage. Store in dry, non-combustible, well ventilated place without sprinkler protection and exclude possible sources of ignition of acetylene gas. Isolate from other materials. Remarks: See Standard for the Installation and Operation of Gas Systems for Welding and Cutting (NFPA No. 51) and Chemical Safety Data Sheet SD-23 (Manufacturing Chemists' Association, Inc.).
- Lithium: Description: Silvery, soft (about as hard as lead) metal, tarnishes to grayish-white on exposure to air. Fire and Explosion Hazards: While massive lithium will not normally ignite spontaneously in air unless heated above its melting point, ignition of the finely divided metal may take place at room temperatures. Unless physically confined, the burning metal has a pronounced tendency to melt and flow out. Combustion is accomplished by the emission of dense, white and opaque fumes, which tend to mask the seat of the fire. Lithium will burn in air, oxygen, nitrogen and carbon dioxide. The susceptibility of molten lithium surfaces to spontaneous ignition is increased by presence of lithium oxides or nitrides. Metallic lithium is sometimes handled at

high temperatures (e.g. 1,000° F.) in closed systems with an inert gas (usually dry argon) in the void space. Leaks of molten lithium from such systems are normally accompanied by spontaneous ignition if the metal is permitted to contact air. At high temperatures lithium may react violently with concrete or other materials containing moisture. Lithium reacts with water to form hydrogen and lithium hydroxide. The reaction rate at room temperature is far less rapid than the corresponding reaction with sodium or potassium. Hydrogen resulting from the reaction may present an explosion hazard when mixed with air. Use of improper extinguishants for lithium fire control may introduce explosion or violent reaction hazards. Life Hazard: Solid material may cause skin and eye burns since it reacts with moisture to form caustic. Similarly the fumes from burning lithium are highly irritating to skin, eyes and mucous membranes. Personal Protection: Wear self-contained breathing apparatus; wear goggles if eye protection not provided. Fire Fighting Phases: Do not use water or halogenated hydrocarbons, such as carbon tetrachloride. Carbon dioxide and dry chemicals are ineffective. Use dry graphite or appropriate metal fire extinguishing powder. Usual Shipping Containers: Hermetically sealed cartridges, cans and drums. Storage: Protect containers against physical damage. Keep away from water or locations where water may be needed for fire in other storage or in the building itself. Avoid high temperatures. Store under kerosene or other neutral oil. Never store under halogenated hydrocarbons. A detached fire resistive building is recommended for quantity storage.

- Magnesium (including alloys) Description: Silvery metal; looks like aluminum but is one-third lighter. Fire and Explosion Hazards: Combustible metal. Fine powder, thin sheets, chips and turnings, are easily ignited and burn with intense heat and brilliant white flame. Powders form explosive mixtures with air which may be ignited by a spark. Pieces having thickness over 1/8-inch are difficult to ignite or to maintain flame as heat is conducted away so rapidly. However, thick pieces can be ignited when enough heat is applied. In finely divided form, will react with water and acids to release hydrogen; also hazardous in such form with chlorine, bromine, iodine, oxidizing agents and acids. Life Hazard: Dust is a slight irritant. Personal Protection:

> In fire conditions protect eyes and skin against flying particles. Avoid direct viewing of magnesium fires as eye injury may result. Fire Fighting Phases: Smother with dry graphite or other suitable dry powders. Do not use foam, Carbon tetrachloride or carbon dioxide. Manual application of water should be conducted with care to prevent contact with burning or molten magnesium. Protect eyes and skin against flying particles. Avoid direct viewing of magnesium fires as eye injury may result. Usual Shipping Containers: Shavings or powder shipped in tightly closed metal or fiber containers. Shipped without containers as ingots, billets, castings, forgings, extruded shapes, rolled sheet and plates. Storage: Protect against physical damage. Store finely divided chips or shavings in detached fire-resistive building, protected from moisture and away from chlorine, bromine, iodine, acids and all possible sources of ignition. Heavier sections may be stored in the open like steel or aluminum.

For all work environments it is important to conduct a survey aimed at identifying the combustible metals or other materials that are present. The physical state of the product (e.g., chips, powders, fine dust), the location, and the quantity of product that may be involved or potentially involved in the fire are important factors to consider both in prevention and in combating fires.

One should always obtain material safety data sheets (MSDSs), and consult personnel familiar with the product and its hazards. If there are combustible metal or other products in a work area, it's extremely important to become aware of that product's specific hazards and the appropriate methods for dealing with them in the event of an incident.

Remember that the only way to extinguish a combustible metal fire involving alkali earth metals and transitional metals is to provide an inert atmosphere of argon or helium, which in most fire situations is impracticable. Applying other agents will contain the fire until the metal has oxidized to a point at which heat production is reduced below the ignition point.

The standard extinguishing agents most of us use for ordinary combustible fires can be dangerous or ineffective on combustible metal fires. As a general rule, the best course of action for metal fires is to allow the fire to burn itself out if the fire can be safely isolated. This may necessitate sacrificing the involved structure. Exposures and uninvolved product can be protected if adequate precautions are taken to ensure runoff does not reach the metal fire or, in the case of alkali metals, exposed metal not involved in the fire. It is also important to consider water sources within the structure that could allow water to reach the fire if breached. It may be necessary to isolate the domestic water supply feed to the fire area. Adequate preplanning would determine the valve locations.

In most cases, combustible metal fires will burn quickly and begin to develop an oxide crust that will limit open burning of the product. It is extremely important that the crust not be disturbed until the metal has completely oxidized to the point of extinguishment. Depending on the size of the fire, this may take 24 hours or longer. Thermal imaging can be a great asset in this case. Even though there may not be signs of external burning, the metal can remain extremely hot and continue to present thermal injury risk to responders and risks associated with application of water if the oxidized crust is disrupted prior to complete extinguishment.

National Fire Protection Association (NFPA) 484, *Standard for Combustible Metals*, is an excellent resource for facilities that use or manufacture combustible metals. The 2006 edition of the standard included a new chapter specifically related to fire prevention, fire protection, and emergency response. This will be extremely beneficial to emergency responders dealing with combustible metal incidents.

6.5 Case Studies

6.5.1 Combustible Metal Dust Led to Fatal Flash Fire[8]

Two separate flash fires at a Hoeganaes Corp. plant killed one worker and injured two others. The fires involved accumulated iron dust that became airborne and combusted. The first flash fire occurred at the Hoeganaes Corp. Gallatin, Tenn. plant Jan. 31, 2011 where one worker was killed and another was seriously burned. The plant experienced a second, similar fire March 29, which caused one injury. The Hoeganaes plant manufactures "atomized" iron powder that is sold to the automotive and other industries for the production of metal parts using powder metallurgy.

In the first, fatal incident, a dust collector was reported to have been out of service for the 2 days leading to the incident. In the second fire, a plant engineer apparently dislodged accumulated iron dust during his work replacing igniters on a furnace; the dust combusted and subsequently injured the engineer.

Tests conducted on samples of metal powder – collected from the plant – determined that the material was combustible. The Chemical Safety

---

[8] http://ehstoday.com/fire_emergencyresponse/fire/csb-combustible-metal-dust-fatal-flash-fire-0511

Board determined that the "presence of combustible dust was known by Hoeganaes at the times of the accidents; it appears the risks were not adequately addressed by the company."

6.5.2 Watco Mechanical Services[9]

Three Texas facilities were cited by OSHA after it was determined combustible metal dust was responsible for causing a flash fire at a work site. The incident, which took place on July 14, 2012, occurred at Watco Mechanical Services in Hockley, TX after employees cutting metal with a torch accidentally set nearby combustible dust aflame during routine blasting procedures at a rail car building. The ensuing fire killed two contracted workers employed by Jordan General and left a third injured. The companies at fault — Watco Mechanical Services, Jordan General Contractors, Inc. and JP Electric — now face penalties collectively totaling $119,840:

Watco Mechanical Services, a Kansas-based company, was dealt 14 serious violations, including "failing to adequately control fugitive emissions of combustible dust; keep work areas clean of combustible dust; provide warning signs to alert employees of combustible dust hazards; and keep emergency cylinder respirators fully charged." Confined space violations were also discovered, with proposed penalties totaling $91,300.

Jordan General Contractors was served seven serious violations, with those relating to combustible dust being their failure to provide proper training on the dangers of working with combustible dust and ensuring cutting operations were put on hold while combustible dust was present. Proposed penalties levied by OSHA were $20,240.

JP Electric, a company contracted to assist with demolition services, was cited with one serious safety violation for failing to prohibit cutting operations in the presence of combustible dust, amounting in a proposed fine of $2,800.

6.5.3 Metal Recycling Facility Fire - California[10]

On July 13, 2010, seven fire fighters were injured while fighting a fire at a large commercial structure containing recyclable combustible metals. At approximately 40 minutes into the response incident, a large explosion propelled burning shrapnel into the air, causing small fires north and south

[9] http://dustincase.ruwac.com/texas-metal-dust-accident-justifies-need-for-immersion-separators/

[10] NIOSH Case Study, http://www.cdc.gov/niosh/fire/reports/face201030.html

of structure, injuring fire fighters, and damaging apparatus and equipment. Realizing that combustible metals may be present, the incident commander ordered fire fighters to fight the fire with unmanned ladder pipes while directing the water away from burning metals. Approximately 2 ½ hours later, two small concentrated areas remained burning and a second explosion occurred when water contacted the burning combustible metals. During the second explosion no fire fighters were injured.

The incident involved a 45,000 square foot multiple business commercial structure that measured approximately 300' x 150' and was built in 1939. The commercial structure was divided into 3 sections with both Type III and Type V (metal clad) construction. The A-side (west) of the structure measured 60' x 100' under a heavy timber bowstring truss roof and exterior block walls covered with a stucco finish. This section of the structure contained denim fabric altering machinery. The larger 210' x 150' open warehouse middle section of the structure was under a metal sawtooth roof (a roof composed of a series of small parallel roofs of triangular cross section, usually asymmetrical with the vertical slope glazed or windowed to allow for light) with concrete reinforced metal beam exterior walls covered with an exterior stucco finish. This section of the structure contained bins, bales, and piles of recyclable metals. The C-side of the structure was an office area that measured approximately 30' x 150'. It was comprised of two stories with a conventional flat roof, wood framed interior walls, and concrete reinforced metal beam exterior walls covered with an exterior stucco finish.

Per the fire department's arson investigation report, the fire originated on the north side of the one-story commercial building and traveled throughout the entire structure over a period of several hours. Titanium products were involved with 2 major explosions. The cause of the fire was undetermined. Occupational injuries and fatalities are often the result of one or more contributing factors or key events in a larger sequence of events that ultimately result in the injury or fatality. NIOSH investigators identified the following items as key contributing factors in this incident that ultimately led to the injuries:

- Unknown building contents
- Unrecognized presence of combustible metals
- Use of traditional fire suppression tactics
- Darkness

6.5.4 Other Case Studies

The Chemical Safety Board has assembled a number of case studies from its own and NFPA investigations over the years. Table 6.2 provides a sampling

Table 6.2 Case studies involving metals (data assembled and reported by the Chemical Safety Board on its website).

Date	City and State	Fatality	Injury	Description	Fuel	Equipment Involved	General Facility Type	Ignition Source	Work Activity Involved
1984	Fairfield, IA	1	1	A fire started in the dust collector which is connected to the wheelabrator. The manufacturer of the machine believes gas from the fire built up in the wheelabrator and caused the third and biggest explosion; two smaller explosions took place in the dust collector. A maintenance worker on the third shift was found about 7 ft from where he was standing before the explosion. A witness saw his head hit the furnace. Another employee was thrown to the floor about 5 ft from where he was standing, sustaining a cut on his upper left arm. The fire marshal officer listed the cause of the fire as static electricity. Neither the dust collector nor the wheelabrator had been used for over a week due to repairs. A stone grinder, which is attached to the dust collector by a duct, was run twice for 8/10 of an hour. A spark from the stone grinder, or heat build-up from newly collected dust, may have started the fire. There was no evidence of fire in any of the ducts.	Aluminum dust	Dust collection system, wheelabrator (Goff airless shot blast equipment for polishing aluminum)	Primary metal industries	Spark	Inspection

(*Continued*)

Date	City and State	Fatality	Injury	Description	Fuel	Equipment Involved	General Facility Type	Ignition Source	Work Activity Involved
1984	Concord, NH	0	1	An employee was burned while standing at the feed end of a time saver sanding machine. Apparently a fire in the dust collector connected to the time saver machine blew back through the ductwork of the machine exhaust hoods and onto the employee's body. The time saver sanding machine collects steel and aluminum dust from the surface sanding of steel or aluminum parts processed through the machine. At the time of this accident the time saver machine operator was feeding steel parts into the sander. This process normally generates visible sparks at the point of sanding. In this instance the sparks became plugged at the local exhaust hood, near the sanding belt on the machine, producing a glowing moss of metal. An employee came into the room to watch the sanding machine operator try to dislodge the glowing moss with a rod while standing on the side of the machine. The employee saw the flexible rubber ductwork behind the sanding machine operator begin to smoke. The employee, now standing at the feed end of the sanding machine, attempted to turn off the electric current to the dust collection system. At this time, the flash fire occurred and burned the front of his body from the waist up. The aluminum dust apparently caught fire from the steel sparks feeding into a model 20.5 fb-55 torit dust collection system (dry type with 4 fabric filter bags). This system was located outdoors. The entire operation was not in compliance with NFPA #65 on aluminum processing, involving the grinding, sanding and polishing of aluminum materials.	Aluminum dust	Time-saver sanding machine	Equipment manufacturer - industrial	Spark	Sanding/ grinding

Date	City and State	Fatality	Injury	Description	Fuel	Equipment Involved	General Facility Type	Ignition Source	Work Activity Involved
1986	Grafton, OH	1	1	Two employees were mixing and bagging the following three compounds: aluminum powder (metallic al), iron oxide, and zirconium silicate. Another employee was bringing in drums of aluminum powder and placing them on a metal deck. One of the drums fell 2 to 3 feet, striking the metal floor and causing a spark. The spark ignited the metal dust, causing an explosion and fire.	Aluminum dust	mixing tank; drum of aluminum powder	Chemical manufacturer - Miscellaneous chemical products	Metal strike	
1987	Marion, NY	1	5	An employee was removing (grinding) Teflon coatings from aluminum fuser rolls when explosion occurred.	Aluminum dust	grinder, aluminum fuser rolls	Fabricated metal products	Unknown	
1987	Salt Lake City, UT	0	6	Employees were present when a surface dresser belt sander metal finishing machine had been used to surface aluminum plate parts. The particulate matter was drawn into a bag dust collector and the process was then changed to steel. A leadman saw smoke coming from the machine, but ignored it. After about an hour, profuse smoke came out of the dust collector and when the residue tray was opened a sparking fire was observed. When the shag door was opened, the fire level increased. An employee used an ABC fire extinguisher to put out the fire, causing the dust to explode into a fireball.	Aluminum dust	surface dresser belt sander metal finishing machine; dust collection system	Fabricated metal products	Unknown	Sanding/ grinding
1989	Robesonia, PA	1	0	Employee working on metal line brushing aluminum dust into mixing drum with hand brush; the static electricity ignited the dust, causing an explosion and fire.	Aluminum dust	Hand brush; mixing drum	Primary metal industries	Static	

(*Continued*)

Date	City and State	Fatality	Injury	Description	Fuel	Equipment Involved	General Facility Type	Ignition Source	Work Activity Involved
1989	Wilkes Barre, PA	1	2	Three employees were in the area of a dry-type dust collection device that was being used to collect aluminum and steel dust from a grinder/polisher. The steel dust created heat and sparks, which ignited the aluminum dust in the dust collector. One of the employees noticed smoke and flames and attempted to extinguish the fire using an H2O fire extinguisher. The fire increased rapidly in intensity, burning all three employees.	Aluminum dust	Dust collection system; grinder/polisher	Fabricated metal products	Spark	Sanding/grinding
1990	Newburgh, IN	0	7	An employee started welding cable tray brackets and conduit pipe supports while standing on top of an exhaust duct leading to the the outside of the building. The duct came from a wet type dust collection machine, and carried aluminum fines and dust. Due to a gap in the duct, aluminum fines were dispersed into open air. Welding sparks ignited the dust and fines, which created a fireball.	Aluminum dust	Lincoln arc welder #6251; Type N Roto-Clone dust collection machine	Electric Services	Spark	Repair/modification
1990	Hot Springs, AR	2	0	Witnessing paint peeling off the lower part of a hopper in a newly-installed aluminum dust collection system, two employees recognized this as an indication that the dust was on fire. The system was closed off and the dampers were shut so that the fire would be smothered from lack of oxygen. The fire appeared to have gone out, when an explosion occurred in the hopper.	Aluminum dust	Dust collection system	Primary metal industries	Unknown	Inspection

Date	City and State	Fatality	Injury	Description	Fuel	Equipment Involved	General Facility Type	Ignition Source	Work Activity Involved
1993	Swissvale, PA	1	0	An employee poured approximately 21 lbs of highly explosive aluminum powder into a rotating metal bucket that contained other powdered metal pigments. During the pouring, a dust explosion and fire occurred. Possible explanation: a buildup of static charge occurred during the pouring of the powder in combination with dust conditions in the metal rotating bucket.	Aluminum dust	Metal bucket	Stone, clay, glass, & concrete products	Static	Loading/ unloading
1993	Abingdon, MD	0	1	A barrel of aluminum dust stored adjacent to electrical panels somehow became contaminated with water, which exothermically reacted with the aluminum dust to form hydrogen gas. The lid blew off, and a cloud of the dust exploded. Nearby combustible material ignited.	Aluminum dust	Barrel	Fabricated metal products	Explosion	
1995	Louisville , KY	1	0	An employee was attempting to change an aluminum powder sensor for a storage/ convey tank. The electrical power to the sensor was still present when the unit went into operation and generated an aluminum cloud. The combination of these factors caused two successive explosions and ensuing fires in which the employee was killed. Electrical test instruments belonging to the employee were found in another location. The electrical power had not been eliminated and the pneumatic energy source had not been identified and controlled.	Aluminum dust	storage/conveyor tank	Primary metal industries	Unknown	Repair/modification

(Continued)

Date	City and State	Fatality	Injury	Description	Fuel	Equipment Involved	General Facility Type	Ignition Source	Work Activity Involved
1995	Olney, IL	2	25	27 employees were working inside a building that contained equipment for assembling aluminum alloy bicycle rims, including 3 conveyor-fed booths in which brushing and polishing were done. The booths had exhaust systems that deposited the aluminum dust into an enclosed basement. Heat or sparks from wheels that were being polished or brushed caused a fire that traveled from a booth to the basement, resulting in explosion and additional fire.	Aluminum dust	conveyor booths	Equipment manufacturer - transportation	Metal strike	
1997	Holmesville, OH	0	2	An explosion and fire occurred in the ductwork of the dust collection system associated with the belting of the aluminum alloy fire hose couplings (which also contain magnesium).	Aluminum dust	hardware	Fabricated metal products		
1999	Scottsboro, AL	1	1	A flash fire occurred during the processing of a military grade aluminum powder. The powder was being manually dumped into a bucket elevator and a shovel was in use to break up the lumps. The shovel lodged in the elevator and caused the elevator to choke; as dust levels built up, ignition occurred.	Aluminum dust	bucket elevator and shovel	Primary metal industries	Unknown	Loading/ Unloading
2002	Sevierville, TN	1	0	Explosion originated from inside a dry-cloth baghouse. Mechanical agitation and/or a static discharge appear to be the likely caused a spark that ignited the dust.	Aluminum dust	baghouse	Chemical manufacturer - Industrial inorganic chemicals	Static	

Date	City and State	Fatality	Injury	Description	Fuel	Equipment Involved	General Facility Type	Ignition Source	Work Activity Involved
2003	Twinsburg Twp, OH	0	0	An explosion in a dust collector at Commercial Alloy Sales on Highland Road ignited a fire about 5:20 p.m. Tuesday. No major injuries were reported, and a damage estimate was not available Tuesday night. The Twinsburg Fire Department reported the explosion came from duct work in the building and from the dust collector unit outside the building. The dust collector was blown apart, with heavy fire and flames coming from the unit, according to the Fire Department.	Aluminum dust				
2003	Champaign, IL	1	1	Two employees were using a torch to remove components of a dust collector. A small, smoldering fire started and one of the employees, using a Halon fire extinguisher, extinguished the fire. While extinguishing the fire, aluminum dust was dispersed into the air, causing an explosion. Both employees were injured in the blast and taken to the burn unit. One of the employees died from the injuries sustained in the blast. The other employee remains in critical condition.	Aluminum dust	torch	Fabricated metal products	Flame	

(Continued)

Date	City and State	Fatality	Injury	Description	Fuel	Equipment Involved	General Facility Type	Ignition Source	Work Activity Involved
2003	Huntington, IN	1	2	A major explosion and fire near a smelting furnace that killed one employee and severely burned another. 5 others were transported to emergency care facilities and released after treatmen. At about 3:00 P M a fire was reported in the number 5 furnace chip melt side well exhaust duct. The maintenance department was sent to evaluated the fire and shut the exhaust ventilation system down to allow the fire in the pipe to self-extinguish as this was the plant policy for aluminum dust fires in the pipe. About 8:10 or 8:15 the employees started the system up after the pipe had cooled and been cleaned out by the employees. Employees cleaned the ash out of the pipe started multiclone blower and checked for proper operation; turned on the dry chip blower #3; then the auger checked multiclone for proper operation; they were starting to leave when a big flash came out of the well where the chips feed in this moved the dust off of the ceiling & ignited it causing a ball of fire to move through the room it also destroyed equipment outside the building & started many small fires inside the building in that area.	Aluminum dust	multi-cone pipe, multi-cone blower, dry chip blower	Primary metal industries		

Date	City and State	Fatality	Injury	Description	Fuel	Equipment Involved	General Facility Type	Ignition Source	Work Activity Involved
2004	West Bend, WI	0	4	Employees smelled smoke coming from the air scrubber machine. The air scrubber collects and directs filings and air from several polishing or buffing stations. As the equipment was turned on, air was introduced to a possible smoldering fire contained to the duct work. An explosion occurred. A coffee urn buffing machine is suspected to have send sparks into duct work and dust collector to initiate the incident	Aluminum dust	Buffing machine, Air scrubber, duct work	Fabricated metal products	Spark	Buffing/ polishing
2005	Hartford, WI	0	2	Employees at Midas performing operations associated with an aluminum spray booth suspected that flow was blocked to the dust collector adjoining the booth. During inspection, employees found the vent pipe to the dust collector glowing. One of the employees shook the pipe causing aluminum dust to dislodge lead to a dust explosion.	Aluminum dust	Dust collection system	Manufacturing		
1990	Steele, AL			Aluminum dust explosion involving scrap processing equipment.	Aluminum dust				

(*Continued*)

Date	City and State	Fatality	Injury	Description	Fuel	Equipment Involved	General Facility Type	Ignition Source	Work Activity Involved
2002	Hempfield, PA	0	3	A spark from a nearby machine apparently got into a dust collector, setting fire to cobalt and tungsten dust that was being collected to be recycled.	cobalt and tungsten dust				
1992	Baltimore, MD	0	1	On Thursday July 30th a panel door blew off the dust hopper and the unit was shut down. Metal dust was still smoldering within the equipment on the first day when two maintenance employees attempted to replace collection bags within the equipment. Opening the hopper caused air to ignite the metal dust that was smoldering. A flash fire occurred.	Ferro-manganese	hopper	Primary metal industries		Maintenance
1987	East Chicago, IN	1	2	Employees were removing dust bags from a Pangborn dust collector located in the 4-A roll shop at the company's plant #2. The dust collector had two rooms: a motor room for the exhaust fan, and a bag room, in which bags hung from the ceiling over a hopper that was protected by a grate. a contractor hired to remove the dust with a vacuum truck, was in the motor room. His head and shoulders were through a 14 by 34 inch porthole leading into the bag room. He was taking bags from one employee and throwing them out of the dust collector. The vacuum truck was connected to the hopper and was in operation when an explosion occurred in the hopper and flames shot upward to where the three employees were working.	Iron dust	Dust collection system	Primary metal industries	Unknown	Repair/modification

Date	City and State	Fatality	Injury	Description	Fuel	Equipment Involved	General Facility Type	Ignition Source	Work Activity Involved
1996	Pittsburgh, PA	1	0	Employee was using a skip hoist and monorail system to raise a 1,000 lb (55 gallon) drum of metal iron dust. The drum overturned and dust spilled throughout the work area, including an operating fork lift, and a dust explosion occurred.	Iron dust	skip hoist, monorail system, 55 gallon drum	Primary metal industries	Unknown	Loading/ Unloading
1998	Garland, TX	1	1	Two employees of a steel recycling crew were removing an overhead crane and beam hoisting assembly. One of the employees was in a manlift (3' x 10' x 4.6') about 8 feet above the ground. He was standing on the lift above a storage bin filled with highly combustible magnesium shavings, when he momentarily lit the cutting torches. Sparks from the torch ignited the shavings, which burned in excess of 500°F. Both employees were burned, one fatally.	magnesium	manlift; storage bin	wholesale trade-durable goods	hot surface	Cutting
1990	Avon Lake, OH	2	1	Employees were mixing magnesium powder (metallic), calcium carbide, and sodium chloride. They were using a steel pipe to loosen a blockage in a magnesium batching hopper. The pipe caused a spark to ignite the magnesium, causing a flash fire and explosion.	Magnesium dust	Magnesium batching hopper	Chemical manufacturer - Miscellaneous chemical products	Metal strike	
1990	Tamaqua, PA	0	2	One employee was manually dumping magnesium powder into a pneumatic feed that would feed larger particles to a hammer mill. Another employee was sweeping. The magnesium dust ignited and caused a flash fire.	Magnesium dust	Pneumatic feed, hammer mill	Primary metal industries	Unknown	

(Continued)

Date	City and State	Fatality	Injury	Description	Fuel	Equipment Involved	General Facility Type	Ignition Source	Work Activity Involved
1992	Gary, IN	1	0	An employee was dumping powdered magnesium into a mixer on a platform. A flash fire occurred when a static spark from the dust collector hose ignited a suspension of magnesium above the mixer opening. The magnesium suspension was created when the second drum was dumped. The employee either jumped or fell to the concrete floor, gashing his head. He suffered third-degree burns over more than 90 percent of his body. He was taken to the Chicago Burn Center, where he died later that day.	Magnesium dust	mixer, dust collector hose	Equipment manufacturer - industrial	Static	Loading/ unloading
1996	Lindon, UT	0	2	Employees were blending magnesium and lime in a large roller drum and hopper. The mixture was then taken from the hopper, up an auger, to a shaker unit, and then dumped onto a belt that carried it into a 3-compartment "pig trailer". Dust levels were controlled by a local exhaust ventilation system. An explosion and flash fire occurred within the hopper, traveled to the roller drum, the shaker unit, the pig trailer, and the open drums of magnesium. Fire put out using talc powder.	Magnesium dust	hopper, roller drum, shaker unit, "pig trailer"	Chemical manufacturer - Miscellaneous chemical products	Unknown	Loading/ Unloading

Date	City and State	Fatality	Injury	Description	Fuel	Equipment Involved	General Facility Type	Ignition Source	Work Activity Involved
1999	Quincy, IL	1	0	Employee was using an electric sander, creating magnesium dust. The sander caught on fire, and the employee dropped it onto the table, which was covered in the dust. A flash fire occurred.	Magnesium dust	Black & Decker electric sander, model 7441 (with 80 grit sandpaper)	Fabricated metal products	Fire	Sanding/ grinding
2000	Fort Wayne, IN	1	0	Employee was grinding magnesium parts on a type N Roto Clone grinding machine, when light fixtures fell from the ceilings, causing a spark that ignited the magnesium dust in the air and on the grinding machine.	Magnesium dust	Grinding machine	Primary metal industries	Hot surface	
2004	Pampa, TX	1	2	The company manufactures oilfield equipment. An explosion and flash fire occurred in a processing area where igniter material is mixed. The mixing process is for the manufacturing of igniters for oil field setting tools used to set oil well plugs. The incident occurred as employees were grinding magnesium powder and potassium perchlorate.	Magnesium dust	Grinding		unknown	Grinding
1995	Elmira, NY			Magnesium dust explosion fire involving a dust collector.	Magnesium dust				

(*Continued*)

Date	City and State	Fatality	Injury	Description	Fuel	Equipment Involved	General Facility Type	Ignition Source	Work Activity Involved
1988	Cleveland, OH	1	8	Three production operations were functioning: the #3 screening system, where an inert material is ground, sized, and screened; the dryer, where ferromanganese is stabilized; and the nugget system, where manganese and iron powders are mixed with a "glue" to make "nuggets." A series of explosions occurred, probably originating at the nugget machine where sparks/fire entered the outside bag dust collector through the duct work, causing a violent explosion within the collector. The top of the collector blew, and flaming pieces of metal flew onto a wooden portion of the plant roof. The fire spread rapidly because of the wood roof, the combustible dust/powders in the plant and, possibly, because of ruptured gas lines.	Manganese dust	Dust collection system	Equipment manufacturer - industrial	Unknown	Loading/ unloading
1988	Lima, OH	1	2	A 30-ton ladle containing 50,000 lbs of molten iron snapped and spilled an unknown quantity of iron onto the floor at the south end of a 20ft pit in an iron foundry. A flash fire was immediately created by the combustion of the molten iron and foundry dust.	Metal dust	30-ton ladle, Alliance overhead cab crane	Primary metal industries	Hot surface	Loading/ unloading
1990	Baltimore, MD	0	2	After a maintenance shutdown, an employee restarted a nitride line. An explosion occurred within the elevator of the line, and hot dust escaped through the vent panel.	Metal dust	Elevator	Primary metal Industries	Unknown	

Date	City and State	Fatality	Injury	Description	Fuel	Equipment Involved	General Facility Type	Ignition Source	Work Activity Involved
1990	Baltimore, MD	0	1	Employee was working with welding contractor, attaching guardrails to a fixed platform and ladder. Sparks ignited a garbage bag, which was removed; however, a spark entered the dust collector and ignited the cloud inside. The employee used fire extinguisher; then an explosion occurred.	Metal dust	Welding device; dust collector	Primary metal industries	Fire	
1991	Gardena, CA	0	2	An employee was cutting the metal rod of a machine press with an acetylene torch while another employee was on a forklift holding the machine press steady. As the metal rod was totally cut off, the forklift jerked and a cloud of dust that had accumulated on top of the lighting fixture drifted downward. As the dust came in contact with the acetylene torch flame, it ignited and formed a fireball that burned both employees.	Metal dust	Machine press, acetylene torch	Rubber & plastic products	Flame	Repair/ modification
1996	Ukiah, CA	1	0	An employee was blending dry metal powder in an open blender in the powder testing/training area of the plant. He was using an equipment prototype when the dry metal powder suspended in the air resulted in a flash fire.	Metal dust	Open blender/ mixer	Equipment manufacturer - industrial	Unknown	Testing
1996	Gallatin, TN	0	1	Sparks from an abrasive disc cutting operation were drawn into the dust collection piping system. Sparks then entered the dust collector itself, where they ignited the steel powder inside.	Metal dust	Cutting device, dust collection system	Primary metal industries	Spark	Cutting

(*Continued*)

Date	City and State	Fatality	Injury	Description	Fuel	Equipment Involved	General Facility Type	Ignition Source	Work Activity Involved
1997	Brooklyn, OH	0	2	A possible fire or spark was produced by a buffing machine(s), where it apparently got into the dust exhaust system. Once this spark or fire got into the exhaust system, it got into the baghouse, caused a small explosion, whereupon the dust in the baghouse was shaken loose and another much larger explosion occurred, sending the force of the explosion and a resultant fire back through the exhaust system. Two employees were severely burned.	Metal dust	Electroplating, plating, polishing, Anodizing, and coloring	Fabricated metal products		
1998	Latrobe, PA	0	0	Firefighters used dry chemicals and sand to put out a particle dust fire, which was contained inside a dust collector (round, about 10-ft in diameter and 15-ft tall).	Metal dust	Dust collection system	Primary metal industries		
2000	Kelseyville, CA	0	1	Employee was using vacuum system to extract metal discs from chemical reaction chamber at steam-powered electrical generation plant. Explosion occurred in the reaction chamber. Explosion caused by ignition of finely divided metal particles; ignition source probably a static spark -- system was not grounded or bonded.	Metal dust	Stretford style chemical reactor, used to Clean steam and reduce SO2	Electric Services	Static	
2003	Blendon Township, MI	0	0	A fire at a steel polishing plant destroyed the building and its contents after dust in a machine sparked flames that couldn't be knocked down by employees. The fire investigation points the source of the flames to a polishing and buffing machine that turns raw metal into polished steel plates, the fire chief said.	Metal dust				

Date	City and State	Fatality	Injury	Description	Fuel	Equipment Involved	General Facility Type	Ignition Source	Work Activity Involved
2003	Hanover, PA	0	0	A fire originated in a dust collection system and ignited a metal dust bin and several filters. The dust collector vacuums dust away from the stations where workers were polishing metal parts for motorcycles.	Metal dust	Dust collection system	Fabricated metal products		Polishing
2003	Rapid City, SD	0	0	The fire broke out in a dust-containment system at the plant and was limited to that area.	Metal dust				
2004	New Washington, OH	0	1	An explosion occurred when a powder coating gun ignited at the plant. Four people were in the building at the time. The only injury was to a worker who got the hair on his arm singed. The filtering system had extensive damage.	Metal dust		Primary metal industries		
1999	Albany , NY	0	0	A dust-collector fines fire at the Albany Casting Plant was caused by housekeeping and a bad dust collector at a cost of $250,000.	Metal dust				
2004	Auburn Hills, MI	0	0	A furnace in the foundry malfunctioned. It sent excessive heat to the system of overhead vent ducts that serve as a large vacuum cleaner, sucking up dust and airborne metal shavings. The excessive heat ignited the metal dust in the system and resulting flash fire spread to the bag house on the roof of the facility.	Metal dust	Dust collection system	Manufacturing	Excessive heat	None

of some of the case studies reported. It is worthwhile for the reader to read through these to gain perspective on the nature and extent of the types of explosion and fire incidents that have occurred with metals and then consider the vulnerability of the operations they have to deal with.

6.6 Good Industry Practices for Prevention and Risk Mitigation

6.6.1 General Good Practices

The following are generally considered by OSHA, NFPA, NIOSH, fire experts and the insurance industry as good industry practices and guidelines for minimizing the risks of dust explosions and fires. The practices discussed below are applicable to all types of explosible dusts. The reader should also refer to National Fire Protection Association Standard 484.

Assume and treat all equipment that handle combustible dusts, as well as any rooms or buildings where combustible dusts can be present and might be put into suspension, as having a dust explosion hazard.

- Implement a management-of-change process in all facilities handling combustible dusts to be certain that no changes occur that could increase the severity or consequence of an existing dust hazard or introduce a dust hazard where none previously existed. Examples of such changes include:
  - Adding new equipment such as blenders, grinders, cutting tools, dust collectors, cyclones, etc.;
  - Increasing temperatures in the process that could result in drier material being handled;
  - Adding new materials;
  - Changing product formulation by adding combustible materials or reducing the proportion of inert materials;
  - Making process changes that reduce the particle size of in-process materials
- In situations where process, equipment, raw material, or product changes are planned that could significantly change the dust properties, one should retest the dust for its explosibility.
- Ensure that the management-of-change process has the following characteristics at a minimum:
  - Provides a method for identification of changes that should be subject to the management-of-change process
  - Provides documentation of the proposed change

- Provides a formal analysis of the loss prevention considerations involved in the proposed change
- Identifies the need for updated personnel training
- Provides for communication of the change and the loss prevention consequences to appropriate personnel such as maintenance, operators, safety, and emergency responders
- Establishes any administrative procedures needed (documentation, checklists that cover hazards, training, etc.)
- Identifies any required authorizations

Where potential for a dust explosion exists, eliminate the potential or minimize the consequences using one of the following methods:

- Control fugitive dust releases using enclosures, collection systems and equipment design.
- Locate dust producing operations in areas separated from different hazard occupancies by construction (dust-tight and explosion-resistant barriers) or distance.
- Minimize chances for dust accumulation by arranging building elements and equipment to reduce the likelihood of dust accumulations. Employ features such as smooth, easily cleaned walls, boxed in or covered horizontal surfaces (beams, joists, etc.), and surfaces sloped a minimum of 60° from the horizontal.
- Where fugitive dust release and accumulation exist in buildings, design the structure to safely vent the potential explosion using damage-limiting construction.
- Locate dust collection and transfer equipment outside, away from important buildings and utilities.
- Construct equipment that processes or transfers combustible particles to contain or safely vent a potential explosion.
- Where explosion venting or containment in equipment is not possible, eliminate the oxygen in the system by inerting, or install an explosion-suppression system.

Practice effective maintenance of production and protection equipment. An effective maintenance program should address the following:

- Identify and eliminate fugitive dust sources continually.
- Test and maintain spark detection and extinguishing systems, explosion isolation devices, and relief vents to ensure

they are in working order per manufacturer's guidelines, or at least monthly.
- Test and maintain metal and non-metal detection and extraction equipment to ensure they are in working order, at least quarterly.
- Check belts and rotating equipment for alignment at least quarterly to prevent these becoming a source of friction heating.
- Lubricate bearings and rotating equipment (fans, blowers, size-reduction equipment) in accordance with manufacturer's guidelines, or at least quarterly.
- Remove accumulated dust on rotating equipment bearings and components to insure free movement and prevent friction heating, at least quarterly.
- Assign accountability and keep accurate records.

The workplace should have practices and protocols that ensure a comprehensive dust fire and explosion awareness program exists at all sites where combustible dust exists either within closed processing systems or as fugitive dust within buildings. The following are important elements to include in the practices:

- Basic education to promote awareness and understanding of the hazards of combustible dusts
- Instruction of new employees on the particular hazards and on precautions relevant to their departments
- A minimum of annual instruction, drill, and familiarization of the local public fire service and/or internal firefighting teams
- Periodic refresher training for all facility personnel

Strict programs should be implemented to control potential dust ignition sources where combustible dusts may be present. As examples:

- The facility should ensure that all electrical equipment is rated Class II, Division 1 or 2, or Zone 20, 21, or 22 per NFPA 70, the National Electric Code, Articles 500, 502, and 506, as appropriate, or an international equivalent.
- Rely on a hot work permit system to manage all hot work operations.
- Prohibit smoking and open flames.

- Provide grounding and bonding of metal components with a resistance of less than 1 x 106 ohms to ground. At a minimum, annually, check for continuity of the metal components and subject all electrical equipment to an initial infrared (IR).

Facilities should prohibit the recycling of air material separator exhaust to buildings or rooms, except where either "a" or all of "b through h" apply:

a. The return air duct discharges into an area that does not contain fugitive dust, combustible equipment or storage, combustible construction, high-value equipment, or equipment that is critical to production, OR
b. Install a filter downstream of the dust air separators that prevents return of dust to the enclosure with a minimum efficiency of 99.9% at 10 microns AND
c. Install a device to measure pressure-drop across the filter with an alarm to indicate when the filter needs to be cleaned or replaced AND
d. Provide support for the filter with a wire mesh screen or other method that allows the filter to withstand a pressure equal to or exceeding the value of P_{red} for the piece of equipment directly upstream from it AND
e. Provide explosion isolation between the building and the last dust collector in the system (the one furthest downstream) AND
f. On explosion-isolation system activation, shut down any connected dust- collection equipment AND
g. Ignitable vapors, gases, or hybrid mixtures are not involved AND
h. The dust-collection system meets the protection requirements in other sections of this data sheet.

Where these features are present, the recycling of air material separator exhaust would not cause the building/room to require explosion protection features such as venting, etc. (other factors present in the building/room could create that need, however).

Careful considerations should be given to room/building explosion hazards. Isolate areas handling combustible dusts from other less hazardous occupancies by separation with construction or distance to minimize damage from the potential explosion or fire. Typical areas needing isolation would be where fugitive dust is not readily controlled for example, grinding, sanding, sawing, open conveying, filing open bins, etc. Excluded would be for example, rooms containing properly vented dust collectors, spray dryers, fluid bed dryers, etc.

Isolate areas handling combustible dusts using the methods listed below in order of preference:

- Detached outside location at least 50 ft (15 m) away from an important building or facility.
- Along an exterior wall of an important building, preferably at a corner to limit exposure.
- Inside an important building on the first floor, either at an exterior corner or along an exterior wall. Avoid locations on upper floors of multistory buildings. Where above-grade locations are unavoidable, ensure the floor and ceiling of the room have the same pressure resistance as the walls.

Construct buildings in which a combustible dust hazard exists of noncombustible Class 1 materials. Ensure physical barriers that isolate dust hazard areas are sealed dust tight using noncombustible materials. Where the barriers have a fire resistance, ensure the seals maintain that rating. Use door seals, window seals, positive room pressurization, etc. to prevent dust from entering and accumulating in adjacent areas that otherwise do not have their own sources of fugitive dust. Ensure physical barriers that isolate dust hazard areas have explosion resistance sufficient to prevent failure before the pressure can be safely vented. Also do not allow openings in explosion-resistant walls. Where openings cannot be avoided, keep all doors in these walls normally closed and make sure they have the same explosion-resistance as the walls themselves.

In buildings where some fugitive dust is likely despite process design and equipment upkeep, arrange building elements and equipment to reduce the likelihood and amount of dust accumulation by any or all of the following, as appropriate:

- Provide smooth interior walls with minimal ledges
- To the extent practical, provide horizontal surfaces such as girders, beams, ledges, and equipment tops with a sloped cover having a smooth finish, to shed dust settling out of the air
- Slope covers at an angle of 60° from horizontal, unless a lesser slope is known to be sufficient

- Enclose or box in overhead structural steel that is out of the reach of normal vacuuming or sweeping operations and that has horizontal ledges (such as I-beams or unshaped channels in the up or sideways position) with a noncombustible material to eliminate pockets for dust accumulation

Buildings can be considered to not have a combustible dust hazard if they contain combustible dust-handling equipment that is designed and protected to control or safely vent an internal explosion. The following are considered prudent precautions where quantities of fugitive dust could be expected in new operations or are excessive in existing operations:

- Survey process equipment to identify the sources of dust release.
- Modify, repair, or replace equipment to eliminate or at least reduce dust escape.
- Provide permanent vacuum pickup points at the locations that release dust, such as grinding, buffing, bag dumping, open transfer points in conveying systems, and other equipment/locations where large quantities of dust are liberated frequently. In some cases this may require construction of a ventilated hood or containment enclosure for existing equipment.
- Operate closed dust-handling systems under a slight negative pressure to reduce dust escape.
- Conduct extra housekeeping in existing operations while equipment is being modified.

For existing buildings containing small, localized amounts (less than 5% of the building area and in no case exceeding 1,000 ft^2 of fugitive combustible dust) can be considered tolerable without explosion damage-limiting construction, if the following conditions exist:

- The fugitive dust escape and accumulation rate is very low, i.e., less than 1⁄16 in. (1.6 mm) for a dust with an approximate bulk density of about 36 lb/ft3 (580 kg/m3) per three month period AND
- The cleaning frequency is high enough to permit one scheduled cleaning to be missed without allowing dust accumulations to reach the unacceptable level of 1⁄16 in. (1.6 mm) noted above.

Facilities should follow good engineering practice in sizing and constructing pressure-relieving and resistant walls. Proper design of the explosion vent relief pressure (P_{stat}) should be based on as low as the wind-resistance design will permit. In a low-wind area, P_{stat} can be as low as 20 psf (0.01 bar), whereas in higher wind areas 30 to 40 psf (0.015 bar) is more typical.

Where a thorough engineering study shows that explosion-venting walls alone cannot provide the needed explosion venting area, roof vents may be used to provide a portion of the needed vent area if snow and ice are not allowed to accumulate on the vent. Any of the following are considered acceptable methods:

- Position the explosion vents at a minimum 60° angle, either on a roof pitched at that angle or as a projection above the roof line. For vents projected above the roof line, the effective vent relief area to be used in vent sizing calculations is the smallest cross sectional area the combustion gases would have to flow through.
- Provide heat tracing along the perimeter and across the surface area of the explosion vent. Leave heat tracing on permanently, or automatically actuate the system whenever the outside temperature drops to 0°C (32°F) or lower. Ensure the wiring for the heat trace cabling incorporates enough slack to allow the explosion vent to deploy as intended.
- Provide explosion venting panels without insulation to allow building heat to melt away snow or ice.
- Expect condensation under the explosion vent and take measures to ensure condensation does not cause problems.

In situations where explosion venting devices swing out of the way rather than rupture, use gravity or mechanical devices to ensure they cannot reclose. To prevent the uncontrolled release of explosion vent panels where they could become missile hazards, take the following steps:

- Provide tethering cables to limit the vent movement.
- Attach tethering cables to no more than two corners, making sure the tethered corners are adjacent.
- To prevent the tethered panels from swinging back into the vent opening after the explosion, make connections at the side or bottom of the panel rather than the top. This allows the panels to completely swing out of the way of the vent opening, leaving an unobstructed path for the

combustion gases to flow along during the explosion, and for fresh air to flow into the protected enclosure after the explosion.

- Consider the explosion vent as a hinged panel when performing calculations to determine the vent area and vented explosion pressure. This will account for the hindrance to the venting process created by the short tethers.

It is also good practice not to attach sprinkler system piping to any wall, ceiling, or roof that could be displaced by the pressure of a room or building explosion.

6.6.2 Considerations for Operations and Maintenance

In buildings where fugitive dust is released despite process design and equipment upkeep, the following actions should be incorporated into a fugitive-dust-control program:

- Assign accountability and keep records
- Commit time, money and resources regularly
- Create a maintenance schedule and review it periodically to ensure it is adequate
- Pay attention to eliminating accumulations above floor level, such as equipment tops and building structural members because this dust is more likely to become suspended (airborne) and create an explosible cloud if it is disturbed
- Use vacuum removal wherever possible using a central, portable, or truck-mounted vacuum system as appropriate
- Where vacuuming is impractical, sweeping or water washdown are other acceptable options

Where compressed air blow-down is the only practical method for cleanup, use the following precautions:

- Perform air blown-down frequently enough to prevent hazardous accumulations of dust
- Limit the extent of blow-down to small areas at a time with as low a volume and pressure of air as possible
- Shut down electrical equipment not suitable for Class II, Division 2 hazardous locations
- Prohibit open flames and hot work, and ensure no hot surfaces exist

Any accumulations of fugitive dust present a potential for a secondary explosion and must be eliminated in buildings without damage-limiting construction (DLC), and controlled in those with DLC. Good practices call for:

- Base dust-cleaning frequency on preventing accumulations of more than 1⁄16 in. (1.6 mm) thickness over more than 5% of the room floor area
- Use fugitive dust exceeding the above, either in thickness or area, as a trigger to initiate cleanup
- Ensure the total area of accumulation does not exceed 1,000 ft^2

Note that the thickness of 1⁄16 in. (1.6 mm) is based on a typical wood or agricultural dust with a bulk density of about 36 lbs/ft^3. The thickness of dust of different density can be based on a ratio of the bulk densities.

A dedicated program for inspecting explosion venting devices periodically should be implemented to prevent their condition or mobility from being impaired. Possible impairments include corrosion, improper painting of movable parts or rupture membranes, icing, snow accumulation on or in front of vents, obstruction by pipes, wires, or other utilities, permanent or temporary equipment located next to the venting device. Any of these can increase the effective vent-relief pressure (P_{stat}) and cause explosion-resisting walls to fail and/or damage the building structure during an explosion.

6.6.3 Assessing and Mitigating Equipment Explosion Hazards

The following is a list of the import Dust Explosion Standards. European standards are available from agencies throughout Europe, such as the British Standards Institution. U.S. standards are those embodied in NFPA publications.

- EN 14491:2006 Dust explosion venting protective systems - This European standard specifies the basic requirements of design for the selection of a dust explosion venting protective system. It covers vent sizing to protect an enclosure against the internal pressure effects of a dust explosion, flame and pressure effects outside the enclosure, recoil forces and the influence of vent ducts.
- EN 14373:2005 Explosion suppression systems - General requirements for explosion suppression components,

evaluating the effectiveness of an explosion suppression system, evaluating the scale up of an explosion suppression system, evaluation and development of design tools for explosion suppression systems, instructions for installation of an explosion suppression system and maintenance instructions for an explosion suppression system.

- EN 14797:2006 Explosion venting devices - This European standard specifies the requirements for venting devices used to protect enclosures against the major effects of internal explosions arising from the rapid burning of suspended dust, vapor or gas contained within. It includes the requirements for the design, inspection, testing, marking, documentation and packaging.
- EN 1127-1:2007 Explosive atmospheres, basic concepts and methodology - Specifies methods for the identification and assessment of hazardous situations leading to explosion and the design and construction measures appropriate for the required safety. This is achieved by hazard identification, risk assessment, reduction of risk and information for use.
- EN 15089:2009 Explosion isolation systems - This standard sets out the criteria for alternative test methods and interpretation means to validate the efficacy of explosion isolations. It covers general requirements for the explosion isolation components, evaluating the effectiveness of an explosion isolation system and evaluating design tools for explosion isolation systems.
- NFPA 68:2007 - NFPA 68 applies to the design, location, installation, maintenance, and use of devices and systems that vent the combustion gases and pressures resulting from a dust deflagration within an enclosure so that structural and mechanical damage is minimized. As part of its revision, the 2007 standard includes a performance-based option, including two new chapters supporting the performance-based aspect.
- NFPA 69:2008 - NFPA 69 applies to the systems and equipment used for the prevention of explosions by the prevention or control of deflagrations, including two new chapters to incorporate a performance- based approach for compliance, in addition to the more traditional specification-based option. The concept of safety integrity levels (SIL) for safety instrumented systems was added to the standard to address

aspects of system reliability and chapters on Detection and Ignition Control and Suppression were updated and expanded. Requirements for explosion prevention control by isolation have been expanded in two chapters – one on passive isolation techniques and another on active techniques.

- NFPA 654:2008 - This standard provides requirements for safety to life and property from fire and explosion associated with the manufacturing, processing, blending, pneumatic conveying, repackaging and handling of combustible particulate solids or hybrid mixtures.

The following are considered good industry practices for assessing and mitigating dust explosion risks.

First, locate all equipment having an explosion hazard outdoors whenever practical. In situations where equipment has been identified as having an explosion hazard, take one of the following steps. Either eliminate the explosion or, mitigate the explosion hazard using explosion venting, suppression, containment, or vacuum operation, or a suitable combination of these.

In situations where equipment handles coarse material, prevent dust from being generated during material handling by suitable methods or pre-clean the coarse material to remove all fine particulate matter from most of the process equipment and surfaces.

Where equipment handles coarse material, prevent dust from being generated during material handling by eliminating airborne dust inside the processing equipment using a liquid mist (water or other compatible liquid). This can be accomplished as follows:

- Apply the liquid suppressant at a point in the process that involves substantial turbulence, e.g., at a discharge spout, to ensure the suppressant is thoroughly mixed in with the material stream.
- Install an interlock that will shut down the solid process stream on suppressant spray-system malfunction if this system is being relied upon as the sole means of hazard control.
- Confirm the effectiveness of the system to eliminate dust in the equipment by visually checking equipment while it is running, for example, by opening access or inspection ports.
- Develop a maintenance and inspection program with written procedures and records of completion to ensure the

dust-suppressant spray system is functioning properly whenever the solid process stream is being handled.

A good practice is to inert the combustible dust by mixing it with a non-combustible dust. This is known as "phlegmatization". Phlegmatization can be accomplished first performing a test to determine whether the mixture is non-explosible per ASTM E1515, *Standard Test Method for Minimum Explosible Concentration for Combustible Dust* or international equivalent. Ensure the noncombustible dust does not separate from the combustible dust during further material handling.

One can also reduce the oxygen level in the process with an inert gas (see next chapter on inerting practices). Where an inert atmosphere is to be used to safeguard pneumatic conveying of freshly ground or freshly manufactured light metal powder (aluminum, magnesium, titanium, zirconium):

- Provide a minimum oxygen concentration of at least 1% to ensure a stable oxide layer can be formed to passivate the dust.
- Where the metal powder is not to be exposed to air, or where testing has shown a lower oxygen level is acceptable, concentrations lower than 1% are acceptable.

The metal powders titanium, magnesium, zirconium, uranium, and thorium, will ignite in pure carbon dioxide. Use argon, helium, or nitrogen (except for titanium, which can be ignited in pure nitrogen) for inerting processes handling these materials.

As an alternative to an inert atmosphere for handling freshly manufactured light metal powder (aluminum, magnesium, titanium, zirconium) take the following steps:

- Employ wet collectors (water wash) rather than dry type
- Remove the accumulated sludge and dispose of it in a safe manner before shutting down the collection system
- Continue collector ventilation at all times until the metal/water sludge is removed from the collector

Aluminum in particular has a very low MIE (as low as 0.1 mJ) making it easily ignitable. Metal powders in general can have high K_{st} and can be very difficult to provide adequate explosion venting for. Aluminum powder when damp with water can form small amounts of hydrogen so it needs to be kept wet at all times.

6.6.3.1 *Explosion Venting*

Explosion venting offers a simple and practical method of protecting plant which is not strong enough to withstand the pressure developed by an explosion. If the relief panel is of a suitable size and design, it will safely vent the explosion without significant damage to the vessel or injury to nearby plant personnel.

Care must be taken in vent sizing and placement and should be in accordance with the appropriate national safety standard (e.g. EN14491 or NFPA68). Explosion reliefs for facilities containing flammable dusts must also be dust tight and have sufficient mechanical strength to resist the wear to which they are subjected and internal pressure fluctuations. On dust collectors in particular, the stress imposed by reverse jet cleaning must be considered in the vent design. Reliefs usually take the form of bursting panels, although explosion doors are common on older plants. Bursting panels provide good, dust-tight seals but may allow the entry of air into plant after an explosion, which may give rise to a subsequent fire. Explosion doors can usually be engineered to shut after operation, but are more difficult to make dust tight and have considerably lower venting efficiency due to their weight. Proprietary panels and doors, explosion tested to the appropriate standard (e.g. EN14797) are recommended. Extensive research over the past two decades has provided soundly based calculation methods to determine the vent area required. Design to is based upon:

- The volume of the vessel to be protected
- The shape of the vessel to be protected
- An estimate of the vessel strength
- The explosion indices (K_{st} & P_{max}) of the dust hazard
- The opening pressure of the vent relief
- The size of the vent relief
- The vent relief efficiency
- The length and shape of any associated vent ducting

When an explosion vent opens as a result of a dust explosion, a fireball of considerable volume, typically ten times the volume of the vessel, will be ejected over a considerable distance. If the vent opens inside a building, anyone in close proximity is likely to be seriously injured. In addition, any secondary dust (i.e. dust lying as a layer on the tops of bins, ledges, etc.) may become involved in a much larger secondary explosion. It is therefore normal practice to duct the vent to outside through to atmosphere via a short length of purpose built ducting. Consideration must be given to flame travel outside and the back pressure effect of the ducting. Alternatively, use flameless explosion vents where a flame arrestor element stops the passage of burning material but still allows the explosion pressure to be abated.

Where indoor equipment with an explosion hazard cannot be relocated outside, provide explosion venting in one of the following ways:

- Locate the vessel next to an exterior wall and vent the explosion to the outdoors via a short vent duct
- Vent the explosion to the surrounding area through an approved explosion quench pipe

Note that a vent area increase (compared to an open or unobstructed vent) will be needed to accommodate reduced venting efficiency caused by the quench pipe.

Install vent ducts that redirect the combustion products from the vent to a safe area in accordance with the following practices:

- Route the vent to a safe outdoor area
- Permit no bends in the duct
- Limit the length to diameter ratio (L/D*) of the duct to 1
- Ensure the vent duct is at least as strong as the P_{red} design of the vessel
- Permit no closures on the discharge end of the duct that obstruct the free venting of the discharged material

Do not provide explosion venting if the dust or its combustion products are poisonous, radioactive, corrosive to nearby equipment or structures, or for any other reason should not be released from an otherwise closed system. Use alternatives to venting, such as explosion hazard elimination, explosion suppression, explosion containment, or high vacuum operation.

Where a vessel requiring explosion venting contains significant obstructions to the free flow of gases, provide several distributed vents at different places on the vessel rather than one large vent of the same area. Where a vessel contains both suspended combustible dust and ignitable liquid vapors or flammable gases (a hybrid mixture), the reactivity of the mixture will require larger vent areas than the combustible dust on its own.

The presence of a flammable vapor or gas can be ignored if the concentration is at or below 5% of its lower explosive limit (LEL). Where the gas concentration exceeds 5% of its LEL, determine the reactivity of the mixture by testing.

6.6.3.2 *Isolation*

One should avoid multiple pieces of interconnected equipment containing a dust explosion hazard. Properly protected equipment can fail when an explosion propagates from another piece of equipment. Isolation systems can prevent that condition.

Provide explosion isolation on all connections of vessels (or vessel groups) designed to contain the explosion pressure (explosion-resistant design). Also provide explosion isolation on all connections between vessels (or vessel groups) individually protected by explosion venting (or other mitigation method) but subject to unacceptable property damage or business interruption by a propagating event.

There are various forms of isolation systems which can be applied as effective flame barriers. One simple example is a rotary air lock (see figure 6.1).

These devices should be installed as follows:

- Ensure the angle between adjacent vanes and the shape of the housing allows two vanes per side to be engaged (near the housing wall) at all times.
- Ensure the vanes (including the tips) are made of metal and have a thickness of at least 1⁄8 in. (3 mm).
- Ensure the gap between the tips of the vanes and the housing between 0.2 and 0.25 mm (less or equal to 0.1 mm for aluminum dust).
- Interlock the rotary air lock to automatically stop in the event of an explosion, to prevent the passing of burning matter. An interlock is not required if burning matter would not cause a second fire or explosion of any significance.

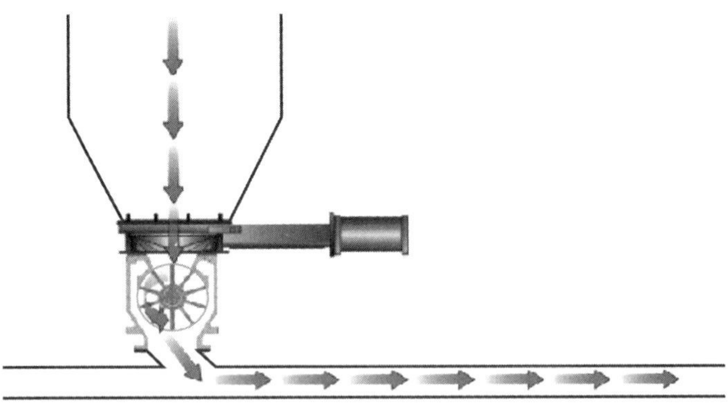

Figure 6.1 Example of a rotary air lock.

Isolation can also be achieved by the use of chokes. Chokes provide explosion isolation by providing an accumulation of powder through which an explosion cannot propagate. In many situations rotary air locks can serve as a choke if the height of powder above the air lock is equivalent to the diameter of the discharge opening and is always present.

Screw conveyors can also serve as chokes where part of the screw is removed or a baffle is in place that ensures a plug of material is always in the screw.

Rapid Action Valves, Gate- or Butterfly-Type Valves can provide isolation. One must ensure the distance between the explosion detection device and the rapid action valve is sufficiently long to allow the valve to fully close before the arrival of the dust flame front.

A flame front diverter or explosion diverter is yet another option to achieve isolation. Do not install this device upstream of an air-moving fan because an explosion originating upstream of the diverter will propagate past the diverter. Also do not use explosion diverters in air streams that have a significant loading of abrasive dust that would eventually erode through the pressure-relieving diverter cover. Explosion diverters should not be used for hybrid mixtures where the flammable vapors exceed the LEL. In using such a device it is important to note that the use of a rupture disk used for the pressure-venting opening should have a release pressure equal to or less than 0.1 barg (1.5 psig). If the diverter is constructed in house and a prefabricated rupture disk is not used, ensure the device meets the following criteria:

- When providing a solid cover, ensure it will quickly and easily release during an explosion.
- When providing a rupture membrane, use a weak weather-resistant material, such as aluminum foil or tar paper.
- Confirm the rupture pressure is at or below 0.1 barg (1.5 psig) by calculations or pressure testing.

A high speed abort gate is also an option. One can actuate the high-speed abort gate by either pressure detection or infrared explosion detection in the upstream vessel expected to experience the explosion. Be sure to ensure the detection system and abort-gate responses are fast enough to completely close the abort gate before the dust flame front reaches it.

A double-dump valve can provide an interlock to ensure both valves do not open simultaneously.

A back-blast damper, yet another option, is essentially a check valve that can be effective in stopping explosion propagation in the opposite direction of normal flow. Ensure that the device is provided with an explosion vent downstream from the normal flow in the system.

The alternative to explosion vent relief is explosion suppression. In their simplest forms, such systems comprise a rapid-acting pressure sensor, control system and a number of pressurized containers of suppressant designed for extremely rapid injection into the protected vessel. The suppressant is commonly sodium bicarbonate. The design and installation of suppression systems should be left to specialist suppliers, but compliance with the appropriate standard (e.g. EN14373) should be specified in the purchase contract.

6.6.3.3 *Explosion Containment*

Since dust explosions generally develop maximum pressures in the range 8–12 bar, it is not usually practical to produce a plant capable of withstanding such pressures unless it is of small volume and round shape. However, certain pieces of equipment, such as hammer mills, are sufficiently robust to contain an explosion. In that instance, it is important to consider the protection of adjoining equipment as containment will increase the explosion severity in what is known as the 'pressure piling' effect. Pressure piling is outside the scope of this discussion of the subject and specialist advice should be sought.

Where explosion containment is the chosen explosion hazard mitigation method, use the explosion resistant-design method. Existing, shock-resistant vessels are tolerable for providing explosion containment. a) Explosion-resistant design is any vessel with a design pressure of 6 barg (87 psig) or more that would not be deformed by a dust explosion that occurs at an initial (pre-explosion) pressure of less than 0.1 barg (1.5 psig). Shock-resistant design is any vessel having 3 barg (43 psig) design pressure or more when designed in accordance with ASME Boiler and Pressure Vessel Code, Section VIII, or yield strength of 6 barg (87 psig) or more according to other codes. With this design, the vessel may deform but not rupture in the event of a dust explosion that occurs at an initial (pre-explosion) pressure of less than 0.1 barg (1.5 psig). To prevent damage to other equipment connected upstream and downstream of the vessels protected by explosion containment, provide explosion isolation.

6.6.3.4 *High Vacuum Option*

Where high vacuum is the chosen explosion hazard mitigation method, use the following criteria:

- Operate the dust-handling system at a sub-atmospheric pressure of less than 0.1 bar absolute (1.5psi absolute)
- Provide an interlock to automatically inert or shut down the process upon a loss of vacuum

One should use only vacuum cleaner systems approved for handling dusts.

6.6.3.5 *Ignition Source Control*

Where a process or system has a high frequency of dust explosions implement the following:

- Operate the equipment that is the source of sparks or hot/glowing particles in an inert atmosphere, OR
- Install a spark-detection system combined with either a spark-extinguishing system or a high-speed abort gate, AND
- Locate the spark-extinguishing system or high-speed abort gate upstream of the first piece of equipment having an equipment explosion hazard, AND
- On activation of the spark-detection system shut down any connected dust collection equipment.

The installation of spark control as described above doesn't eliminate the need for explosion hazard mitigation features.

6.6.3.6 *Addressing Poor Housekeeping Practices*

One should always ask the question – Is there a problem with housekeeping? This is a fundamental question that should be asked of all facilities handling combustible dusts. Dusts on the floors of shop areas, on the surfaces of equipment and walls, in the overhead beams of manufacturing areas, and from equipment exhausting materials into open bins are all suspect areas that deserve close scrutiny. See as examples figures 6.2 and 6.3.

Any dust accumulations outside of equipment poses a potential for a secondary explosion and should be removed and the source of the releases eliminated. As a rule of thumb, 1⁄16 in. (0.0625 in., 1.6 mm) of dust is a cause for cleanup (note, this assumes wood dust with approximate bulk density of about 36 lb/ft^3). This is about the thickness of a US quarter. Where dusts of different bulk density are used, the thickness of the equivalent mass of

Figure 6.2 Dust generated from the discharge of a cyclone into an open container in a room.

Figure 6.3 Shows dust accumulation on overhead beams in a manufacturing area.

1⁄16 in. (1.6 mm) wood dust would be inversely proportional to their bulk densities. For example, for a dust with a bulk density of about 60 lb/ft^3 (950 kg/m3), a layer of about 0.04 in. (1 mm) thick would be equivalent.

Equivalent thickness = 1.6 mm x (580 kg/m3/950 kg/m3) or 1 mm

Any dust accumulations in a small room (200 ft^2) would present a severe secondary explosion hazard. In a larger room, for example 20,000 ft^2, dust covering an area of 200 ft^2 would be a less-severe event.

For practical purposes, some dust accumulations in small parts of large areas can be considered tolerable with frequent cleanup and actions to remove the source of the dust. As a rule of thumb, consider 5% of the area with 1⁄16 in. (1.6 mm) dust capable of being thrown into suspension as the limit of "tolerable." In any building of ordinary construction, any area of dust accumulations exceeding about 1,000 ft^2 (93 m2) is considered unacceptable.

Any dust that is elevated above floor level in joists, beams, equipment tops, etc., should be considered capable of being thrown into suspension. Dust that has agglomerated by age, heat, moisture, etc., should not be considered just as capable of being thrown into suspension but may suggest an ongoing housekeeping problem or source of uncontrolled release. Dust on the floor could be suspended but is less of a risk than elevated accumulations.

Close attention should be paid to dust adhering to walls since this is easily dislodged. Also consider other projections, such as light fixtures, that can provide surfaces for dust accumulation. The available surface area for dust deposits on bar joists or steel beams can be roughly estimated at 5% of floor area. However, some steel beams might have an equivalent surface area as high as 10% of floor area, such as when spans between columns are longer than average for a given geographical area, or when a large building elevation difference requires a stronger roof structure due to anticipated snow drift loadings.

6.6.3.7 Phlegmatization

Phlegmatization is discussed in the next chapter. Combustible and noncombustible dusts must be intimately mixed to truly create a safe non-explosible product. Typically, inert dust ranging from 50% to 75% by weight of the total is required. Carbonates, phosphates, and salts are sometimes used for inerting. In the coal-mining industry, rock dust is used because it is in abundant supply. Unfortunately, no mathematical correlations can predict the required limiting inert powder concentration. All mixtures must be tested to establish the inertness of the mixture.

6.7 Risk Screening Guidelines and Resources

OSHA has developed a generalized approach to performing a fire and explosion risk assessment for dust handling operations.

The **1st step** is to determine whether the dusts generated in a work environment are combustible. Most organic and metallic dusts tend to exhibit some flammability characteristics. Many, such as coal dust, cellulose ester powders, finely ground aluminum and cornstarch are well recognized as being combustible. But one should be cognizant that oftentimes in manufacturing or R&D settings, there can be new or intermediate products developed, or chemicals manufactured under contract to outside customers by toll converters whose combustibility characteristics may be unknown. A literature search supplemented by information supplied by a vendor may provide the information necessary to determine dust combustibility. If such information cannot be readily obtained, then a lab test which involves exposing a small sample of finely ground material to an open flame should be performed to demonstrate whether it is combustible. If the dust is not combustible, then the risk of a dust explosion is non-existent, and the evaluation is complete. But if the dust is combustible, then the next step in the risk assessment process should be initiated. For many processes (such as pneumatic conveying processes using nitrogen or other inert gases), testing may not be done because the inert atmosphere adequately controls the hazard. However, even in these situations, there may be value in knowing the explosive potential of the dust in order to properly classify electrical equipment and determine the level of safety instrumentation needed.

Once a material in dust form has been determined to be potentially combustible, then the **2nd step** should be to assess whether the dust is unstable or explosive. Materials like organic peroxides are unstable and shock sensitive as solids, regardless of their physical form. The risk assessment for assessing such materials needs to be much more extensive than a standard dust explosion screening, since these materials are hazardous even without a high concentration of finely divided particulate matter. References at the end of this chapter can assist the reader in identifying the appropriate tests which can be performed by qualified laboratories.

Once a determination of the dust material stability has been established, then the **3rd step** should focus on a determination of the particle size and particle size distribution. The reader may wish to revisit chapter 1 to place the relevance of size and distribution into proper context. Dusts are generated from solid or granular materials and can exist over a wide range of particle sizes depending on the material handling and processing operation. They may also form through the processes of sublimation and thermal oxidation as well as from combustion-related processes. Particles that are too large to remain airborne settle out due to gravity, while the smallest particles can remain suspended in air almost indefinitely as colloidal suspensions. The unit of measure used to characterize dust particle size is microns or μm. The micrometer is a unit of length equal to 10^{-4} (0.0001) centimeter or approximately 1/25,000 of an inch; or another way of stating this – there are *25,400 microns* in one inch. In metric units a micron represents one-millionth of a meter.

Unlike combustible or flammable liquids, solids must have a certain surface-to-volume ratio in order to sustain combustion. Above this critical size, an accelerating flame front cannot be sustained, so there is no risk of generating dangerous levels of explosive overpressure (although a fire may still be sustained). Different compounds exhibit different maximum particle size behavior; however, industry testing and experience indicates that, for all normal (non-reactive) dusts, there is no explosion hazard above either 420 microns (see NFPA Standards 68 and 654) or 500 microns (see ISO standard). Thus, even if the dust is combustible, if there is no place in the manufacturing process where material exists below the 500 microns particle size (the more conservative of the two sizes), then our risk assessment may be considered to be nearly complete, with one exception that is outlined in step 4 below.

If the particle size is not known, simple screen analyses, which can be performed on-site with a minimum amount of sample, should be run to determine the size distribution and weight fraction for a representative sample of material.

A **4th step** must be to assess whether dust particles are wet with flammable or combustible solvents, or whether the dust itself is likely to give off flammable or combustible solvent vapors or VOCs (Volatile Organic Compounds). Combinations of solvents and solids are referred to as hybrid mixtures. These present unique hazards. For hybrid mixtures, there is a synergistic effect of the dust and vapors that can allow an explosion even if neither the solvent nor the dusts by themselves are in a combustible state. Thus, things like particle size (for solids) and lower flammability limit (LFL for vapors) are not accurate indicators unto themselves of the level of risk. Hybrid mixtures should be evaluated on a case-by-case basis, and generally involve extensive testing, implementation of other risk reduction measures (such as inerting), or both. Expert consultation should be obtained if a hybrid mixture can be present.

A **5th step** in the risk assessment should be a determination of whether a portion of the dust is smaller than 500 microns. A careful review of the particle size data, especially the particle size distribution for sampled dusts (see step 3), should be made in order to assess the existence of fines and also to see whether bimodal particle size distributions exist which extend into the particle size range that poses a potential hazard.

We should ask this point in the assessment (**step 6**), whether there are fines concentrated anywhere in the process that the risk assessment is being performed on. Steps 5 and 6 are generally performed together. Once the size distribution is determined in step 3, step 5 is a simple matter of looking at the results and determining whether small particles are present. Even if there is a fraction of material that is smaller than 500 microns, if this material is never concentrated or segregated, an explosion hazard may not exist. For these cases, a representative sample of the material should be tested in a 20 liter apparatus (standard test setup for dust explosion evaluation) to determine if a hazard exists. A 20L test (actually these are a series of tests) costs between $5,000 and $10,000, and my require takes 4–6 weeks lead time to complete (such tests may be accelerated by paying a premium for faster testing) and is generally done by an outside lab. Usually, somewhere between 1 and 2 pounds (0.5 to 1.0 kg) of material is needed to run a full test spectrum. If only smaller quantities are available, alternative testing methods may have to be used.

Often a more common situation is one where fine material is segregated from the bulk material for processing reasons. This can be accomplished by means of a centrifugal separation device (e.g., cyclone), a filter of some kind, or other type of separator (electrostatic precipitator). Segregation may occur unintentionally, via pneumatic conveying or other processing steps. In any event, if segregation occurs, it increases the risk of a dust explosion in the region of higher fines concentration. Testing on a representative sample of the concentrated fines should be done to determine the level of hazard involved. For processes in the development stage, an artificial sample may need to be generated to perform the required testing.

We now come to a critical step – **step 7**, which is to determine the worst-case parameters. Any risk assessment should be based on a worst case scenario. Remember that we are not dealing with an exact science and set of rules. The limits of science and engineering calculations restrict the analysis from being precise and as such adequate safety margins need to be applied in risk mitigation programs. As such, attention should now focus on determining the four main parameters that define worst case situations for explosion and file. Testing is highly recommended for quantifying the K_{st} value, P_{max}, the minimum ignition energy (MIE), and the minimum explosive concentration (MEC). The first two parameters are considered the *Primary Test Values* and the latter two are the *Secondary Test Values.*

The **K_{st} value** - Dusts are classified by their maximum energy release potential, as measured by the maximum rate of pressure rise in a given test vessel. This parameter, called K_{st}, is defined as follows:

$$K_{st} = (dP/dt)_{max} * V^{1/3}$$

Where P = Pressure

t=time

V= test vessel volume

In metric terms, K_{st} has units of bar-m/sec. This value is determined by a series of tests in a standard 20-liter test sphere, as described in NFPA #68. Once determined, it provides a measure of the speed at which a particular dust combusts and generates pressure. The higher the K_{st} value, the more dangerous the dust, since a faster rate of pressure rise means less time to detect and react (either through relief or suppression) to keep a system safe.

The **P_{max} value** - The second parameter of interest is the maximum pressure obtained from combustion of the dust, known as P_{max}. This is also determined experimentally, and has units of pressure (bar-gauge or bar_g in metric units). The higher the value, the more dangerous the dust, since P_{max} is a measure of how much stress a given vessel might be placed under during an event.

St Values - For simplicity, K_{st} and P_{max} values are grouped into three hazard classes, known as St values. While this has been discussed earlier, the values are restated in the table below:

Hazard Class	K_{st} values, bar-m/sec	P_{max}, bar_g
St-1	< 200	< 10
St-2	201–300	< 10
St-3	> 300	< 12

Using these values and the methods in NFPA #68, deflagration-venting devices can be sized for a variety of physical configurations encountered in the plant.

Now let us turn attention to the secondary test values - minimum ignition energy (MIE) and the minimum explosive concentration (MEC). These two parameters are analogous to those developed for flammable and combustible vapors (although for vapors, the term lower flammability limit is more commonly used than explosive concentration). While venting systems can be successfully sized without these parameters, knowing what the values of MIE and MEC are for a given dust may prevent the need for venting and/or suppression systems, if the particular system of interest does not exceed either of these values.

Minimum Ignition Energy (MIE) - MIE represents the lowest energy level that will ignite an optimum dust/air mixture. Ignition energy increases rather dramatically as the dust concentration moves away from the optimum mixture in the same manner as vapor/air mixtures. However, as a general rule, MIEs for dust are significantly higher than those for solvent vapors. If a dust's MIE is greater than 50 mJ, it is generally not considered a significant explosion hazard. Under these conditions, simple preventive measures, such as grounding and bonding control are usually considered sufficient, although in some circumstances, additional protective measures may be needed.

Minimum Explosive Concentration (MEC) - MEC is the lowest concentration of dust in air that will sustain combustion, and is analogous to the lower flammability limit (LFL) for vapors. Unlike vapors, dusts generally do not exhibit an upper flammable limit or upper explosive concentration that is of any practical use in explosion prevention. Typical values for organic dusts which produce the fastest rate of pressure rise are in the 300–1000 g/m^3 (0.3–1.0 oz ft^3) range, which is considerably higher than comparable numbers for flammable vapors.

Steps 8, 9 and 10 represent risk assessment steps for determining whether the processing steps involving combustible dusts can be safely conducted without the need for explosion venting or designing the vessels to contain a deflagration. These methods, as well as use of reaction explosion suppression systems, are discussed in more detail in NFPA #69. The steps address the following critical questions:

- **Step 8: Will the process under evaluation exceed 25% MEC?**
- **Step 9: Is the dust MIE < 50 mJ?**
- **Step 10: Will inerting the dust working environment be necessary?**

If it can be determined that the dust concentration will never exceed 25% of the MEC, then the risk of propagating an explosion is minimal, and neither inerting nor venting will be necessary. In practice, this is very difficult to determine accurately enough to relieve the need for some risk reduction measures, especially since in the real manufacturing world there are transient operating conditions leading to process upsets, startups or other non-steady state conditions. The nature of these transient operating conditions can lead to higher than expected concentrations of dust.

Similarly, simply having a dust MIE greater than 50 mJ will probably ensure that a static discharge will not be able to ignite the dust/air mixture. However, more energetic ignition sources (such as electrical arcs from improperly installed or maintained electrical devices) can still prove enough of a risk so as to warrant additional action. The resistivity of the solid material (as measured in Mohm-m) is also critical to the proper evaluation of risk as related to ignition energy.

As explained in the next chapter in greater depth, the use of an inert gas, such as nitrogen, is commonly used and often provides an inexpensive risk reduction technique. It may in fact be applied in lieu of running tests to determine MIE and MEC. If inerting is applied, the need for explosion venting should be looked at on the basis of a cost-benefit economic analysis, including both equipment damage and business interruption values. For non-inerted systems where the risk is determined to be above acceptable levels, a properly designed venting system, per NFPA #68, should be used.

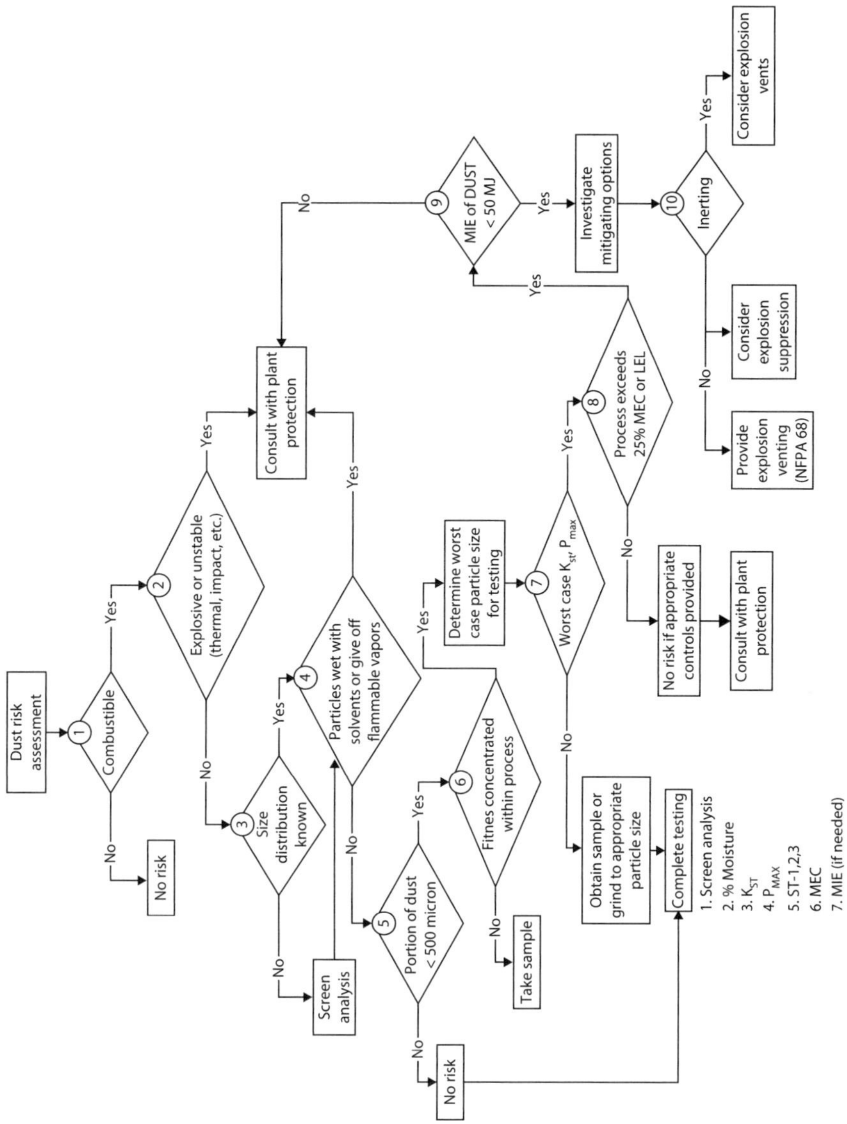

Figure 6.4 Logic diagram of dust explosion risk screening.

Figure 6.4 provides a logic diagram for the reader to follow in performing the risk assessment. It provides a quick reference to the methodology and critical issues that should be considered in identifying the hazards, levels of risk, and in establishing the criteria for implementing risk mitigation measures. The keys to any successful combustible dust program are:

1. Awareness of the Hazard
2. Appropriate Engineering Design of Key Dust Handling Facilities and Equipment, to prevent dust explosions inside of processing equipment
3. Good Housekeeping, to minimize accumulation and prevent dust explosions outside of processing equipment
4. Commitment to making safety the highest priority on the part of the organization

There are no excuses for not placing full and proper resources into risk management. Risk management practices need to be thorough and based on a careful consideration of worst case scenarios. The reasons for applying good practices that are specifically tailored to the facility's operational conditions are to protect lives, prevent injuries, to insulate the public from harm, and to protect valuable assets and the business itself. Half-hearted attempts, poor practices and ineffective policies are irresponsible and can lead to negligence with devastating outcomes. Additional information and resources on the prevention of dust explosions may be obtained from NFPA #654 (Ref 5), numerous publications of the American Institute of Chemical Engineers, and the recommended references at the end of this chapter.

Recommended References

US Chemical Hazard and Investigation Board Website, www.csb.gov

US Department of Labor Directive Number CPL 03-00-008, effective 3/11/08, "Combustible Dust National Emphasis Program (Revised)"

NFPA 68, "Standard on Explosion Protection by Deflagration Venting", 2007 Edition

NFPA 69, "Standard on Explosion Prevention Systems", 2008 Edition

NFPA 654, "Standard for the Prevention of Fire and Dust Explosions from the Manufacturing, Processing, and Handling of Combustible Particulate Solids", 2006 Edition

"Preventing Dust Explosions", Chemical Engineering Progress, Volume 104, No. 3, March 2008, pp 8–14

Bartknecht, W. Dust Explosions: Course, Prevention, Protection. Springer-Verlag, New York, NY, 1989.

Alfert, F., and K. Fuhre. Flame and Dust Free Venting of Dust Explosions by Means of a Quenching Pipe, CMI-Report No. 89-25820-1, Chr. Michelsen Institute (Norway), 1989.

Eckoff, R. K. *Dust Explosions in the Process Industries*, Butterworth-Heinemann, Oxford, 1991.

van Wingerden, K. "Prediction of pressure and flame effects in the direct surroundings of installations protected by dust explosion venting", Journal of Loss Prevention in the Process Industries (Vol. 6, No. 4, 1993): 241–249.

WirknerBott, I., et.al., "Dust Explosion Venting: Investigation of the Secondary Explosion", Paper presented at the Seventh International Symposium on Loss Prevention and Safety Promotion in the Process Industries, Taormina, Italy, 1992.

Lunn, G. A., "Dust Explosion Prevention and Protection Part 3 — Venting of Weak Explosions and the Effect of Vent Ducts", Rugby: Institute of Chemical Engineers, 1988.

Cooper, M. G., et.al., "On the Mechanisms of Pressure Generation in Vented Explosions, Combustion and Flame (1986) 65: 1–14.

Cubbage, P. A., and W. A. Simmonds, "An Investigation of Explosion Reliefs for Industrial Drying Ovens, -Top Reliefs in Box Ovens", Transactions of the Institute of Gas Engineers (1955) 105: 470–526.

Pineau, J. P., "Protection Against Fire and Explosion in Milk Powder Plants", Europex First International Symposium, Antwerp, Belgium, April 16–19, 1984.

Hurlimann, H., "Results from Real Scale Explosion Tests", Third International Symposium on Loss Prevention and Safety Promotion in the Process Industries, Basel, Switzerland, 1980.

van Wingerden, C. J. M., and H. J. Pasman, "Explosion Venting of Partially Filled Enclosures", Conference on Flammable Dust Explosions, St. Louis, Missouri, USA, November 2–4, 1988.

Zeeuwen, J. P., and G. G. M. van Laar, "Explosion Venting of Enclosures Partially Filled with Flammable Dust-Air Mixtures, "International Symposium on Control of Risks in Handling and Storage of Granular Foods, Paris, France, April 2426, 1985.

Siwek, R., "New Knowledge About Rotary Air Locks in Preventing Dust Ignition Breakthrough", Plant Operations Progress (July 1989) Vol. 8, No. 3: 165–176.

Green, D. W., and R. H. Perry, *Perry's Chemical Engineers' Handbook*, Eighth Edition (Chemical Engineers Handbook). New York: McGraw-Hill, 2007.

DeGood, R., "Isolation: Another Way to Take the Bang out of Explosions", CPI Equipment Reporter (Jan-Feb 1988).

Siwek, R., "A Review of Explosion Isolating Techniques," Europex International Seminar, March 1996.

Richmond, J. K., et.al., "A Physical Description of Coal Mine Explosions, Part II." Seventeenth International Combustion Symposium 1978) 1257–1268.

7

Phlegmatization, Diluent Dusts, and the Use of Inert Gases

7.1 Introduction

The term *phlegmatized* is technically applied to an explosive that has had an agent added to stabilize or desensitize it. Sometimes this is desirable either to enhance the handling properties of an explosive material (e.g. when munitions are filled in factories) or to reduce its sensitivity or detonation velocity. As an example, TNT explosive can itself be used to phlegmatize more sensitive explosives such as RDX (to form Cyclotol), HMX (to form Octol) or PETN (to form Pentolite). Other common phlegmatizing agents include paraffin wax, paper or even water (used in water gel explosives). Such agents are often flammable themselves (therefore adding fuel to the blast) or will at least boil off easily. These techniques which have been found useful in ammunition manufacturing have been applied with similar benefits to in dust handling operations.

Explosible dusts can be rendered non-explosible by diluting them with an inert dust as discussed in chapter 5. Further comments on this practice are provided in this chapter.

This chapter also covers the subject of inerting. Partial or complete substitution of the air contained in a process equipment's environment by an inert gas constitutes an effective practice of inerting. Dust explosions can be eliminated when oxygen is completely excluded or is reduced to a concentration below which the flame propagation in the dust cloud, and consequently an explosion cannot occur. This concentration is called the limiting oxygen concentration of LOC.

7.2 Phlegmatization

The practice of phlegmatization is generally applied in the manufacture of explosive materials. An explosive material is capable of producing an explosion by its own energy. An explosive perfectly capable of producing an explosion may liberate its energy without producing one. Black powder, for example, may burn in the open air. An explosion may occur without an explosive, that is, without any material which contains intrinsically the energy needful to produce the explosion. Consider a steam boiler which may explode because of the heat energy which has been put into the water which it contains. But in the case of a steam boiler, the energy is not intrinsic to water, and water is not an explosive. Also, there are explosives which do not themselves explode. Rather, the explosions consist in the sudden ruptures of the containers that confine them, an example being the Chinese firecracker. In the case of a firecracker, fire, traveling along the fuse reaches the black powder – a mixture of potassium nitrate, sulfur, and charcoal – which is wrapped tightly within many layers of paper, the powder burns rapidly and produces gas. It burns very rapidly, for the heat resulting from the burning of the first portion cannot get away, but raises the temperature of the next portion of powder, and a rise of temperature of 10°C more than doubles the velocity of a chemical reaction. The temperature mounts rapidly, gas is produced suddenly, an explosion ensues. The powder burns and the cracker explodes. And in still other cases we have materials which themselves explode. The molecules undergo such a sudden transformation with the liberation of heat, or of heat and gas, that the effect is an explosion.

There are three general classes of explosives:

- Propellants
- Primary explosives
- High explosives

Propellants (also referred to as *low explosives*) are combustible materials, containing within themselves all oxygen needful for their combustion, which burn but do not explode, and function by producing gas which produces an explosion. Examples of low explosives are black powder and smokeless powder (made from colloided nitrocellulose). Explosives of this class differ widely among themselves in the rate at which they deliver their energy.

There are slow powders and fast powders for different uses. The kick of a shotgun is quite different from the persistent push against the shoulder of a high-powered military rifle in which a slower burning and more powerful powder is used.

Primary explosives or *initiators* explode or detonate when they are heated or subjected to shock. They generally do not burn and essentially do not contain the elements necessary for combustion. Rather, the materials themselves explode, and the explosion results whether they are confined or not. They differ in their sensitivity to heat, in the amount of heat which they give off, and in their *brisance,* that is, in the shock which they produce when they explode. Not all of them are brisant enough to initiate the explosion of a high explosive. Examples include mercury fulminate, lead azide, the lead salts of picric acid and trinitroresorcinol, m-nitrophenyldiazonium perchlorate, tetracene, nitrogen sulfide, copper acetylide, fulminating gold, nitrosoguanidine, mixtures of potassium chlorate with red phosphorus or with various other substances, the tartarates and oxalates of mercury and silver.

High explosives *detonate* under the influence of the shock of the explosion of a suitable primary explosive. They do not function by burning and not all of them are combustible; however, many or most of them can be ignited by a flame and in small amounts generally burn tranquilly and can be readily extinguished. When heated to a high temperature by external heat or by their own combustion, they sometimes explode. These differ from primary explosives in not being exploded readily by heat or by shock, and generally in being more brisant and powerful. They exert a mechanical effect upon whatever is near them when they explode, whether they are confined or not. Examples of high explosives include dynamite, trinitrotoluene, tetryl, picric acid, nitrocellulose, nitroglycerin, liquid oxygen mixed with wood pulp, fuming nitric acid mixed with nitrobenzene, compressed acetylene and cyanogen, ammonium nitrate and perchlorate, nitroguanidine.

There are a wide variety of ingredients that are added to explosive compositions which in themselves are not explosive but can enhance the power of explosives, reduce the sensitivity, and aid processing. Aluminum powder is frequently added to explosive and propellant compositions to improve their efficiency. Ammonium nitrate is used extensively in commercial explosives and propellants. It is the most important raw material in the manufacture of commercial explosives and it also provides oxygen in rocket propellant compositions. Commercial blasting explosives contain ammonium nitrate, wood meal, oil and TNT. A mixture of ammonium nitrate, water and oily fuels produces an emulsion slurry which is also used in commercial blasting explosives. Small glass or plastic spheres containing oxygen can also be added to emulsion slurries to increase its sensitivity to detonation. Polymeric materials can be added to secondary explosives to produce polymer bonded explosives (PBXs). The polymers are generally used in conjunction with compatible plasticizers to produce insensitive PBXs. The polymers and plasticizers can be in the nitrated form which will increase the power of the explosive. These nitrated forms are known as energetic polymers and energetic plasticizers.

Phlegmatizers are added to explosives to aid processing and reduce impact and friction sensitivity of highly sensitive explosives. Phlegmatizers can be waxes which lubricate the explosive crystals and act as a binder. The phlegmatizer acts as barrier between molecules of the explosive to slow down the heat transfer and any resultant reaction and reduce impact and friction sensitivity. Typical phlegmatizing agents include wax, paper, water, alcohol, oils such as petroleum jelly and paraffin as well as polymers such as chlorofluoropolymers.

In the shipment of dangerous goods which are explosive or shock sensitive by characterization, lactose, glucose or similar materials may be used as a phlagmetizer provides that the substance contains not less than 90%, by mass, of phlagmetizer. The competent authority may in fact authorize these mixtures to be classified in Division 4.1 (DOT shipping designation) on the basis of a test.

Phlagmetizers are also commonly used for the safe handling of plastics. Commercial polypropylene resins that are polymerized by Ziegler-Natta catalyst systems and have high molecular weight and broad molecular weight distribution use these materials. The chemical structure of polypropylene is influenced by the polymerization system employed during production. Oftentimes peroxides are used in the manufacturing process. The resins produced with a linear aliphatic peroxide produce excessive quantities of VOCs (volatile organic compounds). Other formulations employ cyclic peroxides. In either case, the liquid formulations comprise a liquid phlagmetizer, a liquid plasticizer, and the peroxide. The phlagmetizer stabilizes the peroxide. In order to achieve a safe composition, the phlagmetizer should have a certain minimum flash point and boiling point relative to the decomposition temperature of the peroxide such that the phlagmetizer cannot be removed by boiling/flashing or separated out as a concentrate, thus leaving behind the unstable peroxide composition.

Examples of liquid phlagmetizers for use with cyclic peroxides include glycols and ketones (e.g., acetone) as well as many other chemical compounds. Antioxidants, acid scavengers and various other ingredients can be used a further stabilizing additives.

Products containing phlagmetizers generally reduce the risks of a dust explosion; however, this should not be considered a form of control. The presence and concentration levels of these ingredients may be viewed as safeguards, but cannot be relied upon as substitution for the good industry practices and risk assessment tools explained in earlier chapters.

7.3 Addition of Diluents

Explosible dusts can be rendered non-explosible by diluting them with an inert dust such as calcium sulfate, limestone, sodium bicarbonate, common salt, silicates and even stone dust (see chapter 5). Although the mechanisms by which the effectiveness of these additives are not entirely clear, they are generally thought to serve as heat sinks and or otherwise interfere with flame propagation.

Generally, at least 60% diluent dust is needed to be effective. The diluent dust must be intimately mixed with the explosive dust. The quantity of diluent dust needed to be effective for an intended application has to be determined from a series of explosibility tests. With the exception of coal mining operations, the application of diluent dust inerting is rarely employed simply because large quantities are required and the concern for product contamination.

7.4 Application of Inert Gases

Partial or complete substitution of the air contained in a process equipment's environment by an inert gas constitutes an effective practice of inerting. Dust explosions can be eliminated when oxygen is completely excluded or is reduced to a concentration below which the flame propagation in the dust cloud, and consequently an explosion cannot occur. This concentration is called the limiting oxygen concentration of LOC.

The practice of inerting by application of an inert gas is applicable only to an enclosed system. Any working environment that is open to the atmosphere cannot be effectively interted because of the existence of significant fluctuations in oxygen concentrations. In addition, the loss of inert gas would be uneconomic and may lead to the build-up of inert gas in places which are accessible to personnel, thus creating an inhalation risk due to oxygen deficiency.

The general classes of equipment which are candidates for inerting include processing equipment such as enclosed reactor vessels, mixers, pulverizers, hammer and other types of mills, dryers, ovens, filters, screens, cyclone separators, various dust collectors; storage equipment such as hoppers and silos; and conveyors and bucket elevators including pneumatic conveyors, screw conveyors, en masse conveyors and others.

The practice of inerting is highly recommended for strongly explosible dust (i.e., Kst > 600 bar-m-s^{-1}). Inerting is also useful when dealing with flammable solvent vapors and VOCs. As discussed further on, a critical consideration when designing and applying inerting systems is the need for continuous monitoring of oxygen levels. Estimates of leakage rates from plant process equipment and enclosures and of the efficiency of gas mixing are critical considerations.

The most common inert gases used are nitrogen, carbon dioxide, helium, flue gases and steam. The selection of the gas depends on various factors including cost, availability, reliability of gas supply, the potential of contamination of the dust by the inert gas constituents, including moisture, and the volume effectiveness in reducing explosibility. A few comments on each of the common inert gases are provided below. In considering these options, the reader should bear in mind the principles discussed in chapter 2, particularly with regard to the Limiting Oxygen Concentration (LOC).

Recall that the LOC is defined as the concentration of inert gas that is required to prevent explosion, and must be determined by testing and measurement. As a rough rule of thumb, the LOC for preventing explosion of many dusts is between 8 and 16% with carbon dioxide, and about 5 to 13% with nitrogen. Generally, a greater volume concentration of nitrogen is required than carbon dioxide. Despite these general guidelines, testing is still advised.

Further, where flammable VOCs are suspected, steps must be taken to create a test atmosphere corresponding to the most probable conditions expected in the plant in order to obtain LOC measurements that are representative of plant conditions. Also, when the dust/air mixture is likely to be present at temperatures above room conditions (atmospheric), the LOC test should be modified to simulate plant conditions as close as feasible.

The application of safety margins to the measured LOC is also recommended. The safety margin chosen really depends on the size of the operations, the explosibility of the dust and the level, accuracy and dependability of the oxygen monitors employed. The likelihood of hot surfaces, product smoldering and the possible transport of smoldering products being transported to downstream operations should be taken into consideration when establishing safety margins.

Carbon dioxide is readily available in compressed form generally from proprietary inert gas generators and also from waste gas from on-site processors. Effective, higher oxygen levels on a percent by volume basis are considered permissible compared with nitrogen. Costs for supply are generally moderate. Caution must be exercised because some metal dusts react violently with carbon dioxide – one example being aluminum.

Nitrogen is readily available in compressed or cryogenic forms, and in some instances as a waste gas from on-site process operations. Costs for supply are generally moderate. Nitrogen is generally considered less effective in volume/volume terms than carbon dioxide. Caution must be exercised because some metal dusts react violently with nitrogen at high temperature – one example being magnesium.

Flue gases tend to be readily available as a waste stream from on-site processes or from inert gas generators. These supplies are typically available at low cost. Costs can escalate with this option because additional capital equipment may be needed to cool the gas, remove contaminants, to monitor and/or remove flammable vapors and to remove incandescent materials. Caution in selection is highly recommended because flue gases may react with certain dusts. Also, storage of flue gas may not be economical or practical; and hence adequate supplies may not always be available. If there is a furnace shutdown, then the loss of flue gas supply will result in the loss of the inert gas supply.

Argon or helium are two gases that are unlikely to contaminate products or react with them; however, these represent rather expensive options for inerting.

Steam is generally available in plant operations; however a concern is that during transient operations such as turnarounds, startups and shutdowns, supplies may be interrupted. In addition, steam will condense when temperatures drop leading to a loss of the inert atmosphere. Various types of industrial dryers use the steam that is generated by evaporation from a wet feed to maintain inert conditions within plant environments. This is actually a reasonable pollution prevention practice, but while this may be effective during normal, steady state operations, the inert atmosphere needs to be maintained during the transient periods of operations such as startup and shutdown. One approach to overcome this deficiency is to supply steam or even another inert gas to the system during situations when the operations are not self-inerting. It is advisable to perform continuous monitoring of oxygen levels using sensors that are robust and suitable for operating over a wide range of temperature and humidity conditions, and are unaffected by steam. Provisions need to be made in order to prevent condensation from masking sensors.

The LOC is normally measured at room temperature conditions. It is important to recognize that LOC values obtained under room temperature conditions are not appropriate for environments where steam is used to inert systems at temperatures in excess of 100°C. Rather, the LOC needs to be measured under conditions that are representative of the actual processing temperatures. In addition, it is important to bear in mind that the LOC that is required to prevent an explosion risks may not necessarily prevent the onset of smoldering ignition. If for example, a hot product or intermediate is held in a dryer for an extended period of time, there is a risk that smoldering material may be carried forward into downstream parts of the process. It is therefore important that when performing a risk assessment for fire and explosions, that attention not simply be paid to one or a small group of equipment; rather the integrated parts of the operation need to be taken into consideration.

7.4.1 Best Practices

The best industry practices for the application of inert gases are defined in NFPA Standard 69 (Standard on Explosion Prevention Systems, 2008 Edition, Quincy, MA). The following is a summary of key elements defined in the standard.

Materials other than oxygen can act as oxidants. The Limiting Oxidant Concentration (LOC) depends on the temperature, pressure, and fuel concentration as well as the type of diluent that is relied on. LOC generally tends to decrease as the pressure or temperature prior to ignition increases. Best practice is to test the LOC at the appropriate temperature and pressure. NFPA 69 notes that deviations from the test fuel composition and temperature might possibly be accounted for by using appropriate techniques. For dusts, an appropriate test apparatus should be used in conjunction with a strong ignition source, such as described in the draft of standard ASTM E 27, *Determination of Explosion Characteristics of Dust Clouds*, or in CEN EN 14034-4, *Determination of Explosion Characteristics of Dust Clouds*, Part 4.

NFPA 69 emphasizes that inspection, maintenance and operator training are all necessary requirements of any explosion prevention system. Reliability of the system and its instrumentation "is only as good as the inspection and periodic preventive maintenance they receive. Operator response and action to correct adverse conditions, as indicated by instrumentation or other means, are only as good as the frequency and thoroughness of training provided."

Analyzers and system instrumentation can require frequent periodic inspections. The operation of a system with an oxidant concentration low enough to prevent a deflagration does not mean that the risk of incipient fires are prevented. Smoldering can occur in fibrous materials or dust layers at very low oxidant concentrations, which can ultimately result in a fire or explosion when exposed to higher oxidant concentrations. Extreme caution should therefore be exercised when such systems are opened to the air.

NFPA 69 also notes that purge gases generated by any of the acceptable methods described in the standard might not necessarily be compatible for all applications. The physical and chemical properties of the combustible materials involved establish the type and required purity of the purge gas required.

Personnel should not enter enclosures where the atmosphere is oxygen deficient. If it is necessary to enter such an enclosure, personnel should use self-contained breathing apparatus, preferably the positive-pressure type. Canister-type gas masks should not be used; they do not supply oxygen and do not offer any protection. Also it is essential that the toxicity of certain purge gases be recognized and warnings well posted. The potential for accidental release of purge gases into normally occupied areas should be recognized and the necessary precautions taken to prevent this.

It is also essential to note that the LOC values for dusts of a particular chemical composition could also differ with variations of physical properties such as particle size, shape, and surface characteristics. A particular dust could have combustion properties that differ from those reported in both the general literature and NFPA 69.

Under certain conditions of reducing atmospheres in the presence of sulfur compounds, pyrophoric iron sulfides could form in air-starved atmospheres. When air is admitted into such an atmosphere, the iron sulfides could ignite. A recognized procedure for controlling such ignition is to thoroughly wet the iron sulfide deposits with water and maintain a wetted surface until all deposits are removed and disposed of safely and properly. Another method is to maintain an inert atmosphere in the tank or system containing pyrophoric iron sulfides. The reader is referred to API 2016, *Guidelines and Procedures for Entering and Cleaning Petroleum Storage Tanks*, which provides information covering the control and removal of pyrophoric iron sulfide deposits. Rapid oxidation tends to occur when the deposits dry out. Hence, even though air is admitted slowly, no reactions will occur until the deposits dry out, a process that could take more time than used to admit air. A common practice in industries that deal with such deposits is to keep them wet until they can be removed to a safe location.

Iron sulfide deposits can be thick or become insulated from air by layers of nonreactive materials. When the layers are disturbed, the deposits contact air and can ignite. Furthermore, although procedures are often used to neutralize or remove such deposits before admitting air, it is often difficult to remove all traces of pyrophoric materials.

According to NFPA 69 the rate of application for steam inerting should be sufficient to maintain a steam concentration of at least 2.5 lb/min for every 100 ft^3. This requirement is intended to provide for a sufficient number of isolation points to facilitate maintenance, while holding the number of isolation valves to a manageable number so that accidental shutoff is minimized. The standard also calls for providing consideration to providing a positive means of preventing the backflow of purge gas into other systems where such flow would present a hazard.

One must bear in mind that the objective is to maintain operation outside of the flammable region. This can be achieved by adding either enrichment gas (natural gas or methane) or an inert gas such as nitrogen. In either case, it is important to apply a safety factor. Instrumentation should have redundancy, depending on the criticality of the operation.

NFPA 69 further notes that a calculation of the LOC can result in an overestimation of up to at least 2 volume percent oxygen relative to measured values. This potential error should be taken into account when applying the safety margin.

Products with relatively high vapor pressures can, by themselves, maintain an atmosphere above the upper flammability limit of the vapor. Where flammable atmospheres are estimated, it is good practice to use a padding gas to maintain the oxygen content below the LOC. Such maintenance typically involves almost complete replacement of air, and as such the oxygen analysis of the vapor space is not generally needed. It should be ensured that padding gas capacity maintains padding under adverse conditions, such as simultaneous pump-out of several tanks connected to the same padding supply, possibly with a contraction of vapor volume caused by a sudden summer rainstorm. Such conditions may cause air to be drawn into a container to avoid underpressure damage. Bear in mind also that some monomer tanks need several percent of oxygen to activate dissolved inhibitors. Such tanks may therefore require continuous oxygen monitoring.

The application of enrichment gas (e.g., methane or natural gas) serves the following purposes:

- It elevates the total fuel concentration and can raise it to above the upper flammable limit (UFL).
- It decreases the oxidant concentration in proportion to the concentration of enrichment gas.

- It elevates the LOC due to the better diluent qualities of enrichment gas relative to nitrogen in the air.

In situations where header systems continuously convey vapors to a combustion device such as a flare, operation above the UFL can greatly reduce the quantity of enrichment gas relative to operation below the LOC. The use of oxygen analyzers to control enrichment gas flow is practical only in cases where the nitrogen-to-oxygen ratio is the same as in the air. Where a container has been partly inerted with a diluent such as nitrogen, enrichment gas should be added using flow control, since control via oxygen analyzers would otherwise add insufficient enrichment gas to provide non-flammability. The flow control system can be augmented with gas analyzers to verify correct operation during installation and for periodic performance checks.

Remember that the UFL is a continuous function of oxygen concentration. The greatest UFL corresponds to pure oxygen as the oxidant, and the lowest corresponds to the LOC concentration of oxidant. Systems containing high concentrations of fuel might be safely operated above the LOC, provided that they are nonflammable with respect to the actual UFL envelope. If the oxygen concentration in a system is constrained below a value whose corresponding UFL is U, a safety factor should be applied such that the fuel concentration in the system is maintained at not less than 1.7 U.

A good way to remember the key elements of inerting practices is to review a general flammability chart like the one shown in figure 7.1. When an inert gas is added to a hydrocarbon gas/air mixture, the result is an increase in the lower flammable limit concentration and a decrease in the upper flammable limit concentration. Figure 7.1 illustrates these effects which should be regarded only as a guide to the principles involved.

Any point on the diagram represents a hydrocarbon gas/air/inert gas mixture, specified in terms of its hydrocarbon and oxygen content. Hydrocarbon/air mixtures, without inert gas, lie on the line AB, the slope of which shows the reduction in oxygen content as the hydrocarbon content increases. Points to the left of AB represent mixtures whose oxygen content is further reduced by the addition of inert gas. As indicated in the plot, as inert gas is added to hydrocarbon/air mixtures, the flammable range progressively decreases, until the oxygen content reaches a level generally taken to be about 11 per cent by volume, at which no mixture can burn. The figure of 8 percent by volume inerted gas mixture generally allows some margin beyond this value.

Points C and D represent the lower and upper flammability limit mixtures for hydrocarbon gases in air. As the inert gas content increases, the flammable limit mixtures change; lines CE and DE indicate this, finally

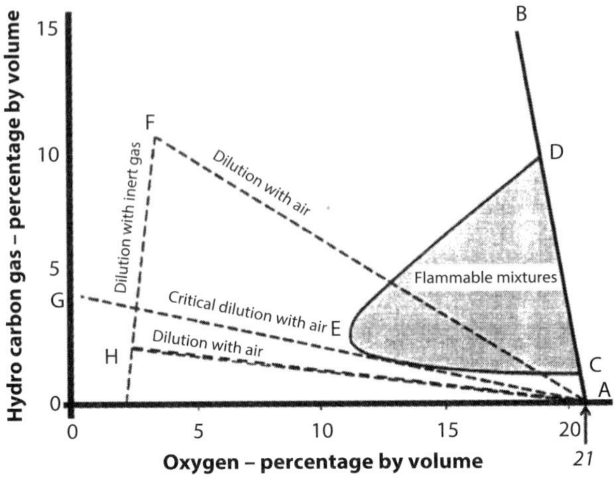

Figure 7.1 Generalized hydrocarbon gas/inert gas mixtures flammability chart.

converging at point E. Only those mixtures represented by points in the shaded area within the loop CED are capable of burning. Changes of composition, due to the addition of either air or inert gas, are represented by movements along straight lines; these lines are directed either towards point A (pure air), or towards a point on the oxygen content axis corresponding to the composition of the added inert gas; such lines are shown for the gas mixture represented by point F. When an inert mixture, such as that represented by point F, is diluted by air, its composition moves along line FA and enters the shaded area of flammable mixtures; this means that all inert mixtures in the region above line GA (critical dilution line) pass through a flammable condition as they are mixed with air (for example, during a gas-freeing operation); those below line GA, such as that represented by point H, do not become flammable on dilution. It will be noted that dilution with additional inert gas, i.e. purging, makes it possible to move from a mixture, such as that represented by F, to one such as that represented by H, by dilution with additional inert gas, i.e. purging.

We should note that there are three general operations that are practiced in the replacement of gas in confined equipment and vessels. They are inerting, purging, and gas freeing. In each of these replacement operations, one of two processes can predominate:

- Dilution, which is a mixing process
- Displacement, which is a layering process

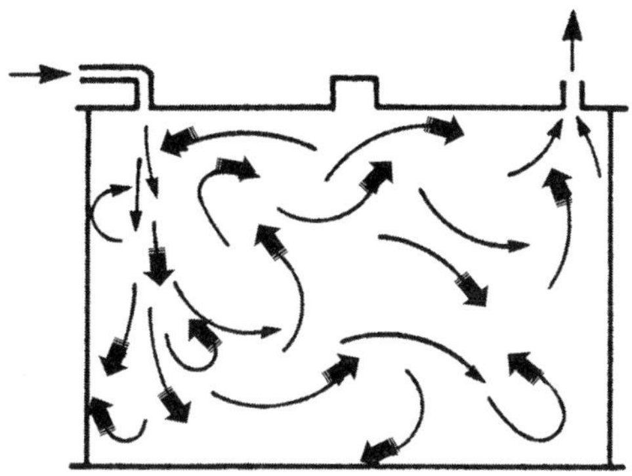

Figure 7.2 Illustrates conceptually turbulent mixing of the dilution gas in a vessel.

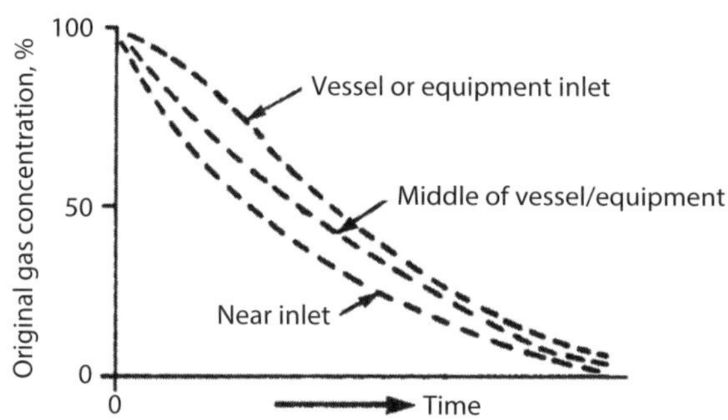

Figure 7.3 Illustrates the sampling location for gas monitoring can vary with sample location.

Figures 7.2 through 7.5 illustrate general principles. Figure 7.2 illustrates an inlet and outlet configuration of a dilution process, highlighting the turbulent nature of the gas flow within the confinement. Figure 7.3 then shows typical curves of gas concentration versus time for different sampling locations within the system.

Now refer to Figures 7.4 and 7.5, which show displacement using an inert gas and the concentrations measured as a function of location within the piece of equipment or vessel, respectively.

These drawings help to illustrate that the two processes have a marked effect on the method of monitoring the process equipment or vessel atmosphere and the interpretation of the results. Figures 7.3 and 7.5 show that the gas replacement process actually taking place within the vessel or equipment must be understood to correctly interpret the reading shown on the appropriate gas-sampling instrument.

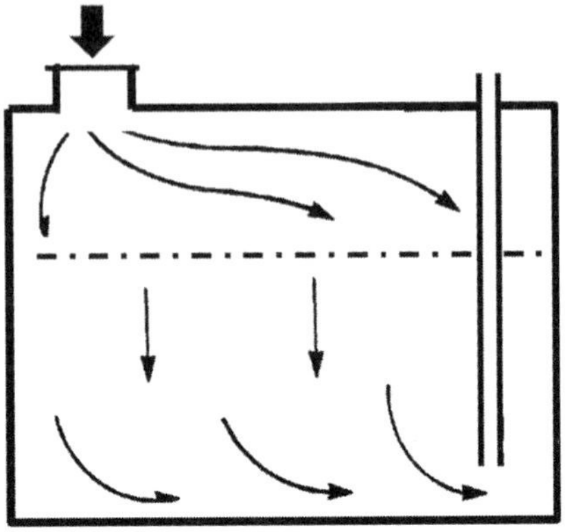

Figure 7.4 Illustrates conceptually displacement of flammable gas in a vessel with an inert gas.

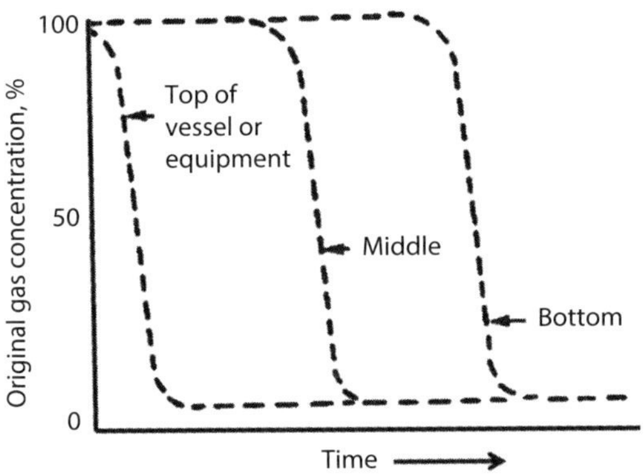

Figure 7.5 Illustrates the sampling location for gas monitoring can vary with sample location.

The dilution theory or viewpoint assumes that the incoming gas mixes with the original gases to form a homogeneous mixture throughout the vessel or piece of equipment; the result is that the concentration of the original gas decreases exponentially. In practice, the actual rate of gas replacement depends upon the volume flow of the incoming gas, its entry velocity and the dimensions of the tank. For complete gas replacement, it is important that the entry velocity of the incoming gas be high enough for the jet to reach the bottom of the tank; it is therefore important to confirm the ability of every installation using this principle to achieve the required degree of gas replacement throughout the vessel.

The following case study will help to illustrate poor practices and good industry remedies.

7.5 Case Study

The following case study examines a fire and explosion incident at a large coke chemical plant, resulting in a fatality, injuries and extensive infrastructure damage. This facility and incident discussed were the subjects of a series of law suits, and as such the facility name is not identified and a number of pertinent facts have been excluded from the discussions.

Coke oven gas (COG) is a by-product of coke manufacturing. It is used as a fuel in coke ovens, boilers, and reheat furnaces. When producing coke by coal carbonization a large volume of gas is produced; this gas is treated in the by-product plant to give a clean fuel gas after removing condensable, corrosive or economically valuable components. The gas is produced over most of the coking period, the composition and rate of evolution changing during this period and being essentially complete by the time the coal charge reaches 700°C. The final yield of clean gas after treatment in the by-product plant depends upon a number of factors, including coal volatile material available and the carbonization conditions.

COG is a complex mixture containing numerous chemical components. COG contains minor components which include ammonia, hydrogen sulfide, hydrogen cyanide, ammonium chloride, benzene, toluene, xylene and naphthalene and other aromatics; tar components; tar acid gases (phenolic gases); tar base gases (pyridine bases); and carbon disulphide. In addition to containing a high concentration of flammable methane gas, it also contains a very high percentage of hydrogen, which is extremely flammable.

Of all these chemicals, the hydrogen, methane, carbon monoxide, paraffinic and unsaturated gases are useful components of the final clean fuel gas. Small amounts of carbon dioxide, oxygen and nitrogen are allowed to remain in the final gas as inert but harmless components. The remainder of the components are removed in the by-product plant as far as is practical. Table 7.1 lists the Flash Point Temperatures, Autoignition Temperatures, and Flammability Limits of COG components in terms of the Lower Explosion Limit, the Upper Explosion Limit and the Flammability Range.

The flash point of a material is the lowest temperature at which the chemical can vaporize to form an ignitable mixture in air. Measuring a flash point requires an ignition source. At the flash point, the vapor may cease to burn when the source of ignition is removed. Table 7.1 shows how dangerous COG is because the chemical components all have a very low flash point temperature or are a flammable gas.

Autoignition temperature is also commonly referred to by engineers as the kindling point of a substance. It is the lowest temperature at which the material will spontaneously ignite in a normal atmosphere without an

Table 7.1 Flammability properties of coke oven gas constituents.

Chemical	Flash Point Temperature, °C	Autoignition Temperature, °C	Flammable Limits, % by Vol.		
			Lower	Upper	Range
Hydrogen	Flammable gas	566	4	74	70
Methane	–187.8	537	5	15	10
Hydrogen Sulfide	Flammable gas	260	4	44	40
Benzene	–11.1	498	1	8	7
Toluene	4.4	480	1	7	6
Xylene	24	464	1	7	6
Naphthalene	88	567	1	6	5
Carbon Monoxide	Flammable gas	609	12	75	63

external source of ignition, such as a flame or spark. This temperature is required to supply the activation energy needed for combustion. The temperature at which a chemical will ignite decreases as the pressure increases or, oxygen concentration increases. It is usually applied to a combustible fuel mixture. The autoignition temperatures for COG constituents are low values. In other words, it does not take a lot of energy or heat for these materials to ignite.

Flammability limits, also called flammable limits, define the proportion of combustible gases in a mixture, between which limits the gas mixture is flammable. Gas mixtures consisting of combustible, oxidizing, and inert gases are only flammable under certain conditions. The lower flammable limit (LFL) describes the leanest mixture that still sustains a flame, i.e. the mixture with the smallest fraction of combustible gas, while the upper flammable limit (UFL) defines the richest flammable mixture. Not only do the constituents of COG have extremely low values for the lower flammability limits, but the major components like methane and hydrogen have very wide limits (the range is the difference between the upper and lower limits). Volatile materials that have very wide flammability limits are the most dangerous because there is a very broad compositional range over which an explosive mixture can form. LFL and UFL are also referred to as the LEL (lower explosion limit) and UEL (upper explosion limit) by engineers, respectively.

For all practical purposes, there is likely no greater hazard at a coke chemical plant than an explosion. Consider as an example one of the major chemical constituents of COG – methane. Methane at room temperature and standard pressure is a colorless, odorless gas. The familiar smell of natural gas as used in homes is a safety precaution achieved by the addition of an odorant, usually blends containing tert-Butylthiol. Methane has a boiling point of −161°C (−257.8°F) at a pressure of one atmosphere. As a gas it is flammable only over a narrow range of concentrations (5–15%) in air. Liquid methane does not burn unless subjected to high pressure (normally 4–5 atmospheres). As a gas, it is flammable and extremely dangerous.

As noted by the U.S. Consumer Product Safety Commission (CPSC)[1] many flammable gases have no smell and their presence cannot be detected by human senses and therefore specific gas detectors must be used. The CPSC notes that the "Accumulation of gas can be produced by leaks which result in the buildup of pockets of gas. These pockets of gas can in turn lead to fire and explosion. Pockets of gas will accumulate at an upper or lower level depending on the gas density and the air currents in the region." The CPSC further notes that the relevant gas density is that of the final gas mixture and not the flammable gas component. The final gas mixture may also be dependent on the degree of mixing in air either by diffusion which is slow or by turbulence.

The CPSC further notes that "Fire results from a flammable gas/air mixture burning after ignition. There are many sources of ignition such as naked flames, sparks, hot surfaces, static electrical sparks, etc. An explosion requires the simultaneous presence of a flammable mixture of gas and oxygen (air) and a source of ignition in a confined space. Such an explosion is usually a deflagration in which the flame front moves with a velocity less than that of sound and the overpressure can exceed 8 bars. In [confined] narrow passages and tubes a deflagration can develop into a detonation in which the flame front travels at a velocity greater than the speed of sound and the overpressure can exceed 50 bars. All flammable gases can cause explosions in the right circumstances."

Further, the CPSC reports that "Once gases have been mixed they will not subsequently separate out into the original constituent gases. However if a mixture is cooled or pressurized the individual components may condense out. Gases diffuse through porous equipment at rates inversely proportional to their molecular weights…"

---

[1] Flammable Gas Safety Code, Annex A: Flammable Gas Safety Manual, U.S.CPS Commission

Because of these factors the coke chemical plant is obligated to manage the large volume of a highly flammable and explosive by-product in a safe manner and to provide all possible safeguards to protect workers and contractors. It is obligated to:

1. Apply OSHA and industry recognized best practices to ensure that its equipment and critical support components are working properly and safely. Critical support components include piping aperture which transport the flammable by-products, pressure relief valves which are intended to prevent the over pressurization of equipment used in processing and transporting the flammable gases, flapper valves which are intended to seal the process and prevent air from entering the system and allowing the flammable gases from becoming flammable mixtures. The consequence of these components failing is to create highly flammable mixtures in confined spaces and as pockets of flammable mixtures in low lying areas, and/or to cause the release of flammable gas streams in atmospheric vents which service the cold boxes. These create highly dangerous environments because when these atmospheres come into contact with energizing sources an explosion and fire can occur.
2. Apply preventive maintenance programs to ensure that the equipment and critical support components it relies on to process, handle and transport the flammable by-products have the highest mechanical integrity possible so that air/oxygen does not mix with the flammable by-products. Oxygen/air is an oxidizer that upon mixing with the hydrocarbon by-products produced by the plant creates an explosive environment. The facility is obligated through both OSHA rules and common sense to exercise diligence and extreme care to ensure that valves, flanges, unions and various component connections along its piping network which transport the gases do not leak the flammable components, and if they unavoidably do, these areas must be isolated from energizing sources. It must inspect and take corrective actions to prevent air from entering into confined spaces where its flammable by-products may leak into in order to prevent these vapors from becoming flammable mixtures when intermingled with air.

3. Apply continuous monitoring for the presence of flammable vapors especially when workers are instructed to perform repairs and modifications to equipment. Because the facility manages such large volumes of a highly flammable by-product it must apply a high degree of diligence to ensure that under no circumstances during operation, shut down, idling, startup, or routine inspections and maintenance turnarounds that flammable vapors are present at conditions which may cause an explosion, especially when personnel are assigned to an area where flammable vapors may potentially be released and there is even the remote possibility of energizing sources being present. A coke chemical plant like any other facility such as a refinery that handles large volumes of flammable materials is obligated to err on the side of safety and therefore apply prudent and precautionary practices such as continuous monitoring for flammable gases in order to reduce the risk of possible incidents.
4. Provide access for inspection and repair of critical components that are intended to provide safety features. Any system sealing device like a flapper valve must be in an accessible location for routine inspection, repairs and replacement; otherwise critical maintenance cannot be implemented and a high risk of failure that can lead to an explosion may occur. Since the 1960s the American Petroleum Institute (API 500) has recommended that facility operators eliminate inaccessible inspection areas for critical components. When infeasible, a facility needs to devise procedures that allow critical components to be remotely monitored and placed on a schedule for servicing and replacement. Corporate representatives testified that quads (described further on) were shut down for regular inspection and turnarounds every 100 days of operation (although records reveal far less frequently). Nothing prevented the facility from gaining access to critical flapper valves (as discussed below, failure of these valves introduced oxygen into an explosive environment) even though they were in a difficult to reach area during a turnaround. The facility simply made repairs to leaking valves, gaskets, and cracked piping when it observed the most obvious leaks and cracks, failing to make inspections and identify failures beforehand. For nearly 50 years it did

not inspect, perform maintenance on, or replace the flapper valves in the cold box atmospheric vent stacks.

5. Inert confined spaces where it is understood that flammable gases may leak into and co-mingle with infiltrating air. The plant's gas recovery process was commissioned in the mid-1960s, making it close to 50 years old. It is both a technology and a hardware that is aging and rapidly approaching the end of its service life. Like all machines and humans, age has its toll. The plant's equipment has been subject to millions of wide pressure and temperature cycling events over normal operations throughout its long history. Subjecting equipment to such conditions as the result of many years of heavy service is compounded by transient operating modes such as upsets, equipment idling, unscheduled shutdowns, startups, and various malfunctions causing severe stresses and metal fatigue. The consequence of these stresses and fatigue are cracks in piping components, vessels, in valves and various critical equipment components and in the floors and walls of the cold box containing the piping aperture which transports the gas. During transient operating modes, metal components associated with piping components and the housing containing these components (i.e., the cold boxes) are subject to thermal stress cracking, thereby making them susceptible to fresh air entering into confined spaces where pockets of flammable gases and hydrocarbon condensates may be present. It has been a long standing industry practice dating back to the turn of the last century to provide proper inerting of these potentially dangerous areas when handling explosive mixtures. Inerting is a standard engineering practice which displaces oxygen/air with a scavenger gas such as nitrogen. Nitrogen will not burn. The displacement of an oxidizer (air/oxygen) with an inert gas, eliminates the possibility of flammable mixtures from igniting when exposed to an energy source.

The operations at the facility are divided into two divisions, one devoted to coke-making and the other devoted to chemicals and energy processing. Within the Chemicals Division, there is an area designated Gas Handling and Light Oil Recovery which consists of two locations referred to as Control Room #1 and Control Room #2. The incident in this matter is an explosion which occurred with the operations and equipment associated with Control Room #2.

At Control Room #2, the plant utilizes a cryogenic process. It consists of a series of regenerators that separate and recover gaseous components from the coke oven gas (COG). The regenerators are vessels which are nearly 50 feet tall and less than 20 feet wide. These units employ ultra-cold temperatures to condense out various components of COG, (e.g., light oils, carbon dioxide, hydrogen sulfide, hydrogen cyanide, ethylene, among others) which are eventually recovered as various "streams" of gas including downriver, underfire and sublimation.

The regenerators are arranged in groups of four that are referred to as "quads." A set of four regenerators sits atop a long rectangular "cold box" which houses the critical equipment components that are comprised of pipes, piping aperture, and related valves connected to the regenerators. Regenerators are clustered into groups of four (called quads), where each quad has its own insulated cold box. The "cold box" consists of an outer shell made of carbon plated and an interior box made of tin sheets. The space between the interior and the exterior box is approximately less than 2 feet, and it is filled with rock wool insulation which is intended to keep the box cold, much like a freezer. Each cold box encloses the bottom of the regenerators and contains the valves and piping. The COG that comes from the regenerator must be kept at low temperatures in order to be safely handled and to efficiently separate chemical components into different recovery streams. The transport piping must therefore be kept at near cryogenic temperatures in order to prevent flashing.

The piping and valves that connect the quads together are housed in the cold boxes at the end(s) of each quad. There are 3 such cold boxes designated by the letters A-C. The lettered cold boxes are distinguished from the quad cold boxes by the fact that they are packed with insulation rather than having an insulated double wall around a hollow space.

Lettered cold boxes (A-C) are connected through the quads and also through the "cross-tie" box that permits gas to bypass the quads and flow between the A and B Cold Boxes. The cold boxes are intended to keep the areas under the Quad 3 generators cold. When properly functioning, this is accomplished by sealing the cold box thus preventing air from being introduced into the box at the ends of the cross-tie (i.e., A-B).

Each quad box is equipped with two atmospheric vent stacks which are approximately ten inches in diameter and extend about ninety feet upward to an end point on the roof of the facility. Near the top of each atmospheric vent stack is a flapper valve. A flapper valve is a check valve in the form of a hinged disk which permits flow in one direction only. A check valve is a device for automatically limiting the flow in a piping system to a single direction (also known as non-return valve). An example of a flapper valve is the part on the bottom of a household toilet that opens to allow water to flow from the tank to the bowl. When closed, it creates a seal between the two spaces. For the homeowner or apartment dweller, a flapper valve is commonly one of the reasons that the water in a toilet may be running constantly. What generally happens is that the valve fails to close all the way. When a flapper valve is damaged and no longer works, it's imperative to fix it immediately, especially when it is employed in a service that may lead to conditions that can contribute to fire and explosion. In the facility's system, the flapper valve is intended to allow any gases which build up in the cold box to flow out through an atmospheric vent stack and to seal against any outside air from entering into the cold box through the atmospheric vent stack.

In addition to the use of flapper valves, the system employs a control valve. This valve serves as a safety relief valve. This valve is opened when the pressure in the cold box becomes excessive. The control valve is housed near the top of the regenerator. The valve is set to automatically open at a certain set point when the pressure in the box becomes too high, thereby depressurizing the cold box by permitting venting through a discharge line. The control valve was designed to relieve the pressure at a set point that is lower than that of the flapper valve (i.e., it opens at a lower pressure than the flapper valve). This aspect is somewhat troublesome because the idea behind an automatic control valve is to protect the box from overpressurization, whereas the flapper valve is intended to allow gases that build up in the system to be evacuated periodically when vapors build up in the cold box.

Prior to the incident in question, one of the atmospheric vent stacks was equipped with a pressure control valve which vents the quad box at a lower pressure than the flapper valve. Also prior to the incident each quad box was equipped with a single point nitrogen injection system which was intended to inert any gas which entered the box and also to help in maintaining a positive pressure within the box. Within each quad cold box, there are various pipes, valves, and expansion joints which serve to connect the regenerator's gas streams to the larger lettered Cold Boxes A, B, and C located within the Control Room #2 area. Unlike the quad boxes, these lettered cold boxes are not in direct contact with the regenerators, but they all form an integrated network of the cryogenic process as they serve as a manifold for a specified set of quads.

The facility's records showed that on the day of the incident various contractors were working in the Control Room #2 area on a project involving the repair and rehabilitation of Cold Box B. Quads 1, 2 , 3 and Cold Box A were in operation at the time. Rehabilitation work was being done on the exterior of Cold Box B which involved, among other things, welding and grinding operations, and the vacuum removal of rock wool insulation from inside the box's walls. At one point the Control Room #2 area was rocked by a large explosion at Quad Box 3 followed by a fire that lasted several hours and was confined to the area around and under the Quad 3 regenerators. Inspections reported that the Quad 3 Cold Box floor had been deformed outward to the point that it impacted the concrete below, and the walls of the Quad 3 Cold Box had become detached and displaced and deformed outwards on both sides of the cold box to the point where they were no longer enclosing the space underneath. The following are major conclusions from the facility's investigation as to the root causes of the explosion:

1. There was a leak of gas from the pipes inside Quad Box 3.
2. The relatively light gas rose up one of the vent stacks and, in a phenomenon described by the investigating team as a "chimney effect," in which ambient air was drawn down into the opposite vent stack through a faulty flapper valve.
3. According to investigators, there may have also been an ingress of air through bad seals and breaches in the integrity of the exterior of the quad box itself.
4. A nitrogen purge system that had been installed in the Cold Box B prior to the incident was not sufficient to inert the environment inside the box. It was concluded that pockets of gas formed from gas leaks and remained inside the box.
5. Although the facility had a monitoring system for flammable vapors, it was not a continuous monitoring system. In other words, it had no means of determining whether an explosive mixture inside Quad 3 existed on a real time basis. As such, the gas/oxygen mixture rose to a dangerous combustible level, but because no-one or no system was used to monitor conditions on a real time basis, the environment within Quad 3 reached dangerous levels and was not detected in time to take corrective actions.
6. Once the gas/oxygen mixture had risen to a combustible level (i.e., within the flammability limits), it was ignited by some unidentified ignition source (possible ones discussed further on).

Figures 7.6 through 7.8 show photos of the extent of damage.

Figure 7.6 Photo showing damage to box from explosion.

Figure 7.7 Photo showing extensive damage to the box from explosion.

Figure 7.8 Photo showing extensive damage piping and insulation from explosion.

The nature of the physical damage described by the Incident Investigation Committee is consistent with a deflagration. The Committee concluded that the large deflagration was precipitated by the unwarranted entry of gas and oxygen into the cold box which was neither dissipated by the nitrogen purge system nor detected by the monitoring system. Once that mixture reached a combustible level, it was susceptible to ignition by a source of energy.

Among the lessons that the facility had to learn the hard way and at the expense of human loss and injury were the following:

The importance of religiously following OSHA standards – Part Number: 1910; Part Title: Occupational Safety and Health Standards, Subpart: H - Subpart Title: Hazardous Materials, Standard Number: 1910.119: Title: Process Safety Management of Highly Hazardous Chemicals. The purpose of this standard is to provide "requirements for preventing or minimizing the consequences of catastrophic releases of toxic, reactive, flammable, or explosive chemicals." As noted by OSHA, these releases may result in toxic, fire or **explosion hazards**.

Paragraph **1910.119(a)(1)(ii)** defines the facilities covered by the standard; namely "A process which involves a flammable liquid or gas (as defined in 1910.1200(c) of this part) on site in one location, in a quantity of 10,000 pounds (4535.9 kg) or more …" The facility in question produces >200 million cubic feet of flammable COG per year. The standard clearly applies.

Paragraph **1910.119(b)** provides the following definition: "Catastrophic release means a major uncontrolled emission, fire, or explosion, involving one or more highly hazardous chemicals, that presents serious danger to employees in the workplace." The same subsection provides the following definition: "Highly hazardous chemical means a substance possessing toxic, reactive, flammable, or explosive properties and specified by paragraph (a)(1) of this section." COG contains high concentrations of flammable gases and chemicals with low flash point temperatures.

The same subsection defines "Normally unoccupied remote facility means a facility which is operated, maintained or serviced by employees who visit the facility only periodically to check its operation and to perform necessary operating or maintenance tasks. No employees are permanently stationed at the facility. Facilities meeting this definition are not contiguous with, and must be geographically remote from all other buildings, processes or persons." The facility's quads meet this definition.

The same subsection states "Process means any activity involving a highly hazardous chemical including any use, storage, manufacturing, handling, or the on-site movement of such chemicals, or combination of these activities. For purposes of this definition, any group of vessels which are interconnected and separate vessels which are located such that a highly hazardous chemical could be involved in a potential release shall be considered a single process." The facility's quads fit this definition.

Paragraph **1910.119(j)** is devoted to "Mechanical integrity of equipment". Paragraph **1910.119(j)(1)(ii)** draws attention to the need for "mechanical integrity of Piping systems (including piping components such as valves)"; and **1910.119(j)(1)(iii)** addresses "Relief and vent systems and devices."

Paragraph **1910.119(j)(2)** requires employers to have "Written procedures - The employer shall establish and implement written procedures to maintain the on-going integrity of process equipment." The facility had no such written procedures for the flapper valves, the piping components housed in its cold boxes or the cold box housing.

Paragraph **1910.119(j)(3)** concerns "Training for process maintenance activities. The employer shall train each employee involved in maintaining the on-going integrity of process equipment in an overview of that process and its hazards and in the procedures applicable to the employee's job tasks to assure that the employee can perform the job tasks in a safe manner." The facility had no training activities because it never implemented any inspection and maintenance program for the flapper valves, the piping aperture housed inside its cold box, nor did it have one concerning the maintenance of the cold box housing. Only after the explosion were such programs implemented.

Paragraph **1910.119(j)(4)** is devoted to inspection and testing. **1910.119(j)(4)(i)** states "Inspections and tests shall be performed on process equipment." **1910.119(j)(4)(ii)** further states "Inspection and testing procedures shall follow recognized and generally accepted good engineering practices."

Paragraph **1910.119(j)(4)(iii)** states "The frequency of inspections and tests of process equipment shall be consistent with applicable manufacturers' recommendations and good engineering practices, and more frequently if determined to be necessary by prior operating experience." This specific paragraph under the OSHA standard alone shows how egregious the facility's derelictions of duties were especially in light of the fact that Quad Box 3 was significantly damaged in a prior explosion a few years earlier. That explosion, like the one described, was attributed to a pocket of gas which had accumulated inside the box and was not purged and then was ignited by a lightning strike. During that earlier investigation, it was also found that there was deterioration in welds on the outside of the cold box which permitted the entry of air. While the box was repaired after the earlier explosion and eventually put back in service, the fact that it was subjected to such a significant physical assault should have been an added reason for the plant to be particularly vigilant in inspecting all aspects of the quad box and its piping system.

Paragraph **1910.119(j)(4)(iv)** states "The employer shall document each inspection and test that has been performed on process equipment. The documentation shall identify the date of the inspection or test, the name of the person who performed the inspection or test, the serial number or other identifier of the equipment on which the inspection or test was performed, a description of the inspection or test performed, and the results of the inspection or test." The facility did not follow this standard because it had no inspection program for any of its cold boxes until after the September 2009 explosion.

Paragraph **1910.119(j)(5)** concerns equipment deficiencies. The standard states "The employer shall correct deficiencies in equipment that are outside acceptable limits (defined by the process safety information in paragraph (d) of this section) before further use or in a safe and timely manner when necessary means are taken to assure safe operation." Over 5 decades of operating its quads the facility never performed any inspections for preventive maintenance purposes. Like the other above standards it was and is obligated to follow, it simply ignored it.

OSHA was established in the 1970s to protect workers. Prior to the implementation and enforcement of OSHA standards many thousands of American workers were injured, contracted disabling occupational diseases, and were killed in catastrophic releases and explosions just like the one that occurred at this facility. Over the past 40 years, responsible industry stakeholders have made safe working conditions a part of the strategic core values of their businesses. This in part has been accomplished by companies investing in proper training of employees to perform detailed inspections and implementing preventive maintenance programs on critical equipment components. For the facility not to take the time and effort to implement a formal inspection and maintenance program on a piece of equipment that is intended to seal its quad box from air which when introduced results in a flammable environment is nothing less that reckless endangerment. For the facility to forego the establishment of formal inspection programs with defined protocols for eliminating gas leaks from the piping aperture contained in the cold box is nothing less than allowing workers to be subjected to ultra-hazardous working conditions. The facility sidestepped its statutory obligations to meet OSHA and its practices are immoral because of the high risks its operations present to its employees and contractors.

Companies need to recognize that simply seeking compliance usually results in mediocre mechanical integrity and process safety. In the hydrocarbon process industry which includes refineries, natural gas processing plants, synthetic fuels plants and chemical manufacturing plants, equipment integrity programs are critical. See for example the presentation made by Reynolds[2] at an inspection summit. Reynolds lists 101 elements that the industry considers essential to maintaining a formal Equipment Integrity Program, among these are:

- Positive Material Identification (PMI)
- Making Temporary Repairs
- Identifying Key and Critical (K/C) Materials Degradation Variables
- Addressing Materials and Corrosion
- Addressing Environmental Stress Cracking Issues for the cold box housing
- Addressing Brittle Fracture Issues for the cold box housing
- Record Keeping of Maintenance and Repairs to Critical Equipment Components
- Training and Certification of Maintenance Personnel especially for control valves, seals and critical piping components
- Formal Inspection Procedures for piping components, flapper valves, control valves and the equipment housing
- Inspection Scheduling
- Remaining Life Calculations for critical equipment
- Equipment Overdue
- Formal Piping Inspection
- Thickness Measurements
- Failure Reporting
- Inspection Recommendation Tracking
- Risk Management
- Corrosion Under Insulation
- External Corrosion Prevention
- Fatigue Failures
- Flange Gaskets
- Fitness for Service evaluations
- Relief Valve Preopping

[2] J. Reynolds, 101 Essential Elements in a Pressure Equipment Integrity Program For the Hydrocarbon Process Industry, Inspection Summit, Galveston, TX, January, 2009

- Knowledge Transfer
- Addressing Localized Corrosion
- Pressure Relief Device Auditing
- Inspection Staffing
- Inspection of Inaccessible Locations
- Process Creep
- Valve Quality Reviews
- Surface Cleaning
- Risk Based T/A Planning
- Failure Analysis

Facilities that do not follow such good industry practices are reckless and irresponsible, and take high risks with their investments, their businesses and the lives and safety of personnel.

The ignition source was not determined in this case. Possible source may have been improper grounding of electrical sources, the generation of electrostatic charges from step and touch activities, nitrogen tars which are known to be explosive, and other various sources considered by the investigators. Two critical issues stand out – one is that improper inerting practices were applied and there was no adequate monitoring for oxygen levels and flammable gases.

The facility relied on a single nitrogen feed which was nothing more than a small pipe with a hole. There were no provisions in the design and operation to ensure that nitrogen intimately mixed in the environment, no engineering computations were applied to estimate pressure and flow requirements, and no considerations were given to intimately mix and or displace gases with the inert gas. Furthermore, there was no constant monitoring inside the environment for oxygen. The concept of LOC was completely ignored. None of the good practices defined and carefully explained in NFPS Standard 69 were followed.

Facilities need to be watchful and attentive to their operations. Good industry practices and standards that are intended to ensure safe operations need to be considered and applied by competent persons with the right engineering and technical skills. Failure to do so can result in a loss of lives, injuries, the loss of major investments and entire businesses. With large, complex and especially older operations, it is impossible to eliminate all possible sources of ignition. Subsequently, practices like inerting, when suitable should be applied but they should be implemented by rigorously following the technical guidance of recognized standards.

8

Augmenting Risk Mitigation with Leak Detection and Repair

8.1 Introduction

Before closing this volume, I wish to emphasize that safety is everyone's business – from CEO, to manager, to secretary and IT support staff, to engineer, technician, contractor and all involved in a business operation. Business is in business to make a profit, but it must be done with a diligent eye on applying good practices, protecting workers and the public. Responsible companies consult safety standards and applying proven practices. In this regard, while dust suppression programs are distinct with specific good practices intended to reduce the risks associated with the handling of combustible dusts, they should not be applied without careful consideration of and integration with other practices which are intended to protect lives and operations. Annex A is a compilation of pertinent data and practices on safe handling of dangerous materials and can be consulted for more useful information. Additionally, this final chapter is intended to provide an overview of good industry practices for leak detection and repair programs (LDAR).

LDAR programs should be considered a critical part of the dust management program in situations where the presence of volatile organic chemicals which are flammable are generated or suspected of being generated within the manufacturing operation. Facilities like pharmaceutical plants handle enormous volumes of diluent solvents as well as manufacture various powdery-like health care products such as tablets and powders. This industry in particular is notorious for major losses of the solvents it relies on in the manufacture of its products, with some solvents estimated to be lost to air in manufacturing operations at levels in excess of 50% (see as an example, USEPA's AP-42 publication, Chapter 13 on Pharmaceuticals which describes solvent losses to air and other media by the industry sector). There are indeed other industry sectors in the chemical industry that operate with large amounts of solvents and generate powders and granular materials. Regardless of the sector, there is a concern for mismanagement of the deadly combination of VOCs (volatile organic compounds) and dusts. As a general warning, the author is of the opinion that it may be virtually impossible to identify and eliminate all possible sources of ignition; this being especially the case in older operations and plants. As such, an extremely high degree of diligence should be applied to eliminating losses of combustible dusts and ignitable vapors into the workroom environments.

LDAR is not only critical for fire and explosion safety, it represents a body of good practices that are intended to reduce and control harmful air discharges. In 1999, EPA estimated that, as a result of noncompliance with federal LDAR guidelines, an additional 40,000 tons of VOCs are emitted annually from valves at petroleum refineries alone.

8.2 Why LDAR Programs are Needed

Apart from the obvious reason of fire protection, fugitive emissions constitute a source of air pollution. EPA has determined that leaking equipment, such as valves, pumps, and connectors, are the largest source of emissions of volatile organic compounds (VOCs) and volatile hazardous air pollutants (VHAPs) from petroleum refineries and chemical manufacturing facilities. EPA[1] has reported that approximately 70,367 tons per year of VOCs and

---

[1] Leak Detection and Repair: A Best Practices Guide, United States Environmental Protection Agency Office of Compliance Office of Enforcement and Compliance Assurance, EPA-305-D-07-001 October 2007

9,357 tons per year of HAPs (Hazardous Air Pollutant) have been emitted from equipment leaks. Emissions from equipment leaks exceed emissions from storage vessels, wastewater, transfer operations, or process vents. EPA has noted that VOCs contribute to the formation of ground-level ozone. Ozone is a major component of smog, and causes or aggravates respiratory disease, particularly in children, asthmatics, and healthy adults who participate in moderate exercise. Many areas of the United States, particularly those areas where refineries and chemical facilities are located, do not meet the National Ambient Air Quality Standard (NAAQS) for ozone. Ozone can be transported in the atmosphere and contribute to nonattainment in downwind areas.

Also, some species of VOCs are also classified as VHAPs (Volatile Hazardous Air Pollutants). Some known or suspected effects of exposure to VHAPs include cancer, reproductive effects, and birth defects. The highest concentrations of VHAPs tend to be closest to the emission source, where the highest public exposure levels are also often detected. Some common VHAPs emitted from refineries and chemical plants include acetaldehyde, benzene, formaldehyde, methylene chloride, naphthalene, toluene, and xylene.

A benefit of a LDAR program is reduction of product losses. Facilities that apply LDAR also increase safety for workers and operators, decrease exposure of the surrounding community, reduce emissions fees, and help facilities avoid enforcement actions.

8.3 Sources of Fugitive Air Discharges

Many emissions are from valves and connectors because these are the most prevalent components and can number in the thousands depending on the nature of the plant. The major cause of emissions from valves and connectors is seal or gasket failure due to normal wear, improper or lack of maintenance. EPA studies have estimated that valves and connectors account for the majority of emissions from leaking equipment with valves being the most significant source; however, other significant sources reported include open-ended lines and sampling connections. The following are typical sources that are associated with leaks:

- Pumps are used to move fluids from one point to another. Two types of pumps extensively used in petroleum refineries and chemical plants are centrifugal pumps and positive displacement, or reciprocating pumps. Leaks from pumps typically occur at the seal.

- Valves are used to either restrict or allow the movement of fluids. Valves come in numerous varieties and with the exception of connectors, are the most common piece of process equipment in industry. Leaks from valves usually occur at the stem or gland area of the valve body and are commonly caused by a failure of the valve packing or O-ring.
- Connectors are components such as flanges and fittings used to join piping and process equipment together. Gaskets and blinds are usually installed between flanges. Leaks from connectors are commonly caused from gasket failure and improperly torqued bolts on flanges.
- Sampling connections are utilized to obtain samples from within a process. Leaks from sampling connections usually occur at the outlet of the sampling valve when the sampling line is purged to obtain the sample.
- Compressors are designed to increase the pressure of a fluid and provide motive force. They can have rotary or reciprocating designs. Leaks from compressors most often occur from the seals.
- Pressure relief devices are safety devices designed to protect equipment from exceeding the maximum allowable working pressure. Pressure relief valves and rupture disks are examples of pressure relief devices. Leaks from pressure relief valves can occur if the valve is not seated properly, operating too close to the set point, or if the seal is worn or damaged. Leaks from rupture disks can occur around the disk gasket if not properly installed.
- Open-ended lines are pipes or hoses open to the atmosphere or surrounding environment. Leaks from open-ended lines occur at the point of the line open to the atmosphere and are usually controlled by using caps, plugs, and flanges. Leaks can also be caused by the incorrect implementation of the block and bleed procedure.

8.4 Good Industry Practices

LDAR is a work practice that is designed to identify leaking equipment so that fugitive emissions can be reduced through repairs. A component that is subject to leaks must be monitored at specified, regular intervals to determine whether or not it is leaking. Any leaking component must then be repaired or replaced within a specified time frame.

Facilities can control emissions from equipment leaks by implementing a LDAR program or by modifying/replacing leaking equipment with "leakless" components. In practice, a combination of both control methods is often applied by responsible companies.

Leaks from open-ended lines, compressors, and sampling connections are usually fixed by modifying the equipment or component. Emissions from pumps and valves can also be reduced through the use of "leakless" valves and "sealless" pumps. Common leakless valves include bellows valves and diaphragm valves. Common sealless pumps are diaphragm pumps, canned motor pumps, and magnetic drive pumps. Leaks from pumps can also be reduced by using dual seals with or without barrier fluid.

Leakless valves and sealless pumps are effective at minimizing or eliminating leaks, but their use may be limited by materials of construction limitations and process operating conditions. Installing leakless and sealless equipment components may be a good choice for replacing individual, chronic leaking components.

When the LDAR requirements were developed, EPA estimated that petroleum refineries could reduce emissions from equipment leaks by 63% by implementing a facility LDAR program. Additionally, EPA estimated that chemical facilities could reduce VOC emissions by 56% by implementing such a program.

Major steps of LDAR are:

- Identification of Components that may leak
- Leak Definition
- Monitoring Components
- Repairing Components
- Recordkeeping

Step 1, Identification – This involves assigning a unique identification (ID) number to each component. One should record each regulated component and its unique ID number in a log. It is important to physically locate each regulated component in the facility, verify its location on the piping and instrumentation diagrams (P&IDs) or process flow diagrams, and update the log if necessary. It is good practice to attach a physical tag on each component that is subject to the LDAR program. Identify each regulated component on a site plot plan or on a continuously updated equipment log. Promptly note in the equipment log when new and replacement pieces of equipment are added and equipment is taken out of service.

Typical problems that are encountered by facilities in establishing programs include not properly identifying all possible or regulated equipment components and not properly documenting components which might be exempt from the program. The basic activities to apply in this step are to:

- Physically tag each regulated equipment component with a unique ID number
- Write the component ID number on piping and instrumentation diagrams
- Set up and institute an electronic data management system for LDAR data and records, possibly including the use of bar coding equipment
- Periodically perform a field audit to ensure lists and diagrams accurately represent equipment installed in the plant

In developing the inventory, avoid improper identification of components as 'unsafe' or 'difficult' to monitor. Components that are identified as being 'unsafe to monitor' or 'difficult to monitor' must be identified as such because there is a safety concern or an accessibility issue that prevents the component from being successfully monitored. All unsafe or difficult-to-monitor components must be included on a log with identification numbers and an explanation of why the component is "unsafe to monitor" or "difficult to monitor." Monitoring can be deferred for all such components, but the facility must maintain a plan that explains the conditions under which the components become safe to monitor or no longer difficult to monitor.

Step 2, Leak Definition –This refers to statutory obligations under the U.S. federal regulations that require a formal LDAR program with Method 21. Method 21 requires VOC emissions from regulated components to be measured in parts per million (ppm). A leak is detected whenever the measured concentration exceeds the threshold standard (i.e., leak definition) for the applicable regulation. Leak definitions vary by regulation, component type, service (e.g., light liquid, heavy liquid, gas/ vapor), and monitoring interval. Most NSPS (New Source Performance Standards) have a leak definition of 10,000 ppm. Many NESHAP use a 500-ppm or 1,000-ppm leak definition. Many equipment leak regulations also define a leak based on visual inspections and observations (such as fluids dripping, spraying, misting or clouding from or around components), sound (such as hissing), and smell. These topics are beyond the scope of this volume. We do note than the LDAR requirements specify weekly visual inspections of pumps, agitators, and compressors for indications of liquids leaking from the seals. Common problems that are encountered in practice include using the wrong leak definition for a particular component due to confusion at facilities where multiple LDAR regulations apply. Best practices to keep in mind are:

- Utilize a leak definition lower than what the regulation requires; especially when consideration of safety.
- Simplify the program by using the lowest leak definition when multiple leak definitions exist.
- Make the lowest leak definition conservative to provide a margin of safety when monitoring components.
- Keep the lowest leak definition consistent among all similar component types. For example, all valves in a facility might have a leak definition of 100 ppm.

Step 3, Monitoring the Components – Here a distinction must be made between monitoring in order to meet statutory reporting requirements for air pollution control and the monitoring that is required for fire and explosion mitigation. First, let us consider the monitoring interval. The monitoring interval is the frequency at which individual component monitoring is conducted. For example, valves are generally required to be monitored once a month using a leak detection instrument, but the monitoring interval may be extended (e.g. to once every quarter for each valve that has not leaked for two successive months for Part 60 Subpart VV, or on a process unit basis of once every quarter for process units that have less than a 2% leak rate for Part 63 Subpart H). However, this level of frequency can be unacceptable from the standpoint of fire and explosion protection. Monitoring for VOCs may need to be on a continuous basis if we are to ensure that process operations and upset conditions do not introduce flammable vapors into the working environment. In contrast, for many NSPS and NESHAP regulations with leak detection provisions, the primary method for monitoring to detect leaking components is EPA Reference Method 21 (40 CFR Part 60, Appendix A); which is a procedure that may be applied at specified intervals with the aim of reducing overall fugitive emissions. Method 21 is a procedure used to detect VOC leaks from process equipment using a portable detecting instrument. Monitoring intervals vary according to the applicable regulation, but are typically weekly, monthly, quarterly, and yearly. For connectors, the monitoring interval can be every 2, 4, or 8 years. The monitoring interval depends on the component type and periodic leak rate for the component type. As one can see, this type of frequency is not reasonable from a safety standpoint. A common problem faced during monitoring practices which can impact both safety and under reporting of fugitives is failing to monitor at the maximum leak location (once the highest reading is obtained by placing the probe on and around the interface, hold the probe at that location approximately two times the response rate of the instrument). Other problems are:

- Not monitoring long enough to identify a leak
- Holding the detection probe too far away from the component interface. The reading must be taken at the interface
- Not monitoring all potential leak interfaces
- Using an incorrect or an expired calibration gas
- Not monitoring all regulated or suspect components
- Not completing monitoring if the first monitoring attempt is unsuccessful due to equipment being temporarily out of service

In some instances, the number of components reported to have been monitored by companies may indicate problems with monitoring procedures. What EPA facility inspectors have found:

- A data logger time stamp showed valves being monitored at the rate of one per second with two valves occasionally being monitored within the same 1-second period.
- At one facility, a person reported monitoring 8,000 components in one day (assuming an 8-hour work day that represents one component every 3.6 seconds).
- Records evaluations showed widely varying component monitoring counts, suggesting equipment might not always be monitored when required.
- Equipment was marked “temporarily out of service” because the initial inspection attempt could not be performed. However, the equipment was in service for most of the period, and no subsequent (or prior) inspection attempts were performed to meet the monitoring requirement.

All of these areas should be examined and the pitfalls avoided in setting up the program. It is a good practice to audit the LDAR program to help ensure that the correct equipment is being monitored and to ensure that the proper frequency of monitoring is being applied. As part of the good practices one should perform QA/QC of LDAR data to ensure accuracy, completeness, and to check for inconsistencies. Also eliminate any obstructions (e.g., grease on the component interface) that would prevent monitoring at the interface. It is important to remember that even when records show a realistic number of components are being monitored, if there are no oversight or accountability checks, then there is no guarantee that components are actually being monitored.

When auditing the program, pay attention to the actual measurement practices and try to identify factors that may prevent the instrument from identifying leaks, such as

- Relying on dirty instrument probes
- Leakage from the instrument probes
- Not zeroing instrument meter
- Incorrect calibration gases used
- Not calibrating the detection instrument

Under EPA enforcement and inspection actions, some facilities were found not following proper monitoring procedures, resulting in a lower number of leaking components being reported. These were most often found associated with unintentional actions such as not allowing for sufficient time to identify a leak. As an example if a worker moves the probe around the component interface so rapidly that the instrument does not have time to properly respond, then a component may never be identified as leaking. Another example is if a worker fails to find the maximum leak location for the component and then does not spend twice the response time at that location, then the monitoring instrument will not measure the correct concentration of hydrocarbons and the leak may go undetected. Optical leak imaging shows the importance of identifying the maximum leak location, as hydrocarbons are quickly dispersed and diluted by air currents around the component.

Other problems identified include holding the probe too far away from the component interface. The probe must be placed at the proper interface of the component being analyzed. Placing the probe even 1 centimeter from the interface can result in a false reading, indicating that the component is not leaking, when in fact it is leaking. It is important to eliminate any issues (e.g., grease on the component interface) that prevent monitoring at the interface (e.g., remove excess grease from the component before monitoring or use a monitor that won't be impacted by the grease and is easy to clean.

For equipment with rotating shafts (pumps and compressors), Method 21 requires the probe be placed within 1 centimeter of the shaft-seal interface. Placing the probe at the surface of the rotating shaft is a safety hazard and should be avoided.

Step 4, Repairing Leaking Components – LDAR programs that are built around statutory requirements require the repair of leaking components as soon as practicable, but not later than a specified number of calendar days (usually 5 days for a first attempt at repair and 15 days for final attempt at repair) after the leak is detected. But this may not be considered acceptable for operations where there is a real risk fire and explosion. Initial attempts at repair include, but are not limited to, the following practices where practicable and appropriate:

- Tightening bonnet bolts
- Replacing bonnet bolts
- Tightening packing gland nuts
- Injecting lubricant into lubricated packing

If the repair of any component is technically infeasible without a process unit shutdown, LDAR programs generally allow the component to be placed on the Delay of Repair list, where the ID number is recorded, and an explanation of why the component cannot be repaired immediately is provided. An estimated date for repairing the component must be included in the facility records. Again, this approach may not be a reasonable practice when there is a concern for fire and explosion and elimination of all sources of energetics have not been considered.

Note that the 'drill and tap' method for repairing leaking valves is generally considered technically feasible without requiring a process. Drill and tap is a repair method where a hole is drilled into the valve packing gland and tapped, so that a small valve and fitting can be attached to the gland. A packing gun is connected to this fitting and the small valve is opened allowing new packing material to be pumped into the packing gland. Many facilities consider this a permanent repair technique, as newer, pumpable packing types are frequently superior to the older packing types they replace. Packing types can be changed and optimized for the specific application over time. Unit shutdown should be tried if the first attempt at repair does not fix the leaking valve. A component is considered to be repaired only after it has been monitored and shown not to be leaking above the applicable leak definition. Some common problems encountered in plants are:

- Not repairing leaking equipment within the required amount of time specified by the applicable regulation or specified plant requirements
- Improperly placing components on the Delay of Repair list

- Not having a justifiable reason for why it is technically infeasible to repair the component without a process unit shutdown
- Not exploring all available repair alternatives before exercising the Delay of Repair exemption (specifically as it pertains to valves and "drill and tap" repairs)

Among some good practices are:

- Develop a plan and timetable for repairing components
- Make a first attempt at repair as soon as possible after a leak is detected
- Monitor components daily and over several days to ensure a leak has been successfully repaired
- Replace problem components with "leakless" or other technologies

Step 5, Recordkeeping – Recordkeeping is an important step not simply for meeting statutory requirements, but for maintaining and implementing a proper preventive maintenance program. The following are good industry practices that are also part of the state and federal LDAR programs:

- Maintain a list of all ID numbers for all equipment subject to an equipment leak regulation.
- For valves designated as "unsafe to monitor," maintain a list of ID numbers and an explanation/review of conditions for the designation.
- Maintain detailed schematics, equipment design specifications (including dates and descriptions of any changes), and piping and instrumentation diagrams.
- Maintain the results of performance testing and leak detection monitoring, including leak monitoring results per the leak frequency, monitoring leakless equipment, and non-periodic event monitoring.

For leaking equipment follow the following protocols:

- Attach ID tags to the equipment
- Maintain records of the equipment ID number, the instrument and operator ID numbers, and the date the leak was detected

- Maintain a list of the dates of each repair attempt and an explanation of the attempted repair method
- Note the dates of successful repairs
- Include the results of monitoring tests to determine if the repair was successful

Some common problems that are encountered and to be avoided are:

- Not keeping detailed and accurate records
- Not updating records to designate new components that are subject to LDAR due to revised regulations or process modifications

More best practices include:

- Performing internal and third-party audits of LDAR records on a regular basis to ensure compliance with both the environmental regulations and the safety program
- Electronically monitor and store LDAR data including regular QA/QC audits
- Perform regular records maintenance
- Continually search for and update regulatory requirements
- Properly record and report first attempts at repair
- Keep the proper records for components on Delay of Repair lists

Generally, placing a leaking component on the Delay of Repair list is permissible only when the component is technically infeasible to repair without a process unit shutdown (e.g., for valves the owner/operator must demonstrate that the emissions from immediate repair will be greater than waiting for unit shutdown). Repair methods may exist, such as "drill and tap" for valves, that allow leaks to be fixed while the component is still in service. Failing to consider such repair methods before exercising the Delay of Repair list may constitute noncompliance with repair requirements (usually 15 days under federal LDAR standards), but also heighten risks for fires. Components placed on the Delay of Repair list must be accompanied by their ID numbers and an explanation of why they have been placed on the list. These components cannot remain on the list indefinitely – they must be repaired by the end of the next process unit shutdown.

Experience has shown that facilities with an effective record of preventing leaks integrate an awareness of the benefits of leak detection and repair into their operating and maintenance program. A strong and well balanced program should stem from careful evaluation of best practices identified at your facility, and also by looking to lessons learned from past incidents and an analysis of the root causes of noncompliance.

Programs should never be ad-hoc – there must be written LDAR program that specifies the regulatory requirements and facility-specific procedures for recordkeeping certifications, monitoring, and repairs. A written program also delineates the roles of each person on the LDAR team as well as documents all the required procedures to be completed and data to be gathered, thus establishing accountability. The plan should identify all process units subject to federal, state, and local LDAR regulations and be updated as necessary to ensure accuracy and continuing compliance. Further, the integration of the program in relationship to its support of dust suppression and risk reduction should be clearly defined in the written program. Some important elements that should be incorporated into the written program include:

- An overall, facility-wide leak rate goal that will be a target on a process-unit-by-process-unit basis
- A list of all equipment in light liquid and/or in gas/vapor service that has the potential to leak VOCs and VHAPs, within process units that are owned and maintained by each facility
- Procedures for identifying leaking equipment within process units
- Procedures for repairing and keeping track of leaking equipment
- A process for evaluating new and replacement equipment to promote the consideration of installing equipment that will minimize leaks or eliminate chronic leakers
- A list of "LDAR Personnel" and a description of their roles and responsibilities, including the person or position for each facility that has the authority to implement improvements to the LDAR program
- Procedures (e.g., a Management of Change program) to ensure that components added to each facility during maintenance and construction are evaluated to determine if they are subject to LDAR requirements, and that affected components are integrated into the LDAR program

A training program will provide LDAR and fire protection personnel the technical understanding to make the written safety programs work. It also will educate members of the LDAR and dust suppression team(s) on their individual responsibilities. These training programs can vary according to the level of involvement and degree of responsibility of LDAR personnel. LDAR training elements may cover:

- Provide and require initial training and annual LDAR refresher training for all facility employees assigned LDAR compliance responsibilities, such as monitoring technicians, database users, QA/QC personnel, and the LDAR Coordinator.
- For other operations and maintenance personnel with responsibilities related to LDAR, provide and require an initial training program that includes instruction on aspects of LDAR that are relevant to their duties (e.g., operators and mechanics performing valve packing and unit supervisors that approve delay of repair work). Provide and require "refresher" training in LDAR for these personnel at least every three years.
- Collect training information and records of contractors, if used.

Appendix A: General Guidelines on Safe Work Practice

Dust explosion and fire safety management programs should be carefully integrated with the overall safe work practices and procedures of the facility. This appendix provides useful general information and good industry practices for safe work ethics and the handling of chemicals.

Appendix A.1 – Glossary

Biological monitoring means the measurement and evaluation of a substance, or its metabolites, in the body tissue, fluids or exhaled air of a person exposed to that substance or blood lead level monitoring.

Class of dangerous goods, means the number assigned to the goods in the ADG Code indicating the hazard, or most predominant hazard, exhibited by the goods.

Combustible substance means a substance that is combustible and includes dust, fibres, fumes, mists or vapours produced by the substance.

Container means anything in or by which a hazardous chemical is, or has been, wholly or partly covered, enclosed or packed, including anything necessary for the container to perform its function as a container.

Correct classification means the set of hazard classes and hazard categories assigned to a hazardous chemical when it is correctly classified.

Division of dangerous goods, means a number, in a class of dangerous goods, to which the dangerous goods are assigned.

Exposure standard represents the airborne concentration of a particular substance or mixture that must not be exceeded. The exposure standard can be of three forms:

- 8-hour time-weighted average
- peak limitation
- short term exposure limit.

Flash point means the lowest temperature (corrected to a standard pressure of 101.3 kPa) at which the application of an ignition source causes the vapours of a liquid to ignite under specified test conditions.

GHS means the *'Globally Harmonized System of Classification and Labelling of Chemicals, 3rd Revised Edition'*, published by the United Nations as modified under Schedule 6 of the WHS Regulations.

Hazard means a situation or thing that has the potential to harm people, property or the environment. The GHS covers physicochemical, health and environmental hazards for hazardous chemicals.

Hazard category means a division of criteria within a hazard class.

Hazard class means the nature of a physical, health or environmental hazard.

Hazard pictogram means a graphical composition, including a symbol plus other graphical elements, that is assigned to a hazard class or hazard category.

Hazard statement means a statement assigned to a hazard class or hazard category describing the nature of the hazards of a hazardous chemical including, if appropriate, the degree of hazard.

Hazardous chemical means a substance, mixture or article that satisfies the criteria for a hazard class, but does not include a substance, mixture or article that satisfies the criteria solely for one of the following hazard classes:

a. acute toxicity—oral—category 5;
b. acute toxicity—dermal—category 5;
c. acute toxicity—inhalation—category 5;
d. skin corrosion/irritation—category 3;
e. serious eye damage/eye irritation— category 2B;
f. aspiration hazard—category 2;
g. flammable gas—category 2;
h. acute hazard to the aquatic environment—category 1, 2 or 3;
i. chronic hazard to the aquatic environment—category 1, 2, 3 or 4;
j. hazardous to the ozone layer.

Label means written, printed or graphical information elements concerning a hazardous chemical that is affixed to, printed on, or attached to the container of a hazardous chemical.

Manufacture includes the activities of packing, repacking, formulating, blending, mixing, making, remaking and synthesizing of the chemical.

Mixture means a combination of, or a solution composed of, two or more substances that do not react with each other.

Placard means a sign or notice:

a. displayed or intended for display in a prominent place, or next to a container or storage area for hazardous chemicals at a workplace
b. that contains information about the hazardous chemical stored in the container or storage area.

Precautionary Statement means a phrase that describes measures that are recommended to be taken to prevent or minimise the adverse effects of exposure to a hazardous chemical or the improper handling of a hazardous chemical.

Substance means a chemical element or compound in its natural state or obtained or generated by a process:

- including any additive necessary to preserve the stability of the element or compound and any impurities deriving from the process, but
- excluding any solvent that may be separated without affecting the stability of the element or compound, or changing its composition.

Supply includes selling or transferring ownership or responsibility for a chemical.

Appendix A.2 – Hazard Classes and Categories

ADG class/category, packing group	Equivalent GHS class and category as classified under the WHS Regulations
Class 1 Explosives	
Unstable explosives (Goods too dangerous to be transported)	Unstable explosives
Division 1.1	Division 1.1
Division 1.2	Division 1.2
Division 1.3	Division 1.3
Division 1.4	Division 1.4
Division 1.5	Division 1.5
Division 1.6	Division 1.6

(*Continued*)

ADG class/category, packing group	Equivalent GHS class and category as classified under the WHS Regulations
Class 2 Gases	Gases under pressure NOTE: The GHS has 4 categories which correspond to the transport condition under the ADG Code. They are: • Gas under pressure – Compressed gas • Gas under pressure – Liquefied gas • Gas under pressure – Refrigerated liquefied gas • Gas under pressure – Dissolved gas
Division 2.1	Flammable gases category 1 Flammable aerosols category 1 and 2
Division 2.2	Oxidising gases category 1 Gases under pressure not otherwise specified
Division 2.3	Acute toxicity: Inhalation categories 1–4 (Note: category 4 only up to LC_{50} of 5000 ppmV) Skin corrosion / irritation categories 1A-C
Class 3 PG I	Flammable liquids category 1
Class 3 PG II	Flammable liquids category 2
Class 3 PG III	Flammable liquids category 3
Division 4.1 Self Reactive substances types A-G [1]	Self reactive substances type A-F Type G are not classified under WHS Regulations as hazardous chemicals.
Division 4.1 PG II	Flammable solids category 1
Division 4.1 PG III	Flammable solids category 2
Division 4.2 PG 1	Pyrophoric liquids category 1 Pyrophoric solids category 1
Division 4.2 PG II	Self heating substances category 1
Division 4.2 PG III	Self heating substances category 2

ADG class/category, packing group	Equivalent GHS class and category as classified under the WHS Regulations
Division 4.3 PG I	Substances and mixtures which in contact with water emit flammable gases, category 1
Division 4.3 PG II	Substances and mixtures which in contact with water emit flammable gases, category 2
Division 4.3 PG III	Substances and mixtures which in contact with water emit flammable gases, category 3
Division 5.1 PG I	Oxidising solids, oxidising liquids, category 1
Division 5.1 PG II	Oxidising solids, oxidising liquids, category 2
Division 5.1 PG III	Oxidising solids, oxidising liquids, category 3
Division 5.2 Organic Peroxides types A-G [1]	Organic peroxides type A-F. Type G are not classified under WHS Regulations as hazardous chemicals.
Division 6.1 PG I	Acute toxicity: Oral category 1 Acute toxicity: Dermal category 1 Acute toxicity: Inhalation category 1 (dusts, mists, vapours)
Division 6.1 PG II	Acute toxicity: Oral category 2 Acute toxicity: Dermal category 2 Acute toxicity: Inhalation category 2 (dusts, mists, vapours)
Division 6.1 PG III	Acute toxicity: Oral category 3 Acute toxicity: Dermal category 3 Acute toxicity: Inhalation category 3 (dusts, mists, vapours)
Division 6.2	No equivalent GHS class and not classified under WHS Regulations as hazardous chemicals.

(Continued)

ADG class/category, packing group	Equivalent GHS class and category as classified under the WHS Regulations
Division 7	No equivalent GHS class and not classified under WHS Regulations as hazardous chemicals.
Class 8 PG I	Skin corrosion category 1A
Class 8 PG II	Skin corrosion category 1B
Class 8 PG III	Skin corrosion category 1C Corrosive to metals category 1
Class 9 [2]	Class 9 dangerous goods are not classified under the WHS Regulations.
Goods too dangerous to be transported	Self reactive substances type A [1] Organic peroxides type A [1] Unstable explosives
C1 combustible liquids (flash point 60–150°C)	Flammable liquids category 4 (flash point 60–93°C)

[1]Depending on packing method, self-reactive substances and organic peroxides type A will either be classified as 'Goods too dangerous to be transported' or their comparative Divisions (4.1 or 5.2)

[2]Class 9 dangerous goods include ecotoxicological hazard classes and categories

Appendix A.3 – Prohibited Carcinogens, Restricted Carcinogens and Restricted Hazardous Chemicals

The table below shows prohibited carcinogens, restricted carcinogens and restricted hazardous chemicals.

The prohibition of the use of carcinogens listed in table C.1, column 2 and the restriction of the use of carcinogens listed in table C.2, column 2 apply to the pure substance and where the substance is present in a mixture at a concentration greater than 0.1%, unless otherwise specified.

Table C.1 Prohibited carcinogens.

Column 1 Item	Column 2 Prohibited carcinogen [CAS number]
1	2-Acetylaminofluorene [53-96-3]
2	Aflatoxins

3	4-Aminodiphenyl [92-67-1]
4	Benzidine [92-87-5] and its salts (including benzidine dihydrochloride [531-85-1])
5	bis(Chloromethyl) ether [542-88-1]
6	Chloromethyl methyl ether [107-30-2] (technical grade which contains bis(chloromethyl) ether)
7	4-Dimethylaminoazobenzene [60-11-7] (Dimethyl Yellow)
8	2-Naphthylamine [91-59-8] and its salts
9	4-Nitrodiphenyl [92-93-3]

Table C.2 Restricted carcinogens.

Column 1 **Item**	**Column 2** **Restricted carcinogen [CAS Number]**	**Column 3** **Restricted use**
1	Acrylonitrile [107-13-1]	All
2	Benzene [71-43-2]	All uses involving benzene as a feedstock containing more than 50% of benzene by volume Genuine research or analysis
3	Cyclophosphamide [50-18-0]	When used in preparation for therapeutic use in hospitals and oncological treatment facilities, and in manufacturing operations Genuine research or analysis
4	3,3'-Dichlorobenzidine [91-94-1] and its salts (including 3,3'-Dichlorobenzidine dihydrochloride [612-83-9])	All
5	Diethyl sulfate [64-67-5]	All
6	Dimethyl sulfate [77-78-1]	All
7	Ethylene dibromide [106-93-4]	When used as a fumigant Genuine research or analysis

(*Continued*)

Table C.2 (*Cont.*)

Column 1 Item	Column 2 Restricted carcinogen [CAS Number]	Column 3 Restricted use
8	4,4'-Methylene bis(2-chloroaniline) [101-14-4] MOCA	All
9	3-Propiolactone [57-57-8] (Beta-propiolactone)	All
10	o-Toluidine [95-53-4] and o-Toluidine hydrochloride [636-21-5]	All
11	Vinyl chloride monomer [75-01-4]	All

Table C.3 Restricted hazardous chemicals.

Column 1 Item	Column 2 Restricted hazardous chemical	Column 3 Restricted use
1	Antimony and its compounds	For abrasive blasting at a concentration of greater than 0·1% as antimony
2	Arsenic and its compounds	For abrasive blasting at a concentration of greater than 0·1% as arsenic For spray painting
3	Benzene (benzol), if the substance contains more than 1% by volume	For spray painting
4	Beryllium and its compounds	For abrasive blasting at a concentration of greater than 0·1% as beryllium
5	Cadmium and its compounds	For abrasive blasting at a concentration of greater than 0·1% as cadmium
6	Carbon disulphide (carbon bisulphide)	For spray painting

7	Chromate	For wet abrasive blasting
8	Chromium and its compounds	For abrasive blasting at a concentration of greater than 0·5% (except as specified for wet blasting) as chromium
9	Cobalt and its compounds	For abrasive blasting at a concentration of greater than 0·1% as cobalt
10	Free silica (crystalline silicon dioxide)	For abrasive blasting at a concentration of greater than 0·1% For spray painting
11	Lead and compounds	For abrasive blasting at a concentration of greater than 0·1% as lead or which would expose the operator to levels in excess of those set in the regulations covering lead
12	Lead carbonate	For spray painting
13	Methanol (methyl alcohol), if the substance contains more than 1% by volume	For spray painting
14	Nickel and its compounds	For abrasive blasting at a concentration of greater than 0·1% as nickel
15	Nitrates	For wet abrasive blasting
16	Nitrites	For wet abrasive blasting
17	Radioactive substance of any kind where the level of radiation exceeds 1 Bq/g	For abrasive blasting, so far as is reasonably practicable
18	Tetrachloroethane	For spray painting
19	Tetrachloromethane (carbon tetrachloride)	For spray painting
20	Tin and its compounds	For abrasive blasting at a concentration of greater than 0·1% as tin
21	Tributyl tin	For spray painting

Appendix A.4 – Requirements for Health Monitoring

The table below provides guidance for health monitoring.

Column 1 Item	Column 2 Hazardous Chemical	Column 3 Type of health monitoring
1	Acrylonitrile	Demographic, medical and occupational history Records of personal exposure Physical examination
2	Arsenic (inorganic)	Demographic, medical and occupational history Records of personal exposure Physical examination with emphasis on the peripheral nervous system and skin Urinary inorganic arsenic
3	Benzene	Demographic, medical and occupational history Records of personal exposure Physical examination Baseline blood sample for haematological profile
4	Cadmium	Demographic, medical and occupational history Records of personal exposure Physical examination with emphasis on the respiratory system Standard respiratory questionnaire to be completed Standardised respiratory function tests including for example, FEV_1, FVC and FEV_1/FVC Urinary cadmium and $\beta_2$microglobulin Health advice, including counselling on the effect of smoking on cadmium exposure

5	Chromium (inorganic)	Demographic, medical and occupational history Physical examination with emphasis on the respiratory system and skin Weekly skin inspection of hands and forearms by a competent person
6	Creosote	Demographic, medical and occupational history Health advice, including recognition of photosensitivity and skin changes Physical examination with emphasis on the neurological system and skin, noting any abnormal lesions and evidence of skin sensitisation Records of personal exposure, including photosensitivity
7	Crystalline silica	Demographic, medical and occupational history Records of personal exposure Standardised respiratory questionnaire to be completed Standardised respiratory function test, for example, FEV_1, FVC and FEV_1/FVC Chest X-ray full size PA view
8	Isocyanates	Demographic, medical and occupational history Completion of a standardised respiratory questionnaire Physical examination of the respiratory system and skin Standardised respiratory function tests, for example, FEV_1, FVC and FEV_1/FVC
9	Mercury (inorganic)	Demographic, medical and occupational history Physical examination with emphasis on dermatological, gastrointestinal, neurological and renal systems Urinary inorganic mercury

(*Continued*)

10	4,4'-Methylene bis (2chloroaniline) (MOCA)	Demographic, medical and occupational history Physical examination Urinary total MOCA Dipstick analysis of urine for haematuria Urine cytology
11	Organophosphate pesticides	Demographic, medical and occupational history including pattern of use Physical examination Baseline estimation of red cell and plasma cholinesterase activity levels by the Ellman or equivalent method Estimation of red cell and plasma cholinesterase activity towards the end of the working day on which organophosphate pesticides have been used
12	Pentachlorophenol (PCP)	Demographic, medical and occupational history Records of personal exposure Physical examination with emphasis on the skin, noting any abnormal lesions or effects of irritancy Urinary total pentachlorophenol Dipstick urinalysis for haematuria and proteinuria
13	Polycyclic aromatic hydrocarbons (PAH)	Demographic, medical and occupational history Physical examination Records of personal exposure, including photosensitivity Health advice, including recognition of photosensitivity and skin changes
14	Thallium	Demographic, medical and occupational history Physical examination Urinary thallium
15	Vinyl chloride	Demographic, medical and occupational history Physical examination Records of personal exposure

Appendix A.5 – Risk Assessment Process Logic Diagram

An overview of the process for the assessment of health risks arising from the use of hazardous chemicals in the workplace is provided below.

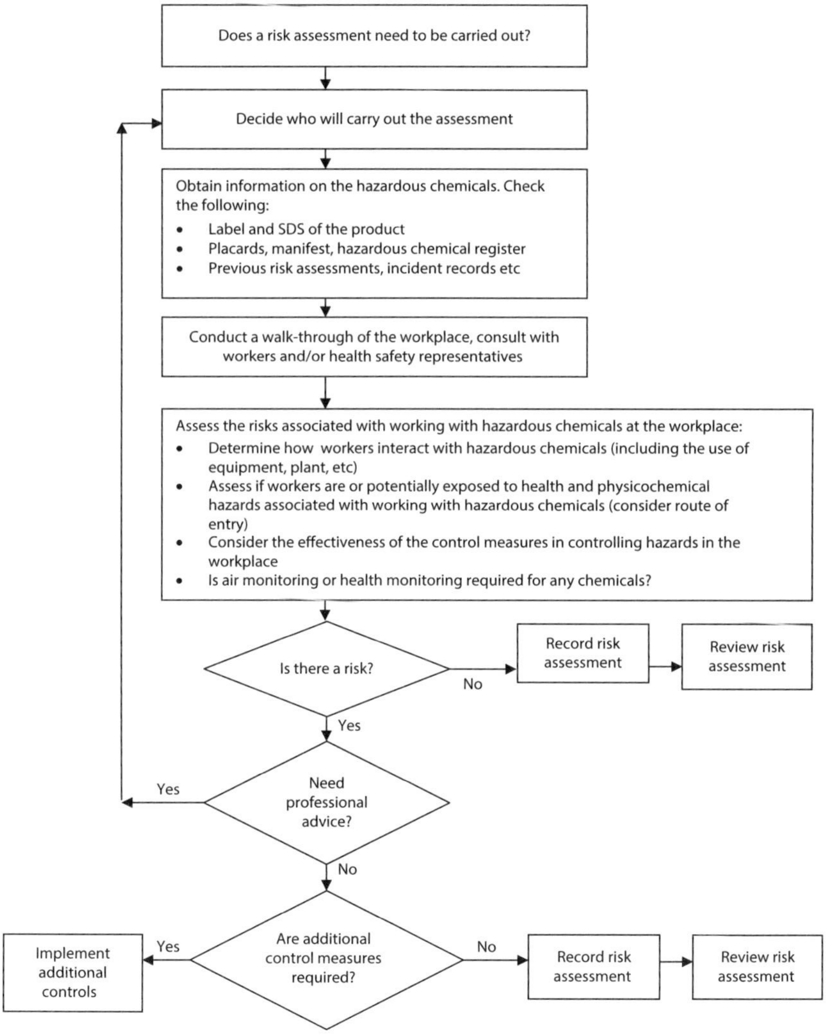

Appendix A.6 – Risk Assessment Audit Checklist

Questions	**Yes**	**No**
1. Does a risk assessment need to be carried out?	☐	☐
2. Has it been decided who should carry out the risk assessment?	☐	☐
3. Have all the hazardous chemicals in the work place been identified?	☐	☐
Has a hazardous chemical register been produced?	☐	☐
4. Has information about the hazardous chemicals been gathered? *(refer to labels, SDS, placards and relevant Australian Standards for the type of hazardous chemical)*	☐	☐
Q. 5 – 9 should be answered for each hazardous chemical or each process where hazardous chemicals are used in the workplace		
5. Have you checked other records associated with the hazardous chemical? *(Consider previous assessments, monitoring records, injury or incident records, induction training, task-specific training etc)*	☐	☐
If 'Yes', are there any hazardous chemical previously assessed as 'high' or as 'significant risk'? Specify the risk(s):	☐	☐
6. Does the chemical have health hazards? *(consider potential acute / chronic health effects and likely route of entry)*	☐	☐
7. Does the hazardous chemical have physicochemical hazards?	☐	☐
8. Does the hazardous chemical have an exposure standard? (refer to the *Workplace Exposure Standards for Airborne Contaminants*)	☐	☐
9. Do workers using the hazardous chemical require health monitoring? (refer to *Part 7.1, Division 6 and Schedule 14 of the WHS Regulations*) If 'Yes', air monitoring may be required.	☐	☐

Questions	Yes	No
10. Are workers, or can workers be potentially, exposed to hazardous chemicals at the workplace, including by-products and waste? For each hazardous chemical or group of hazardous chemicals in the work unit, find out: • Is the substance released or emitted into the work area? • Are persons exposed to the chemical? • How much are the persons exposed to and for how long? *Air monitoring may be required to determine exposure* • Are there any risks associated with the storage and transport of the chemical? Have all hazardous chemicals in the workplace been identified? If not, repeat Q.2 for the next hazardous substance.	☐	☐
11. Are control measures currently in the workplace well maintained and effective in controlling the hazards? If 'No', take appropriate action	☐	☐
12. What are the conclusions about risk? Only answer 'Yes' to one conclusion. • Conclusion 1: Risks are not significant • Conclusion 2: Risks are significant but effectively controlled If you answer Yes to conclusion 1 or 2, go to Q.14. • Conclusion 3: Risks are significant and not adequately controlled • Conclusion 4: Uncertain about risks If you answer 'Yes' to conclusion 3 or 4, go to Q.13.		
13. Have actions resulting from conclusion about risks been identified? • Seek expert advice • Requires appropriate control measure • Requires induction training • Requires on-going monitoring • Requires health monitoring • Requires emergency procedures and first aid		
14. Has the assessment been recorded? ☐ ☐ only a notation in the register.)	☐	☐

Appendix A.7 – Examples of Common Fuel and Oxygen Sources

Fuel type	Examples*	Label elements
Flammable gases	Liquefied petroleum gas (LPG), natural gas, hydrogen, acetylene, hydrogen sulphide, carbon monoxide	
Flammable and combustible liquids	Petrol, mineral turpentine, lighter fluid or 'shellite', kerosene, methylated spirit, acetone, ether, ethanol, hexane, pentane, naphtha, some solvent based paints, diesel, including biodiesel, petroleum based oils, some oil based paints, cotton seed, linseed and eucalyptus oils	 (flammable liquids categories 1-3 only).
Flammable and combustible solids	Bitumen, asphalt, fats and greases, waxes, shellac, acetate and nitrocellulose films, timber and timber products, paper, cardboard, dry grasses, hay, straw, plastics, silk, granulated rubber, metal shavings, filings	 (flammable solids only)
Other fire risk chemicals**	Pyrophoric substances like some barium and calcium alloys, iron sulphide and celluloid scrap	
Dusts	Any dusts that can are generated through other processes, such as metal grinding, filling, etc	none
Chemical reactions***	Water reactive chemicals like calcium carbide, sodium hydride, and some aluminium, lithium, magnesium or zinc powders (which liberate flammable gases like hydrogen on contact with water or acids).	

Notes:

* The form of the substance or material can significantly affect the risk. In general, the smaller the particle size the greater the risk. For example, fine shavings or powders of metals present a much greater risk than metals in the bulk or massive form.

** Pyrophoric substances can react spontaneously in contact with air.

*** Chemical reactions which generate gases can also cause explosions through an increase in the pressure in the container in which the chemical is stored if the gas cannot escape, even if that gas does not itself ignite.

Examples of oxygen sources	Label elements
Oxygen and air cylinders in welding equipment, hospitals for treatment of patients, reticulated gas supplies in a laboratory, air tanks in self contained breathing apparatus (SCBA) equipment	
Nitric acid, nitrates, nitrous oxide, sodium hypochlorite, chlorates, perchlorates, hydrogen peroxide and organic peroxides, potassium permanganate	

Note: Oxygen gas is always present in the air so you should assume that it is present in your workplace. Although oxygen gas itself is not flammable, it will cause a fire to burn with more intensity and at a higher temperature. In oxygen-enriched atmospheres (greater than around 23%) some substances that are not normally flammable can even self ignite.

Appendix A.8 – Fire and Explosion Risks

Industry	Process	Hazards
Agriculture	Grain silos and auger loaders	Combustible particles in the form of husks and fine dusts, dust explosions
	Chaff and hay processing and storage	Combustible particles & dusts and spontaneous combustion of haystacks
	Milling grains and sugars, cellulose, milling fibres - cotton, linen, polyesters, possible peroxide powders	Flammable and combustible materials, dusts and fibres, possible static build-up, oxidising agents
	Processing oil and oil seeds –cottonseed, linseed, other vegetable oils, canola, olives	Combustible oils with possible combustible wastes
	Viticulture and alcoholic spirit manufacture	Flammable and combustible materials & vats or tanks containing flammable vapours
	Drying and processing grains & vegetables e.g. tobacco drying, vegetable preparation	Cellulose fibres, dusts, and other combustible material, rotting vegetable matter produces methane gas

(*Continued*)

Industry	Process	Hazards
Agriculture	Flammable or combustible pesticides	Some pesticides contain flammable or combustible carrier liquids
	Liquid and gaseous ammonia for nitrogen fixing in soils	Flammable gas, toxic gas, corrosive
Automotive industry	Manufacture	Fuels, oils, spray painting, electrical
	Motor mechanics	Fuels, oils, solvents, oxy-acetylene
	Auto electrical	Battery charging, oils and sparks
	Upholstery - vinyls, plastics, glues & solvents, wadding	Flammable and combustible materials
Bakeries	Transferring and pouring flour	Grain flour dusts, heat generation
Battery industry	Recharging wet cells	Hydrogen gas generation and sparks
Bootmaker / Shoe repairs	Gluing, grinding and buffing rubber, leather and plastics	Flammable glues & vapours
Construction industry	Curing agents	Flammable
Chemical industry (manufacturing)	Bulk storage, mixing, blending, aerosol cans - Acetone, ether, polishes, oils, waxes, matches, fire lighters, cigarettes etc	Flammable gases, flammable liquids, flammable or combustible solids and other hydrocarbons, sulphur
	Plastics manufacture and rotomoulding	Flammable and combustible solids, powders, oxidation, heat, static sparks
Drycleaners	Solvent cleaners	Flammable liquids and vapours
Elcctrical industry	Power generation, transformers and transmission lines	Combustible oils, high temperatures and heat, sparks, fires

Industry	Process	Hazards
Explosives industry	Manufacturing, storage, mixing/blending, loading, including auger loaders, nitrates, explosive powders oxidising agents	Potentially explosive metal powders and dusts, mechanical attrition milling, temperature and pressure, flames, heat, incompatible materials
Fibreglass work	Catalysts and resins used contain styrenes and organic peroxides, also use of solvents such as acetone and Methyl ethyl ketones (MEK)	Flammable liquids, oxidising substances and exothermic heat generation capable of causing combustion in other flammable or combustible materials
Film industry	Acetate and nitrocellulose films as well as solvents	Highly flammable and may be liable to spontaneous combustion when exposed to air
Food industry	Grains, flours, sugars, fermentation gases, alcohols	Combustible particles in the form of husks and fine dusts, flammable or combustible gases, and liquids
Gas industry	Manufacturing, storage, transmission, pumping and transport	LPG, methane, hydrogen, acetylene, gas accumulation in tanks, pipes and tankers
Laboratories	Mixing, blending, storage, heating, reactions, acids, alkalis, oxides & peroxides, use of Bunsen burners	Flammable and combustible gases, liquids, solids, dusts, exothermic heat, flames, oxidising agents
Metal production & manufacturing, iron, steel & foundry work, product manufacture	Melting, casting, milling, grinding, welding, electroplating	Molten metals and heat, mechanical attrition milling, metal dusts, shavings, filings, welding gases and sparks, Flammable solvents and electrolysis can produce hydrogen gas bubbles
Mining	Coal mining	Coal dust, methane gas, hydrogen gas, sulphur powder

(*Continued*)

Industry	Process	Hazards
Mining	Metaliferous mines	Iron, aluminium, magnesium, zinc Metal powders and dusts
Paint industry	Oil and solvent based paints, spray painting	Flammable and combustible aerosolised particles, mists vapours, fumes
Paper and cardboard manufacturing	Paper & cardboard processes bleaching fibres and paper - use of peroxides, fibreboard box manufacture	Combustible particles in the form of fibres and dusts, flammable or combustible materials and articles, oxidising agents
Petroleum Industry and other chemical manufacturing	Crude oil & other petroleum products such as petroleum gases, petroleum fuels & oils including diesel & biodiesel bitumen, kerosene, etc	Generation of flammable and combustible hydrocarbons in the form of flammable gases
Pharmaceutical	Bulk storage, mixing & blending	Flammable and combustible materials and articles
Plastics manufacture	Plastics incl. vinyls, ethylene, styrene, vinyl chloride	Flammable and combustible solids, powders, oxidation, heat, static sparks
Printing industry	Inks, dyes, solvents, paper and cardboard	Flammable and combustible materials and articles. For example: paper & cardboard
Recycling and waste disposal	Landfill burial of organic wastes, tyre shredding, paper and cardboard accumulation	Generation of methane gas, combustible rubber particles and dusts, combustible paper products
Road works	Asphalt and bitumen, LPG heating, kerosene & solvents	Flammable and combustible materials and articles
Sewage treatment	Organic waste treatment	Generation of methane and hydrogen sulphide gases
Textile industry	Cotton, linen, silk, synthetics	Fibres

Industry	Process	Hazards
Tyre manufacture	Hot rubber moulding, gluing and grinding rubber	Heat, flammable & combustible glues, combustible dusts and solids
Underground car parks & cellars	Accumulation of heavier than air gases, carbon monoxide	Flammable gas and asphyxiant
Wood working	Milling and processing, Furniture and cabinet making glues, thinners, oils, waxes, plastics, rubber, shellac	Saw dust, and fine wood dusts, flammable & combustible solvents

Appendix A.9 – Practical Examples of Control Measures

The following table illustrates some situations involving risks from hazardous chemicals that may be encountered in the workplace, and provides some examples of controls that may be considered to eliminate or minimise the risks. The conclusions you make in your assessment should be supported by clear and valid evidence.

Examples	Examples of controls
• Use of petrol driven vehicles in poorly ventilated work areas	• Ensure adequate ventilation. Consider use of electric or diesel vehicles.
• Activities which involve prolonged skin contact with hazardous chemicals that are either readily absorbed through the skin or that can directly affect the skin	• Change work practices to avoid skin contact, or select and use appropriate PPE to control exposure
• Handling of caustic or acidic chemicals where there is a potential for splashes onto the skin or eyes	• Consider installing automated systems to dispense or transfer chemicals between containers. Use eye protection. Provide an emergency eyewash facility.
• Dry sweeping of fine particulates	• Use vacuum cleaning as an alternative, or wet cleaning methods.

• Manually cleaning printing screens or large printing rollers with large quantities of volatile solvents.	• Automate or enclose the process and ensure adequate ventilation. Use non-volatile solvents or detergent/water based cleaning solutions
• Processes for which monitoring results approach or exceed exposure standards	• Upgrade ventilation systems so that monitoring results are well below the exposure standard. More efficient ventilation systems may avoid the need for expensive air monitoring in some situations.
• Evidence of significant quantities of fine deposits on workers and surfaces, or processes that generate fine mists or solid particulates (including fumes) within the breathing zones of workers	• Review control measures of the process to minimise release of particles at the source. Examples may include enclosing the process or installing ventilation systems. Review and revise housekeeping procedures to remove dust build up more frequently.
• Application of volatile chemicals over large surface areas	• Substitute less volatile and hazardous solvents.

The following information provides more specific guidance and recommendations on managing the risks for particular types of hazardous chemicals, primarily those hazardous chemicals that are dangerous goods. It gives in more detail some precautions that you should consider to assist in the safe management of higher hazard chemicals like gases under pressure, flammable liquids and solids, self-reactive and oxidising substances as well as advice on how to manage the risks during the abandonment or removal of underground storage tanks.

Gas Cylinders (Gases under pressure)

Used or empty cylinders should be treated with the same precautions as for full cylinders, since residual hazards remain.

Testing and maintenance of gas cylinders

Gas cylinders need to be tested periodically to ensure that they remain safe to use. A poorly maintained gas cylinder can leak, exposing workers to harmful or potentially explosive vapours, or fail catastrophically. In-built safety features may also become inoperable over time. Details of inspection and testing for gas cylinders are provided in AS 2030.1: *Gas cylinders – General requirements.*

As a guide, gas cylinders should be tested every 10 years for dry gases and more frequently for damp or corrosive gases – check with the gas supplier if you need advice. The last test date will be stamped on the cylinder near the valve or on the collar, or on the foot ring of some small cylinders. If the test period has expired, the cylinder may be unsafe to use and it should not be refilled until it is re-tested (and receives a new date stamp). However, it is permissible to use up the cylinder's contents after its test date has expired, prior to testing. Alternatively it could be replaced with a new cylinder. Testing stations can give advice on disposal of a used cylinder if you wish to replace it. Owners of cylinders should keep records of testing and test dates.

Storage and handling of gas cylinders

Cylinders may be stored safely by following these steps:

- Any cap provided for use with a cylinder is kept in place on the cylinder at all times when the cylinder is not being filled and not connected for use
- The cylinder valve is kept securely closed when not in use, including when empty (unless the cylinder is connected by permanent piping to a consuming device)
- Any removable valve protection cap or valve outlet gas tight cap or plug is kept in place on the cylinder at all times (unless the cylinder is being filled or connected for use)
- Keep the cylinder secured against unintended movement by installing chains preventing the cylinder from falling
- Do not lubricate valves or attempt repair of leaks – if the valve is not closing properly, immediately remove the cylinder to a safe area outdoors and seek expert assistance
- Have a water hose or fire extinguisher handy to put out any small fire close to the cylinder – a water spray can also be used to keep the cylinder cool in the event of a fire

To ensure the in-built safety features of a cylinder function correctly, cylinders of liquefied flammable gas need to be positioned so that the safety relief device is in direct contact with the vapour space within the cylinder.

Keep the cylinder upright, unless the design permits other positions – this depends on the position and operation of the relief device. If in doubt check the manufacturer or supplier instructions.

For further guidance on safe storage and handling of gas cylinders, refer to AS 4332: *The storage and handling of gases in cylinders.*

Further advice on storage and handling of specific gases is available from the following Australian Standards:

- AS/NZS 2022: *Anhydrous ammonia – Storage and handling*
- AS 1894: *The storage and handling of non-flammable cryogenic and refrigerated liquids*
- AS/NZS 2927: *The storage and handling of liquefied chlorine gas*
- AS 3961: *The storage and handling of liquefied natural gas*
- AS/NZS 1596: *The storage and handling of Liquefied Petroleum (LP) gas*
- AS 4839: *The safe use of portable and mobile oxy-fuel gas systems for welding, cutting, heating and allied processes*
- AS 4289: *Oxygen and acetylene gas reticulation systems*

Unodourised Liquefied Petroleum Gas (Lp Gas) or Dimethyl Ether

Although the sense of smell should not be relied upon to detect gas leaks and hazardous chemicals, it can often provide some level of warning to nearby workers in some instances. Unodourised LP Gas can be particularly hazardous and, due to the absence of any discernable odour, it cannot be detected by the sense of smell. Dimethyl ether (DME), a highly flammable gas, is often used as a propellant for LP gas.

Using the following control measures can reduce the risks from storing and using unodourised LP Gas:

- Keep the storage and handling of unodourised LP Gas or DME to a minimum, and restrict uses to those for which no less hazardous alternative is available (e.g. aerosol propellant)
- The area where it is stored and handled should be well ventilated, or in a room designed for that purpose fitted with explosion ventilation, or in the open. Access to these areas should be restricted to essential personnel.

- Gas detection equipment should be installed to detect gas where an explosive atmosphere could develop and provide an automatic alarm before dangerous levels of gas are reached so that immediate action may be taken. The gas detector should emit an audible sound and have a visual display.

Flammable Liquids in Packages and in Bulk (Class 3 dangerous goods)

Australian Standard AS 1940: *The storage and handling of flammable and combustible liquids* provides guidance on the safe storage and handling of flammable and combustible liquids, including aspects such as package stores, bulk storage, tank design, pipework and valves.

Abandoning or Removing Underground Tanks of Flammable Liquids

The WHS Regulations require notification to the regulator when an underground, partially underground or fully mounded tank containing flammable liquids or flammable gases is to be abandoned. When the container no longer contains hazardous chemicals, placards and signs should be removed.

Any work on existing or abandoned underground tanks or associated pipework is potentially dangerous where residual levels of the flammable gases, liquids and vapours are present. Introducing an ignition source may cause an explosion or other dangerous occurrence unless suitable procedures are adopted.

Tar-like deposits and sludge may have accumulated in the tank and pipe work. Flushing with water may not remove them and vapour testing may not detect this. Exposure of these deposits to air and sunlight under normal temperatures, or work involving heat (e.g. use of grinders or oxy-acetylene cutting), may release vapours creating a potential explosion hazard.

By following the steps listed below, the likelihood of dangerous occurrences can be minimised or even eliminated:

- Remove the tank from the ground and transport to a disposal area and arrange for the tank to be decommissioned.
- Fill the tank with an inert solid material like concrete or sand.
- If it is intended that the tank be used again (within two years), you can fill the tank with water and a corrosion inhibitor

Further information on removal and disposal of underground tanks is available in Australian Standards, for example AS 4976: *The removal and disposal of underground petroleum storage tanks.*

Self-Reactive Substances, Flammable Solids, Pyrophoric Liquids And Solids, Self-Heating Substances And Mixtures And Substances Which in Contact With Water Emit Flammable Gas (Class 4 dangerous goods)

There are a number of key considerations for controlling the fire risks from storing and handling the above types of hazardous chemicals. These include:

- Ensuring non-combustible materials are used in the construction of buildings and storage areas
- Installing and maintaining appropriate fire protection systems
- Utilising separation distances (or barriers such as fire resistant screen walls)
- Ensuring ignition and heating sources are controlled within the storage and handling areas, for example, electrical equipment used in these areas is intrinsically safe
- Ensuring adequate ventilation and/or extraction is provided to avoid creation of a hazardous atmosphere or hazardous area
- Installation of explosion doors or vents if there is the potential that flammable gases or vapour could be formed or there is the potential to form combustible dusts
- Ensuring that the storage area is moisture free and protected from the elements
- Ensuring that measures are taken to protect light or temperature sensitive materials, for example, by installing temperature controls or protecting from direct sunlight

Tanks to be used for storing or handling these hazardous chemicals should be designed and operated to ensure that:

- Moisture cannot enter the tanks
- Valves and fittings are readily accessible, easily operated and operate as designed

- If practical, remote operation for primary shut off valves at the tank is provided

Flammable Solids (Class 4.1 dangerous goods)

Nitrocellulose film and other nitrocellulose products – handling and storage

Nitrocellulose film and product containing nitrocellulose can represent a significant explosion hazard if the risks are not properly controlled. Risks can be minimised by:

- Reducing the amount of material stored or handled in the work area at any one time
- Ensuring the storage and handling area is constructed from non-combustible materials
- Ensuring there is sufficient means of escape in the event of an emergency. For example, use of outward opening doors, and removing all non-essential furniture and equipment from the work area to allow unimpeded access to the emergency exit
- Eliminating all ignition sources, including
    - Using intrinsically safe electrical wiring and equipment suitable for use in hazardous areas
    - Guard or enclose heating elements and other electrical equipment to prevent ignition or decomposition of any nitrocellulose products
    - Keeping the temperature of any surfaces and equipment (including enclosures) to a suitably safe temperature for the material being used
- Installation of an automatic sprinkler system
- Preventing the accumulations of excessive amounts of waste materials
- Displaying suitable signs warning of hazards and precautions (for example, "No smoking")

Oxidizing Agents (Class 5.1 dangerous goods)

Oxidising substances are hazardous chemicals that are reactive and can support combustion. They can react and are incompatible with a range of other substances including organic materials (wood, paper) and hydrocarbon solvents. You should always refer to the SDS to check for any incompatibilities with the materials you are using, storing or handling.

Unintended dangerous reactions of oxidising agents can be avoided by observing the following precautions:

- Keep away from combustible or readily oxidisable materials, including fuel containers, sulfur and powdered metal and any other incompatible materials. Stores of oxidisers should be a reasonable distance away (for example, at least 5 m)
- Place packages and containers on clean pallets, racks or shelving to allow easier detection of leaks and to prevent contact with other substances. Some oxidising chemicals can ignite on contact with timber, therefore, old and weathered pallets should not be used
- Eliminate sources of heat if practicable. If this is not practicable, ensure that heat sources do not allow the oxidising agents to be heated to within about 15°C of their decomposition temperature
- Keep packages closed when not in use to avoid spillage
- Do not park or drive any vehicles (e.g. forklifts) nearby because heat from the engine or fuel or oil leaks may cause a dangerous occurrence
- Do not store any liquids above oxidizing agents in case leaks cause incompatible materials to spill onto the oxidising substance
- Do not allow accumulation of dust and keep surfaces clean in areas where oxidising substances are handled in the workplace
- Clean up spillages immediately and dispose of waste in accordance with your local regulations. Do not mix substances in the waste bin because they might react or cause a fire

Solid (dry) pool chlorine

If your workplace keeps large quantities of solid (dry) pool chlorine on the premises, avoid dangerous reactions by observing the precautions listed above. You should also ensure that the pool chlorine is kept a safe distance away (e.g. at least 10 m) from any ammonium salt like ammonium sulfate, or be separated from it by suitable bunding.

Organic Peroxides (Class 5.2 dangerous goods)

Organic peroxides are capable of self-reaction and stabilizers are usually necessary. Some are classified as "Goods too dangerous to be transported" and extreme caution is needed when storing or handling these materials.

Like oxidising agents, organic peroxides can be highly reactive with incompatible materials and precautions are necessary to avoid unintended reactions occurring. Risks can be eliminated or minimised by observing the following precautions:

- Keep packages in a specifically designated and designed cabinet, room or external storage building containing explosion vents and/or doors to limit the effects in the event of an explosion
- Keep a suitable safety zone (e.g. 5 m) opposite the cabinet or storeroom doors and blow out panels
- Use cabinet doors with friction or magnetic catches to allow any pressure build up to escape more easily
- Keep nothing else in the organic peroxides store. If this is not practicable, then measures should be taken to ensure that incompatible materials cannot come into contact with the organic peroxides
- Keep the storage area free of waste, dirt, dust or metal filings (these could react with spillages) or any combustible materials
- Eliminate ignition sources inside, or outside within a suitable exclusion zone (e.g. 3m) of the storage area or entrance to the store
- Keep packages on sealed or laminated hardwood or coated metal shelves free from rust or corrosion to avoid a harmful reaction in the event of a spill
- Keep a space of at least 100 mm between the packages and the floor, ceiling, or walls. Fitting a guarding system or raised shelving can assist with this
- Keep suitable spill containment equipment close to the store which can be accessed quickly and used in the event of a spillage
- If opening packages, take them at least 3 m clear of the store. Reseal all packages before returning them to the store

Temperature controls can be important in the safe handling and storage of organic peroxides. To avoid harmful reactions or decomposition of the organic peroxides due to temperature:

- Determine any critical temperatures including any recommended maximum temperature. The label and SDS may provide this information. Otherwise, other sources should be consulted. Keep the store within the recommended temperature range for the different types of organic peroxides present and keep organic peroxides out of direct sunlight
- Do not permit heating to be installed in the storage area

If cooling or refrigeration is required to maintain the desired temperature in the storage area, expert advice should be obtained because air conditioners and unmodified refrigerators are potential ignition sources.

Further information on storage and handling of organic peroxides can be obtained from AS 2714: *The storage and handling of organic peroxides.*

Corrosives (Class 8 dangerous goods)

Corrosive substances and mixtures [class 8 dangerous goods] can be either alkaline or acidic and these two categories are incompatible. Acids should never be stored with alkaline chemicals due to the potential for harmful reactions. Some reactions of acids and alkaline chemicals can be highly exothermic and rapidly generate large amounts of gas, causing an explosion risk.

Risks associated with storage and handling of corrosive substances and mixtures can be eliminated or minimised by observing the guidance in the following Australian Standards:

- AS 3780: *The storage and handling of corrosive substances*
- AS 1940: *The storage and handling of flammable and combustible liquids* (where the corrosive substance or mixture is also a flammable liquid or has a dangerous goods Subsidiary Risk of Class 3 (flammable liquid)
- AS/NZS 3833: *The storage and handling of mixed classes of dangerous goods, in packages and intermediate bulk containers*

Eyewash and safety showers should be readily accessible where corrosives are handled or transferred.

Glossary of Terms

The following are terms that the reader should become familiar with in dealing with dust explosion and fire safety.

Air Material Separator: A broad term for a device designed to separate powders from the air in which it is transported. Most commonly, this would be a cyclone or dust collector.

Barg: bar, gauge (unit of pressure)

Blanketing (or Padding): The technique of maintaining an atmosphere that is either inert or fuel-enriched in the vapor space of a container or vessel.

Bonding: Electrical connection between two electrically-conductive objects that minimizes any difference in electrical potential between them.

Burning to Detonation: This refers to the phenomenon which can take place when an explosive substance is confined in a tube and ignited at one end. The gas generated from the chemical decomposition of the explosive mixture becomes trapped, resulting in an increase in pressure at the burning surface; this in turn raises the linear burning rate. In detonating explosives the linear burning rate is raised so high by pressure pulses generated at the burning surface that it exceeds the velocity of sound, resulting in a detonation.

Chips: Particles produced from a cutting or machining that are not oxidized and that are not diluted by noncombustible materials. Chips vary in ease of ignition and rapidity of burning, depending on their size and geometry.

Combustible Concentration Reduction: According to NFPA #69, the technique of maintaining the concentration of combustible material in a closed space below the lower flammable limit.

Combustible Metal: Any metal, composed of distinct particles or pieces, regardless of size, shape, or chemical composition that will burn.

Combustible Metal Dust: Any finely divided metal 420 μm (microns) or smaller in diameter (that is, material passing a U.S. No. 40 standard sieve) that presents a fire or explosion hazard. Any time a combustible dust is processed or handled, the potential for explosion or fire exists. The degree of hazard will vary depending on the type of combustible dust, conditions, the amount of material present, and processing methods used.

Combustion: A chemical process of oxidation that occurs at a rate fast enough to produce heat and usually light in the form of either a glow or flame.

Damage-Limiting Construction (DLC): Construction designed to minimize the damage from a deflagration (explosion) in equipment or building. This can be pressure resistive, pressure relieving, or some combination of the two. Most commonly this would be vent panels on enclosures (buildings or equipment) releasing at a pressure below the strength of the enclosure.

Deflagration: A substance is classed as a deflagrating material when a small amount of it in an unconfined condition suddenly ignites when subjected to a flame, spark, shock, friction or high temperatures. Deflagrating explosives burn faster and more violently than ordinary combustible materials. The burning of a deflagrating explosive is therefore a surface phenomenon which is similar to other combustible materials, except that explosive materials do not need a supply of oxygen to sustain the burning. The propagation of an explosion reaction through a deflagrating explosive is based on thermal reactions. The explosive material surrounding the initial exploding site is warmed above its decomposition temperature causing it to explode. Explosives such as propellants exhibit this type of explosion mechanism. The transfer of energy by thermal means through a temperature difference is a relatively slow process and depends very much on external conditions such as ambient pressure. The speed of the explosion process in deflagrating explosives is always subsonic; that is, it is always less than the speed of sound.

Deflagration Suppression: The technique of detecting and arresting combustion in a confined space while the combustion is still in its incipient stage, thus preventing the development of pressures that could result in an explosion (NFPA #69).

Deposit Velocity: The minimum air (or other gas) velocity needed to prevent dust-particle fallout during pneumatic-material conveying and to pick up any dust particles deposited during airflow interruption. The velocity varies with particle weight, density, and aerodynamic properties.

Design Strength: Pressure to which a vessel can be exposed without any risk of damage (because a safety factor has been applied to the yield strength).

Detonation: Explosive substances which on initiation decompose via the passage of a shockwave rather than a thermal mechanism are referred to as detonating materials or explosives. The velocity of the shockwave in solid or liquid explosives is between 1500 and 9000 m s-l, an order of magnitude higher than that for the deflagration process. The rate at which the material decomposes is governed by the speed at which the material will transmit the shockwave, not by the rate of heat transfer. Detonation can be achieved either by burning to detonation or by an initial shock.

Differential Scanning Calorimeter: The technique of differential scanning calorimetry (DSC) is very similar to DTA. The peaks in a DTA thermogram represent a difference in temperature between the sample and reference, whereas the peaks in a DSC thermogram represent the amount of electrical energy supplied to the system to keep the sample and reference at the same temperature. The areas under the DSC peaks will be proportional to the enthalpy change of the reaction. DSC is often used for the study of equilibria, heat capacities and kinetics of explosive reactions in the absence of phase changes, whereas DTA combined with TGA is mainly used for thermal analysis.

Differential Thermal Analysis (DTA): This is a test method that involves heating (or cooling) a sample and an inert reference sample under identical conditions and recording any temperature difference which develops between them. Any physical or chemical change occurring to the test sample which involves the evolution of heat will cause its temperature to rise temporarily above that of the reference sample, thus giving rise to an exothermic peak on a DTA plot. Conversely, a process which is accompanied by the absorption of heat will cause the temperature of the test sample to lag behind that of the reference material, leading to an endothermic peak.

Double Dump Valve: An arrangement of two gate or butterfly valves in series. Only one is open at a time. This valve is often used where material discharged from one vessel is gravity fed to another vessel (i.e., not pneumatically conveyed) such as a dust collector discharging into a hopper below it or a material blender or grinder discharging into a pneumatic conveying system.

Dust: Any sample of solid particles with a median size smaller than 500 microns.

Dust, Combustible: Established by tests that expose the material to ignition sources of various intensities, such as a spark, a match flame, a Bunsen burner, or a Meker burner. Any organic material, unoxidized metal particles, or other oxidizable materials (e.g., zinc stearate) should be considered combustible unless testing proves otherwise. A combustible dust is not always an explosible dust.

Dust, Explosible: Established by ASTM test E1226 (Test Method for Pressure and Rate of Pressure Rise for Combustible Dusts) or E1515 (Test Method for Minimum Explosible Concentration of Combustible Dusts), or National/International equivalent (e.g., ISO 6184/1, Explosion Protection Systems – Part 1: Determination of explosion indices of combustible dusts in air). Dust with a median particle size greater than 500 microns can be assumed to be non-explosible as long as particles smaller than 500 microns cannot be segregated during material handling.

Enclosure: A confined or partially confined volume.

Explosion Isolation: System or single device that prevents the propagation of explosion effects from one volume to an adjacent volume.

Explosion Mitigation: Methods used to reduce damage from the explosion after the explosion has started.

Explosion Prevention: Methods used to prevent an explosion by controlling either the air, fuel, ignition source, or a combination of these.

Flame Arrester: A device that prevents the transmission of a flame through a flammable gas/air mixture by quenching the flame on the surfaces of an array of small passages through which the flame must pass.

Flame Burning Velocity: Defined by NFPA (Standard 69) as the burning velocity of a laminar flame under specified conditions of composition, temperature, and pressure for unburned gas.

Flammable Limits (flammability limits): The minimum and maximum concentrations of a combustible material in a homogeneous mixture with a gaseous oxidizer that will propagate a flame.

Flammable Range: The range of concentrations between the lower and upper flammable limits.

Grounding: Electrical connection between a conductive object and the ground that minimizes the difference in the electrical potential between the object and ground.

Heat of Explosion: When an explosive material is initiated either to burning or detonation, its energy is released in the form of heat. The liberation of heat under adiabatic conditions is called the 'heat of explosion'. The heat of explosion provides information about the work capacity of the explosive.

Heats of formation: The heats of formation for a reaction can be described as the total heat evolved when a given quantity of a substance is completely oxidized in an excess amount of oxygen, resulting in the formation of carbon dioxide, water and sulfur dioxide. The energy liberated during the formation of combustion products is known as the 'heat of explosion'. If these products are then isolated and allowed to burn in excess oxygen to form substances like carbon dioxide, water, *etc.*, the heat evolved added to the heat of explosion would be equal to the 'heat of combustion'. Consequently, the value for the heat of combustion is higher than the value for the heat of explosion for substances which have insufficient oxygen for complete oxidation. The value for the heat of formation can be negative or positive. If the value is negative, heat is liberated during the reaction and the reaction is exothermic; whereas, if the value is positive, heat is absorbed during the reaction and the reaction is endothermic. For reactions involving explosive components the reaction is always exothermic. In an exothermic reaction the energy evolved may appear in many forms, but for practical purposes it is usually obtained in the form of heat. The energy liberated when explosives deflagrate is called the 'heat of deflagration', whereas the energy liberated by detonating explosives is called the 'heat of detonation' in kJ mol^{-1} or the 'heat of explosion' in kJ kg^{-1}.

Ignitable Liquid: Any liquid or liquid mixture that is capable of fueling a fire, including flammable liquids, combustible liquids, inflammable liquids, or any other reference to a liquid that will burn. An ignitable liquid must have a fire point.

Inerting: A technique by which a combustible mixture is rendered non-ignitible by adding an inert gas or a noncombustible dust.

Kinetics: Kinetics is the study of the rate of change of chemical reactions. Reactions can be very fast, *i.e.* instantaneous reactions such as detonation, those requiring a few minutes, i.e. dissolving sugar in water, and those requiring several weeks, *i.e.* the rusting of iron. In explosive reactions the rate is very fast and is dependent on the temperature and pressure of the reaction, and on the concentration of the reactants.

Kst: The dust explosibility constant, defined as the maximum rate of pressure rise of a dust explosion in a 1 cubic meter vessel. The units are bar meter per second (bar m/s). The test method used to obtain this constant is standardized worldwide. This value (K_{st}) is used in all modern dust explosion vent sizing to characterize the reactivity (i.e., explosibility) of a particular dust. Only metric units are used for this constant.

Limiting Oxidant Concentration (LOC): The concentration of oxidant in a fuel-oxidant-diluent mixture below which a deflagration cannot occur under specified conditions.

Lower Flammable Limit (LFL): The lowest concentration of a combustible substance in an oxidizing medium that will propagate a flame.

MEC: Minimum explosible concentration, the lowest concentration of dust that can support a self-propagating explosion. (The terms LEL [lower explosible limit] or LFL [lower flammable limit] mean the same, but are not often used in the context of dust explosions.)

Media Type Collector: A device (enclosure) that separates dry, solid material from air by passing the air through a dry filtering medium. Examples are enclosures with bag-type filters, cartridge-type filters (normally a pleated filter arranged in a cylindrical shape, similar to automobile air filters), rotary drum filters, and panel filters.

MIE (Minimum Ignition Energy): The minimum amount of thermal energy released at a point in a combustible mixture to cause indefinite flame propagation under specified test conditions. The lowest value of

MIE, known as LMIE, is found at a certain optimum mixture. It is this value that is usually reported as the MIE.

Oxidant: A gaseous material that can react with a fuel (either gas, dust, or mist) to produce combustion.

Oxidant Concentration Reduction: NFPA #69 defines this as the technique of maintaining the concentration of an oxidant in a closed space below the concentration required for ignition to occur.

Phlegmatization: The process of mixing inert dusts with combustible dusts to reduce or eliminate the explosion hazard.

Pblast, Max: The localized pressure as a result of the fireball and pressure from a vented explosion.

Pmax: The maximum pressure developed in the 20-liter sphere when testing dust for explosibility characteristics by ASTM E1226 method. It is factor used to help size explosion vents.

Pred: Highest explosion pressure in a vessel protected with explosion vents; usual units are bar_g or psig.

Pstat: Explosion vent relief pressure; usual units are bar_g or psig.

Psia: Pounds per square inch, absolute (unit of pressure).

Psig: Pounds per square inch, gauge (unit of pressure).

Pyrophoric Material: Term refers to a substance capable of self-ignition on short exposure to air under ordinary atmospheric conditions.

Ribbon: A piece of metal that is less than in. (3.2 mm) in two dimensions or less than an inch in single dimension. These sizes are generally thought of as a powder.

Shock to Detonation: Explosive substances can be detonated if they are subjected to a high velocity shockwave. Detonation will produce a shockwave which will initiate a secondary explosive if they are in close contact. The shockwave forces the particles to compress, and this gives rise to adiabatic heating which raises the temperature to above the decomposition temperature of the explosive material. Explosive crystals undergo an exothermic chemical decomposition which accelerates the shockwave. If the velocity of the shockwave in the explosive composition exceeds the velocity of sound, detonation will take place, Although initiation to detonation does not take place instantaneously the delay is negligible, being in microseconds.

Strong Ignition Source: A strong ignition can provide more than approximately 100 Joules of energy. Examples of a strong ignition source include open flame, welding arc, gas or dust explosion, and electric arc/short. One should consider the presence of combustible building construction or large amounts of combustible storage along with unacceptable levels of fugitive dust as presenting the possibility of a dust cloud coincident with a flame (strong ignition source). Conversely, examples that would not be considered a strong ignition source include frictional sparks, mechanical impact sparks, static sparks, cigarettes, hot surfaces, overheated electrical components.

Strong Vessel: A vessel that can withstand explosion pressures in excess of 0.2 bar_g (3 psig) without being damaged or destroyed. This includes most process vessels constructed or used in Europe.

Thermogravimetric Analysis (TGA): This is a technique used for the study of explosive or unstable reactions. In TGA the sample is placed on a balance inside an oven and heated at a desired rate and the loss in the weight of the sample is recorded. Such changes in weight can be due to evaporation of moisture, evolution of gases, and chemical decomposition reactions, i.e. oxidation. TGA is generally combined with DTA, and a plot of the loss in weight together with the DTA thermogram is recorded. These plots give information on the physical and chemical processes which are taking place. In an explosive reaction there is a rapid weight loss after ignition due to the production of gaseous substances and at the same time heat is generated.

Tube Sheet: The mounting plate for cartridge-type filters or bag-type filter tubes and cages.

Upper Flammable Limit (UFL): The highest concentration of a combustible substance in a gaseous oxidizer that will propagate a flame.

Ultimate Strength: Pressure at which an enclosure will be torn open (i.e., ruptured).

Weak vessel or Enclosure: A structure that cannot withstand explosion pressures in excess of 0.2 bar_g (3 psig) without being damaged or destroyed. This includes most rooms, buildings, and many North American process vessels.

Yield Strength: Pressure at which an enclosure will be deformed without rupturing.

Index

Also of Interest

Check out these other related titles from Scrivener Publishing

From the Same Author

Industrial Gas Flaring Practices, by Nicholas P. Cheremisinoff, PhD, ISBN 9781118937878. With the consequences of the world's gas flaring practices only just beginning to be understood or even studied, this volume is the first in decades to tackle a very difficult hot-button issue for our time that could significantly reduce CO2 emissions and affect global warming.

Emergency Response Management for Offshore Oil Spills, by Nicholas P. Cheremisinoff, PhD, and Anton Davletshin, ISBN 9780470927120. The first book to examine the Deepwater Horizon disaster and offer processes for safety and environmental protection. *NOW AVAILABLE!*

A Guide to Save Material and Chemical Handling, by Nicholas P. Cheremisinoff, PhD, and Anton Davletshin, ISBN 9780470625828. This handy volume is a ready "go to" reference for the chemical engineer, process engineer, or chemist working with the industrial chemicals and processes involving corrosion protection, high temperature and pressure services. The volume provides an assembly of useful engineering and properties data on materials of selection for process equipment, and the chemical properties, including toxicity of industrial solvents and chemicals. *NOW AVAILABLE!*

Other Related Titles from Scrivener Publishing

Critical Reflections on Nuclear and Renewable Energy, by Way Kuo, ISBN 9 781118773420. Provides a personal yet scientific comparative study of nuclear energy and a spectrum of other energies from a reliability point of view, as well as outlining guidelines for developing energy policies. *NOW AVAILABLE!*

Ethics *in Engineering,* by James Speight and Russell Foote, ISBN 9780470626023. Covers the most thought-provoking ethical questions in engineering. *NOW AVAILABLE!*

Biofuels Production, Edited by Vikash Babu, Ashish Thapliyal, and Girijesh Kumar Patel, ISBN 9781118634509. The most comprehensive and up-to-date treatment of all the possible aspects for biofuels production from biomass or waste material available. *NOW AVAILABLE!*

Biogas Production, Edited by Ackmez Mudhoo, ISBN 9781118062852. This volume covers the most cutting-edge pretreatment processes being used and studied today for the production of biogas during anaerobic digestion processes using different feedstocks, in the most efficient and economical methods possible. *NOW AVAILABLE!*

Bioremediation and Sustainability: Research and Applications, Edited by Romeela Mohee and Ackmez Mudhoo, ISBN 9781118062845. Bioremediation and Sustainability is an up-to-date and comprehensive treatment of research and applications for some of the most important low-cost, "green," emerging technologies in chemical and environmental engineering. *NOW AVAILABLE!*

Coal-Fired Power Generation Handbook, by James G. Speight, ISBN 9781118208465. The most complete and up-to-date handbook on power generation from coal, this book covers all of today's new, cleaner methods for creating electricity from coal, the environmental challenges and concerns involved in its production, and developing technologies. *NOW AVAILABLE!*

Bioremediation of Petroleum and Petroleum Products, by James Speight and Karuna Arjoon, ISBN 9780470938492. With petroleum-related spills, explosions, and health issues in the headlines almost every day, the issue of remediation of petroleum and petroleum products is taking on increasing importance, for the survival of our environment, our planet, and our future. This book is the first of its kind to explore this difficult issue from an engineering and scientific point of view and offer solutions and reasonable courses of action. *NOW AVAILABLE!*

An Introduction to Petroleum Technology, Economics, and Politics, by James Speight, ISBN 9781118012994. The perfect primer for anyone wishing to learn about the petroleum industry, for the layperson or the engineer. *NOW AVAILABLE!*

Sustainable Energy Pricing, by Gary Zatzman, ISBN 9780470901632. In this controversial new volume, the author explores a new science of energy pricing and how it can be done in a way that is sustainable for the world's economy and environment. *NOW AVAILABLE!*

Green Chemistry and Environmental Remediation, Edited by Rashmi Sanghi and Vandana Singh, ISBN 9780470943083. Presents high quality research papers as well as in depth review articles on the new emerging green face of multidimensional environmental chemistry. *NOW AVAILABLE!*

Energy Storage: A New Approach, by Ralph Zito, ISBN 9780470625910. Exploring the potential of reversible concentrations cells, the author of this groundbreaking volume reveals new technologies to solve the global crisis of energy storage. *NOW AVAILABLE!*

Troubleshooting Vacuum Systems: Steam Turbine Surface Condensersand Refinery Vacuum Towers, by Norman P. Lieberman. ISBN 978-1-118-29034-7. Vacuum towers and condensing steam turbines require effective vacuum systems for efficient operation. Norm Lieberman's text describes in easy to understand language, without reference to complex mathematics, how vacuum systems work, what can go wrong, and how to make field observations to pinpoint the particular malfunction causing poor vacuum. *NOW AVAILABLE!*

Fundamentals of the Petrophysics of Oil and Gas Reservoirs, by Buryakovsky, Chilingar, Rieke, and Shin. ISBN 9781118344477. The most comprehensive book ever written on the basics of petrophysics for oil and gas reservoirs. *NOW AVAILABLE!*

Petroleum Accumulation Zones on Continental Margins, by Grigorenko, Chilingar, Sobolev, Andiyeva, and Zhukova. Coming in September 2012, ISBN 9781118385074. Some of the best-known petroleum engineers in the world have come together to produce one of the first comprehensive publications on the detailed (zonal) forecast of offshore petroleum potential, a must-have for any petroleum engineer or engineering student. *NOW AVAILABLE!*

Mechanics of Fluid Flow, by Basniev, Dmitriev, and Chilingar. Coming in September 2012, ISBN 9781118385067. The mechanics of fluid flow is one of the most important fundamental engineering disciplines explaining both natural phenomena and human-induced processes. A group of some of the best-known petroleum engineers in the world give a thorough understanding of this important discipline, central to the operations of the oil and gas industry.

Zero-Waste Engineering, by Rafiqul Islam, ISBN 9780470626047. In this controversial new volume, the author explores the question of zero-waste engineering and how it can be done, efficiently and profitably. *NOW AVAILABLE!*

Formulas and Calculations for Drilling Engineers, by Robello Samuel, ISBN 9780470625996. The most comprehensive coverage of solutions for daily drilling problems ever published. *NOW AVAILABLE!*

Emergency Response Management for Offshore Oil Spills, by Nicholas P. Cheremisinoff, PhD, and Anton Davletshin, ISBN 9780470927120. The first book to examine the Deepwater Horizon disaster and offer processes for safety and environmental protection. *NOW AVAILABLE!*

Advanced Petroleum Reservoir Simulation, by M.R. Islam, S.H. Mousavizadegan, Shabbir Mustafiz, and Jamal H. Abou-Kassem, ISBN 9780470625811. The state of the art in petroleum reservoir simulation. *NOW AVAILABLE!*